NATIONAL INSTITUTE OF
WATER AND ATMOSPHERIC RESEARCH (NIWA)

The Marine Fauna of New Zealand:

King Crabs of New Zealand, Australia and the Ross Sea (Crustacea: Decapoda: Lithodidae)

Shane T. Ahyong

Marine Biodiversity and Biosecurity
National Institute of Water and Atmospheric Research Ltd
Private Bag 14901, Kilbirnie
Wellington, New Zealand

NIWA Biodiversity Memoir 123

2010

Cataloguing in Publication

AHYONG, S. T.
The marine fauna of New Zealand: King crabs of New Zealand, Australia and the Ross Sea (Crustacea: Decapoda: Lithodidae) / by Shane T. Ahyong—Wellington: NIWA (National Institute of Water and Atmospheric Research Ltd), 2010
(NIWA Biodiversity Memoir, ISSN 1174-0043; 123)

Soft cover: ISBN 978-0-478-23285-1
Hard cover: ISBN 978-0-478-23286-8

Series Editor: Dennis P. Gordon

Copy edited by: Tracy Farr

Typeset and indexed by: Geoff Gregory, Word Therapy, Paraparaumu

Printed and bound by: Graphic Press & Packaging Ltd, Levin

Received for publication—15 February 2010

NATIONAL INSTITUTE OF
WATER AND ATMOSPHERIC RESEARCH (NIWA)

The Marine Fauna of New Zealand:

King Crabs of New Zealand, Australia and the Ross Sea (Crustacea: Decapoda: Lithodidae)

Shane T. Ahyong

Marine Biodiversity and Biosecurity
National Institute of Water and Atmospheric Research Ltd
Private Bag 14901, Kilbirnie
Wellington, New Zealand

ABSTRACT

The king crabs, Lithodidae, are among the largest and most conspicuous deepwater crustaceans worldwide, including the polar regions. Prior to the present study, 10 extant species were believed to occur in New Zealand and Australia: *Lithodes murrayi*, *Lithodes longispina*, *Neolithodes brodiei*, *Neolithodes flindersi*, and six species of *Paralomis*. Three species in two genera of lithodid have been recorded from the Ross Sea, Antarctica. Surprisingly, however, the Australian and New Zealand lithodid fauna has never been studied in detail and the identities of most species in the region have never been explicitly scrutinised. The Lithodidae from New Zealand, Australia, and the Ross Sea are thus fully reviewed. A total of three genera and 23 species are now known from the study area: 13 species in New Zealand, 12 in Australia (four shared by both regions) and three in the Ross Sea (of which one species extends onto the Macquarie Ridge). Neither *L. murrayi* nor *L. longispina* occurs in the region. Previous records of *L. longispina* and *L. murrayi* from New Zealand are based on different ontogenetic phases of the same undescribed species, herein named ***L. aotearoa*** sp. nov. Previous records of *L. longispina* from Australia are based on *Lithodes richeri* and an undescribed species, herein named ***L. australiensis*** sp. nov. Records of *Lithodes murrayi* from Macquarie Island are referable to ***L. macquariae*** sp. nov. Thus, *L. longispina* is reliably known only from Japan, and *L. murrayi* is reliably known only from the southwestern Indian Ocean. *Lithodes murrayi* does not exbihit a circumpolar distribution as formerly believed. *Neolithodes brodiei* is known only from New Zealand; reports of the species from Australia are based on the recently described *N. flindersi* Ahyong, 2010. Previous reports of *Paralomis histrix* from New Zealand are based on juvenile *Paralomis zealandica*, and *P.* cf. *histrix* from southeastern Australia are referable to a new species, ***P. echidna*** sp. nov. In total 14 new species are described, five exclusively from New Zealand, five exclusively from Australia, and four shared by both regions. Of the three lithodid species known from the Ross Sea, *Neolithodes yaldwyni* Ahyong & Dawson, 2006 and *Paralomis stevensi* Ahyong & Dawson, 2006 are presently known only from Antarctic waters; the third species, *P. birsteini* Macpherson, 1988 ranges onto the Macquarie Ridge. *Paralomis stevensi* is reported for the first time from the Amundsen Sea. *Paralomis shinkaimaruae* Takeda & Hatanaka, 1984 is synonymised with *P. zealandica* Dawson & Yaldwyn, 1971. Results of the present study demonstrate considerably higher species richness than previously known, essentially doubling the known fauna of the region.

Keywords: Crustacea, Decapoda, Lithodoidea, Lithodidae, king crabs, systematics, taxonomy, new species, Australia, New Zealand, Antarctica

CONTENTS

INTRODUCTION

The king crabs, Lithodoidea, comprise a small group (~100 species) of largely deepwater species, of which more than 60 occur in the Pacific Ocean. They are best known on the basis of several commercially important northern hemisphere species, especially *Paralithodes camtschaticus* (Tilesius, 1815) (Red King Crab) and *Lithodes aequispinus* (Benedict, 1895) (Golden King Crab). In the southern hemisphere, *Lithodes santolla* (Molina, 1782) (Southern King Crab) and several species of *Paralomis* are commercially landed off southern South America, and exploratory fisheries for lithodids are currently being undertaken in New Zealand waters. Although the lithodoids are exceptionally crab-like in appearance, they are not true crabs (infraorder Brachyura), rather being part of the infraorder Anomura, which includes squat lobsters and hermit crabs. Until 2007, the king crabs comprised the family Lithodidae, containing two subfamilies, Lithodinae and Hapalogastrinae, within the superfamily Paguroidea. McLaughlin *et al.* (2007), however, regarded the king crabs as only distant relatives of the hermit crabs, and raised the Lithodidae and its constituent subfamilies to superfamilial and familial rank, respectively — Lithodoidea, containing Lithodidae and Hapalogastridae. Hapalogastrids occur only in the northern Pacific, so the present study deals only with Lithodidae sensu McLaughlin *et al.* (2007).

The lithodoids are the paradigm example of carcinisation, that is, the evolution of a crab-like form from a non-crab-like ancestor. No other decapods more closely approximate the brachyuran form than do lithodoids. The process of attaining crab-like form is typically understood to involve a shortening and broadening of the carapace (from an elongate, often subcylindrical form), and a reduction of the abdomen, which is held folded beneath rather than extended behind the cephalothorax. Lithodoids have long been thought to be derived from within the asymmetrical hermit crabs, the so-called 'hermit to king' hypothesis (Bouvier 1894a, b, 1897; Cunningham *et al.* 1992). Indeed, the close relationship of lithodids to asymmetrical hermit crabs, particularly to pagurids, has long been widely accepted through the placement of lithodoids among the paguroids in most classification schemes of the Anomura (Martin & Davis 2001). The primary evidence used for this argument was the asymmetry of the somites and pleopods found in the female abdomen of lithodoids. Several recent studies, however, have challenged the 'hermit to king' hypothesis, citing apparently implausible transformation pathways (McLaughlin & Lemaitre 1997; McLaughlin *et al.* 2004) and proposed a distant relationship between lithodids and other paguroids (McLaughlin *et al.* 2007). In contrast, most other phylogenetic studies have recovered a close relationship between lithodids and pagurids supporting the derivation of king crabs from shell-dwelling hermit crabs (Richter & Scholtz 1994; Morrison *et al.* 2002; Ahyong & O'Meally 2004; Tsang *et al.* 2008; Ahyong *et al.* 2009; Bracken *et al.* 2009). As such, recognising the superfamily Lithodoidea, distinct from Paguroidea, as proposed by McLaughlin *et al.* (2007) is not compatible with the weight of phylogenetic evidence gathered to date. However, pending ongoing studies of anomuran phylogeny, the classification of McLaughlin *et al.* (2007) is followed here, with Lithodoidea treated as a superfamily, and Lithodidae restricted to non-hapalogastrine king crabs. Irrespective of the phylogenetic position of the king crabs, all workers recognise the validity of the two major clades, corresponding to Lithodidae and Hapalogastridae, respectively.

WORLD STUDIES

Prior to the last three decades, most research into the Lithodidae was focused on the northern hemisphere fauna. Key studies include Bouvier (1895, 1896), Benedict (1895), Hansen (1908), Schmitt (1921), and Makarov (1938). Enrique Macpherson (1988b) revised the Atlantic fauna, covering 30 species, and studied smaller collections from the Indian Ocean and south-western Pacific Ocean (e.g. Macpherson 1988a, c, 1989, 2001, 2003; Macpherson & Chan 2008). Lithodids from Antarctic and sub-Antarctic waters have been described by Macpherson (2004), Spiridonov *et al.* (2006), and Ahyong & Dawson (2006). Most recently, lithodids have been studied by Takeda & Nagai (2004), Takeda & Bussarawit (2007), Hall & Thatje (2009b), Ahyong & Chan (2010), Ahyong (2010) and Ahyong *et al.* (2010) from various Atlantic and Indo-Pacific localities. Dawson (1989) provided a useful world bibliography of the references dealing with Lithodidae, and Zaklan (2002) provided a synoptic overview of the biology and taxonomy of the group.

PREVIOUS REGIONAL STUDIES

Studies of Antipodean decapods date back more than a century (e.g. Heller 1865; Hess 1865; Miers 1876; Haswell 1882; Hutton 1882; Filhol 1883, 1885a), though lithodids were discovered in New Zealand and Australian waters only much more recently. The first lithodid to be recorded in the region, identified as *Lithodes murrayi* Henderson, 1888 by Hale (1941), was collected in 1930 near Macquarie Island (south of Tasmania) by the British, Australian and New Zealand Antarctic Research Expedition (BANZARE). No other lithodids were recorded from the region until 1958, when a juvenile specimen of *Paralomis zealandica* from Cook Strait, New Zealand, identified then as *Lithodes* sp., was figured by King (1958). In 1960 a specimen, identified as *Lithodes murrayi*, was captured by crayfishers in Foveaux Strait, New Zealand (Dell 1963; Yaldwyn & Dawson 1970). Since then, increased deepwater exploration in New Zealand waters by the New Zealand Oceanographic Institute (now NIWA) led to the discovery of not insignificant populations of lithodids, among them *Neolithodes brodiei* Dawson & Yaldwyn, 1970 and *Paralomis zealandica* Dawson & Yaldwyn, 1971. Subsequently, other species were also reported from New Zealand: *Lithodes longispina* Sakai, 1971, *Paralomis histrix* (de Haan, 1844), *Paralomis hirtella* de Saint Laurent & Macpherson, 1997, and *Paralomis dawsoni* Macpherson, 2001 (see Dawson & Yaldwyn 1985; Webber & Naylor 2004a, b; Ahyong *et al.* 2007; Dawson 2008). Additionally, the first and only known fossil lithodid was described from mid-late Miocene formations of New Zealand, *Paralomis debodeorum* Feldmann, 1998. In total, seven names have been applied to extant species of Lithodidae in New Zealand.

In Australia, lithodids were not known from mainland waters until the 1980s when deepwater exploration off New South Wales (by New South Wales Fisheries) and off Victoria and Tasmania (by the Commonwealth Scientific and Industrial Research Organisation, CSIRO) discovered five species of lithodids on seamounts, outer shelf and slope waters. These species were reported as *Lithodes longispina, Neolithodes brodiei, Paralomis* cf. *birsteini, P.* cf. *histrix*, and *P.* cf. *phrixa* (see Poore 2004). Poore *et al.* (2008) subsequently recorded *P.* cf. *phrixa* and *Lithodes* aff. *longispina* from southwestern Australia. Including the record of *L. murrayi* from Macquarie Island, six species have been reported from Australian waters, of which three were apparently shared with New Zealand (Poore 2004). Ahyong (2010) showed that Australian records of *N. brodiei* were based on an undescribed species, named *N. flindersi* Ahyong, 2010. Otherwise, no further progress on the taxonomy of Lithodidae in New Zealand and Australian waters has been made, although significant collections in both countries have been accumulated.

Three species of lithodids are known from the Ross Sea region, Antarctica (Ahyong & Dawson 2006). The first lithodid recorded from the Ross Sea (near Scott Island) was identified as the North Atlantic *Paralomis spectabilis* Hansen, 1908 (Birstein & Vinogradov 1967), an apparently bi-polar species. Macpherson (1988a), however, showed that the southern records of *P. spectabilis* were referable to a distinct species, *Paralomis birsteini* Macpherson, 1988a. More intensive sampling of the Ross Sea over the last half decade by NIWA and the New Zealand Ministry of Fisheries recovered two further species, the known lithodid fauna now comprising *Paralomis birsteini* Macpherson, 1988a, *P. stevensi* Ahyong & Dawson, 2006, and *Neolithodes yaldwyni* Ahyong & Dawson, 2006. Note that Davie (2002) inadvertently listed *Lithodes murrayi* from the Possession Island in the Ross Sea, rather than Possession in the southwestern Indian Ocean.

Thus, a total of 13 nominal lithodid species have been reported from the study area, although only the Ross Sea fauna has been well characterised. This study commenced as a revision of the New Zealand Lithodidae. However, the presumed faunal overlap between New Zealand, Australian and Antarctic species, along with taxonomic ambiguity of several nominal species, required revision of the regional fauna. Thus, the extant Lithodidae of New Zealand, Australia and the Ross Sea are revised below.

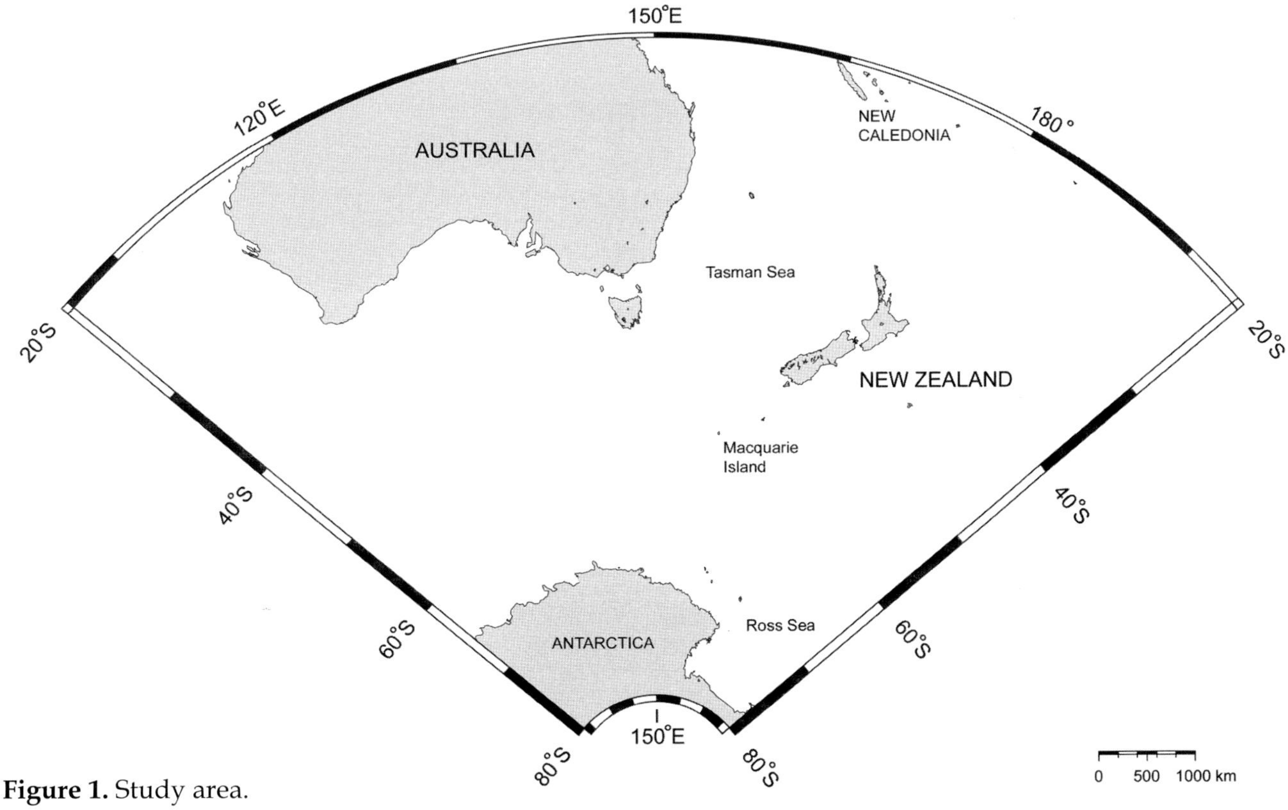

Figure 1. Study area.

MATERIALS AND METHODS

STUDY AREA AND STUDY MATERIAL

The study area encompasses the region bounded by New Zealand Exclusive Economic Zone (EEZ), southern Australia (south of 30°S), including the Macquarie Ridge, and the Ross Sea, Antarctica (Fig. 1). Lithodid collections held in the National Institute of Water and Atmospheric Research, Wellington, New Zealand (NIWA), Museum of New Zealand Te Papa Tongarewa (NMNZ), Australian Museum (AM), Institut de recherche pour le développement, Nouméa (IRD), Museum Victoria (NMV), Tasmanian Museum (TM), and South Australian Museum (SAM) were comprehensively studied, along with type and other specimens in the collections of the Muséum national d'Histoire naturelle, Paris (MNHN), National Science Museum, Tokyo (NSMT), National Taiwan Ocean University, Keelung (NTOU), Raffles Museum of Biodiversity Research, National University of Singapore (ZRC), National Museum of Natural History, Smithsonian Institution, Washington D.C. (USNM), Florida Museum of Natural History (FLMNH), and Zoological Laboratory of the Kyushu University, Fukuoka, Japan (ZLKU). The majority of specimens from the New Zealand region were collected by various benthic surveys conducted by NIWA (formerly NZOI) (RV *Tangaroa*, RV *Kaharoa*) and by numerous commercial fishing vessels. For economy, the vessel name is not explicitly mentioned under Material Examined; *Tangaroa* and *Kaharoa* station numbers are prefaced by TAN and KAH respectively. In other instances, the vessel name is mentioned. Lithodids collected from Australian waters were collected by New South Wales Fisheries (FRV *Kapala*) along the southeast coast, various CSIRO research programmes (RV *Soela*, RV *Southern Surveyor*), especially between Tasmania and Western Australia, and miscellaneous commercial fishing vessels. All images are by the author unless indicated otherwise.

MEASUREMENTS AND TERMINOLOGY

Morphological terminology is discussed below and illustrated in Figs 2–4. Carapace length (cl) is measured along the dorsal midline from the apex of the rostrum to the posterior carapace midline. Postorbital carapace length (pcl) is measured from the posterior margin of the orbit to the posterior margin of the carapace. Carapace width (cw) is greatest carapace width excluding spines. Other measurements are shown in Fig. 4. All measurements are in millimetres (mm). Distribution maps were produced with the aid of *Online Map Creation* (www.aquarius.ifm-geomar.de).

MORPHOLOGY

Characters currently used in lithodid taxonomy are derived from the cephalothorax, especially the antennae and ornamentation of the carapace; the ornamentation and proportions of the chelipeds (pereopod 1) and walking legs (pereopods 2–4); and abdomen. Owing to the considerable allometric changes that typically occur between juvenile and adult stages, diagnoses and descriptions given under taxonomic accounts are based on adults. Juvenile morphology is discussed in the Remarks for each species.

Cephalothorax: The cephalothorax comprises an acron (bearing the eyes), five cephalic somites bearing appendages (antennule, antenna, mandible, maxillule, and maxilla) and eight thoracic somites bearing maxillipeds 1–3 and pereopods 1–5. These somites are fused and entirely covered by the carapace. The cephalothorax is ventrally represented by a series of fused sternal plates. In most lithodids, the cephalothorax is covered by a well-calcified carapace which is generally pentagonal or pyriform. A prominent rostrum is present anteromedially. In *Paralomis* and *Neolithodes*, the rostrum is trifid, consisting of a median spine and a pair of divergent dorsal spines. In *Lithodes*, the rostral structure differs from that of *Paralomis* and *Neolithodes*, having a strong anterior projection with a pair of dorsal spines placed at some distance from the base, demarcating an anterior and posterior portion. The anterior portion is itself distally bifid, though the distal spines are sometimes absent owing to damage or moulting abnormalities. The posterior portion also bears a prominent ventral, anteriorly directed spine, herein termed the ventral rostral spine. The ventral rostral spine in *Lithodes* appears to be homologous with the median rostral spine of *Paralomis* and *Neolithodes*, but different terminology is employed here for taxonomic convenience. In most species of *Lithodes*, the ventral rostral spine is not visible dorsally, but in *L. santolla* (Molina, 1782) and *L. confundens* Macpherson, 1988b the anterior and posterior dorsal portions are much reduced in length, and are distally overreached by the ventral rostral spine. As a result, in dorsal view, the anterior dorsal portion may appear to be distally trifid, rather than bifid as it actually is.

The orbital margin is delimited medially by the rostrum and laterally by the outer orbital spine. The eyes have a unisegmental peduncle (ocular peduncle) and well-pigmented, distoventrally positioned cornea. The ocular peduncle usually bears small spines or granules. The hepatic region bears, at the anterolateral angle, the anterolateral spine; and posteriorly, a long spine adjacent to the junction with the branchial margin, termed the hepatic spine. At any given body size, the hepatic spine is usually the longest of the carapace spines. The branchial margin typically bears several major spines (longer than adjacent spines, especially in juveniles) that serve as useful landmarks: one above pereopod 3 (primary marginal branchial spine 1), and one above pereopod 4 (primary marginal branchial spine 2). These large spines effectively divide the branchial margin into discrete regions, herein termed the anterior branchial margin, lateral branchial margin, and posterior branchial margin. The two branchial regions are separated anteriorly by the gastric and cardiac regions, and posteriorly by the intestinal region, marked by paired intestinal spines in *Lithodes* and *Neolithodes*. The margin between each of these major spines is lined with varying numbers of shorter spines. The number of posterior branchial marginal spines is diagnostic in species of *Lithodes*.

The dorsal surface of the carapace is divided into regions by shallow grooves. The most obvious is the cervical groove, which crosses the carapace and separates the gastric and hepatic regions from the branchial and cardiac regions. The gastric region is higher than the other regions and lies behind the rostrum; it usually bears four major spines (most obvious in *Lithodes* and *Neolithodes* but not always obvious in other genera) in addition to other spinules or granules that may be present. The regions flanking the gastric region are termed hepatic. The cardiac region lies medially behind the gastric region and bears two or four major spines. The cardiac region is followed by the intestinal region and usually bears one or two spines along the midline. The intestinal region is often ill-defined and difficult to distinguish from the branchial or cardiac regions. The branchial regions flank the cardiac and intestinal regions and typically bear three major spines of which the anterior two are longer. The spine pattern is consistent across most lithodids, and homologous spines can be identified across species and genera, except in species in which the carapace is entirely and densely covered with spines or tubercles, or dorsally unarmed as in *Paralomis zealandica*, *P. hirtella*, and *P. dawsoni* (although spines may be present in early juveniles). As is well known in lithodids, the length of carapace spines changes dramatically between juveniles and adults, and must be considered during the identification process. As indicated by Macpherson (1988b), however, although the proportional length of spines varies considerably with size, the number and position of spines is remarkably constant within most species.

Thoracic sternum: The visible thoracic sternites are represented ventrally by the fused sternites 4–7 forming the sternal plastron. Sternal characters are not used in lithodoid taxonomy, although the presence of a deep median fissure on sternite 5 (between the first pair of walking legs) is useful at generic level, and will separate *Lithodes* and *Neolithodes* (fissure present) from *Paralomis* (fissure absent).

Abdomen: The abdomen is short, broad, variously calcified, and folded underneath the cephalothorax. It consists of six somites plus telson. The terga are variously calcified but the sternal surfaces are membranous and uncalcified. Males lack pleopods, but females often have a pair of small first pleopods and uniramous, unpaired left pleopods 2–5. Uropods are absent.

Abdominal somite 1 is reduced and often armed with a pair of submedian spines. Somite 2 is widest and deepest, and bears important diagnostic features. It is plesiomorphically divided into five plates: median, paired submedian, and paired marginals. In *Neolithodes*, all five plates are separate and demarcated by a suture. The median and submedian plates are large and broad whereas the marginal plates are very narrow. In *Paralomis*, all five plates are indistinguishably fused into a single large plate. In *Lithodes*, the median and submedian plates are always indistinguishably fused into a single plate (the 'central' plate), but the marginal plates may or may not be fused with the central plate. Where the marginal plates are fused with the remaining plates, a shallow crease is usually visible indicating the line of fusion. The condition of the marginal plate of somite 2 (fused or unfused) is a useful diagnostic character for adults. The marginal plates are always free in juveniles, even in species in which the plates are fused as adults (e.g. *Lithodes aotearoa*). Tergites 3–5 are symmetrical in males, and asymmetrical in females (via enlargement of the left plates and reduction of the right plates), but are variously calcified between lithodid genera. In *Paralomis*, the terga are composed of median, submedian, and marginal plates, of which the marginals are often subdivided (females with left marginals fused with submedians). In *Lithodes*, the median plates are composed of blunt nodules, the submedians are entire, and the marginals are subdivided (females with left marginals fused with submedians). The terga in *Neolithodes* are membranous and densely covered with spinular or nodular calcified elements; in females, the submedians and marginals are fused into a single plate. Tergite 6 comprises an elongate central plate and narrow, subdivided marginal plates. The telson is usually short and semicircular.

In the Hapalogastridae (not known from the study area), the tergites of somites 3–5 are entirely uncalcified.

Antennule: The antennule is the first cephalic appendage, and consists of a 3-segmented peduncle and paired flagella. The basal article bears the statocyst. The distal two articles are typically cylindrical and unarmed. Two multi-segmented flagella arise apically from the ultimate peduncular article. The upper flagellum is long and bears numerous long aesthetascs along its ventral margin. The lower flagellum is short and small-segmented.

Antenna: The antenna is the second cephalic appendage and consists of a 5-segmented peduncle and long, uniramous flagellum. The proximal articles of the peduncle are usually spinose. The first article bears the urinal or green gland, opening anteromesially. The exopod, termed the scaphocerite, is present in most lithodids including *Paralomis*, but reduced to a small sclerite or spine in *Lithodes* and *Neolithodes*. The scaphocerite, when present, varies from being a simple spine with a few subsidiary spinules, to an elaborate multispinous structure.

Pereopod 1 (chelipeds): The first pereopods are the chelipeds and usually asymmetrical in size, especially in males. In females, the size difference in the chelipeds is much less pronounced. The right cheliped is the larger (rarely vice versa), as in Paguridae. The cheliped consists of dactylus, propodus, carpus, merus, ischiobasis, and coxa. Unlike the walking legs (pereopods 2–4) in which the segments more-or-less articulate in the same plane, the arrangement of the cheliped segments allow articulation in multiple planes, making positional terminology such as extensor and flexor margins difficult to apply. Therefore, in the case of the chelipeds, terminology used here refers to the position of structures when the cheliped is held forwards with the dactylus uppermost.

Pereopods 2–4 (walking legs 1–3): Pereopods 2–4 are the walking legs, and their spination, relative overall lengths and the relative segment proportions are often diagnostic at species level. The walking legs are similar in structure and ornamentation, differing slightly in length and proportions. They consist of dactylus, propodus, carpus, merus, ischiobasis, and coxa. The length of the walking legs is sexually dimorphic, with that of males usually about 10–15% longer than in size-matched females. The lengths and proportions of walking leg segments, especially the merus and propodus, vary allometrically, typically being more slender in juveniles than adults. As in other Anomura, the female

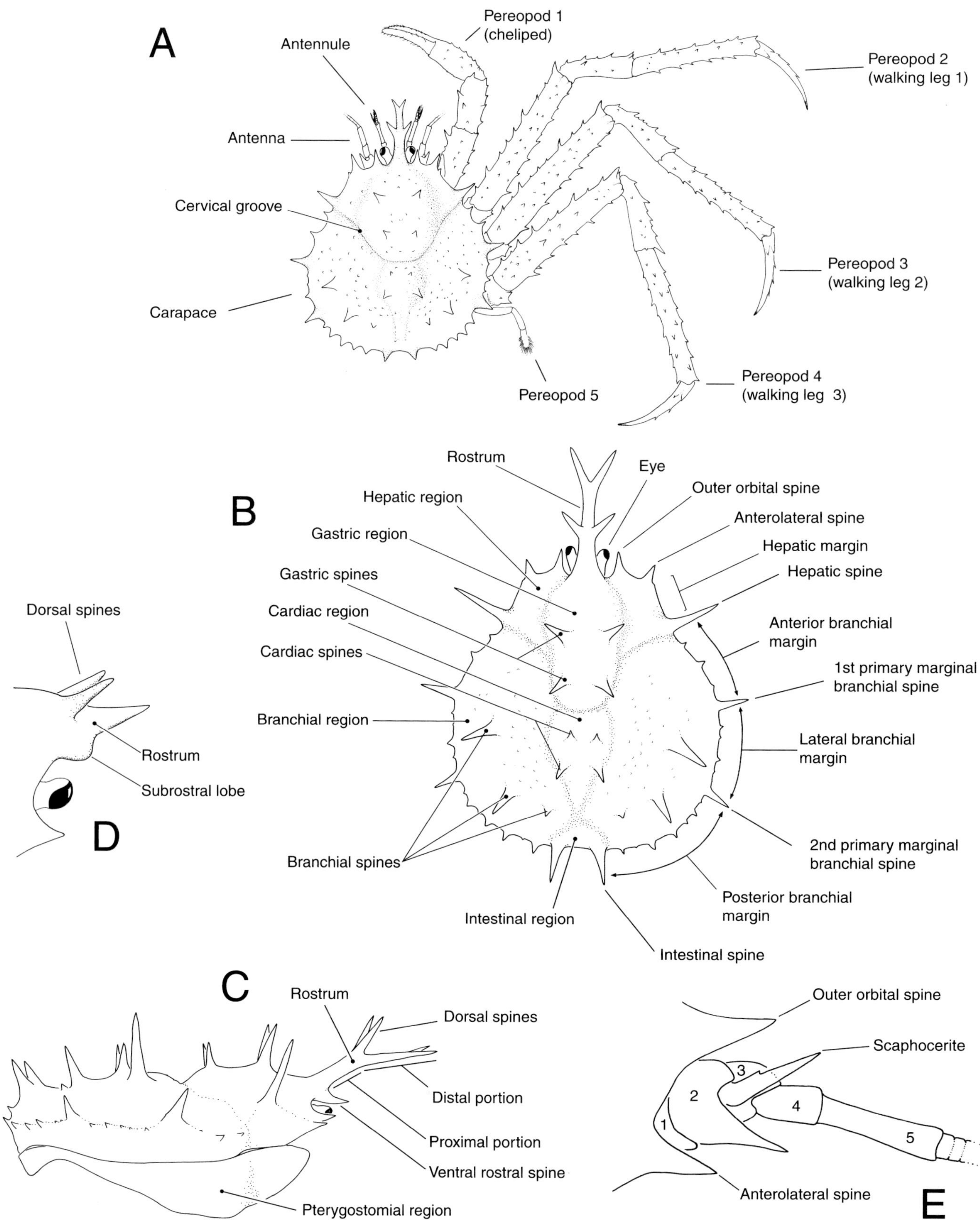

Figure 2. General morphology. A, habitus. B–C, carapace spines and regions. D, rostrum features of *Paralomis* and *Neolithodes*. E, antenna.

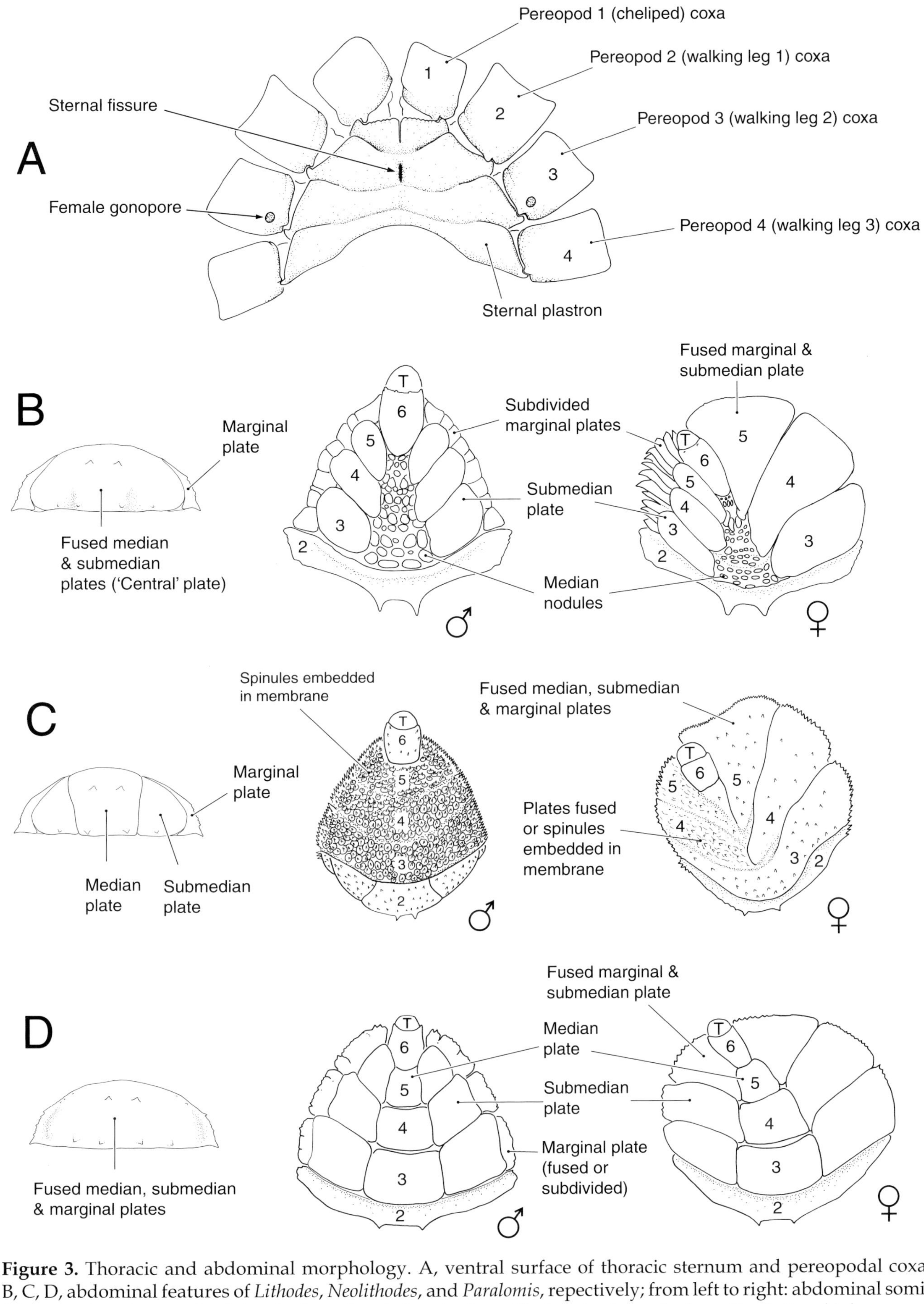

Figure 3. Thoracic and abdominal morphology. A, ventral surface of thoracic sternum and pereopodal coxae. B, C, D, abdominal features of *Lithodes*, *Neolithodes*, and *Paralomis*, repectively; from left to right: abdominal somite 2 in posterior view, male abdominal somites 2–6 and telson (T), female abdominal somites 3–6 and telson.

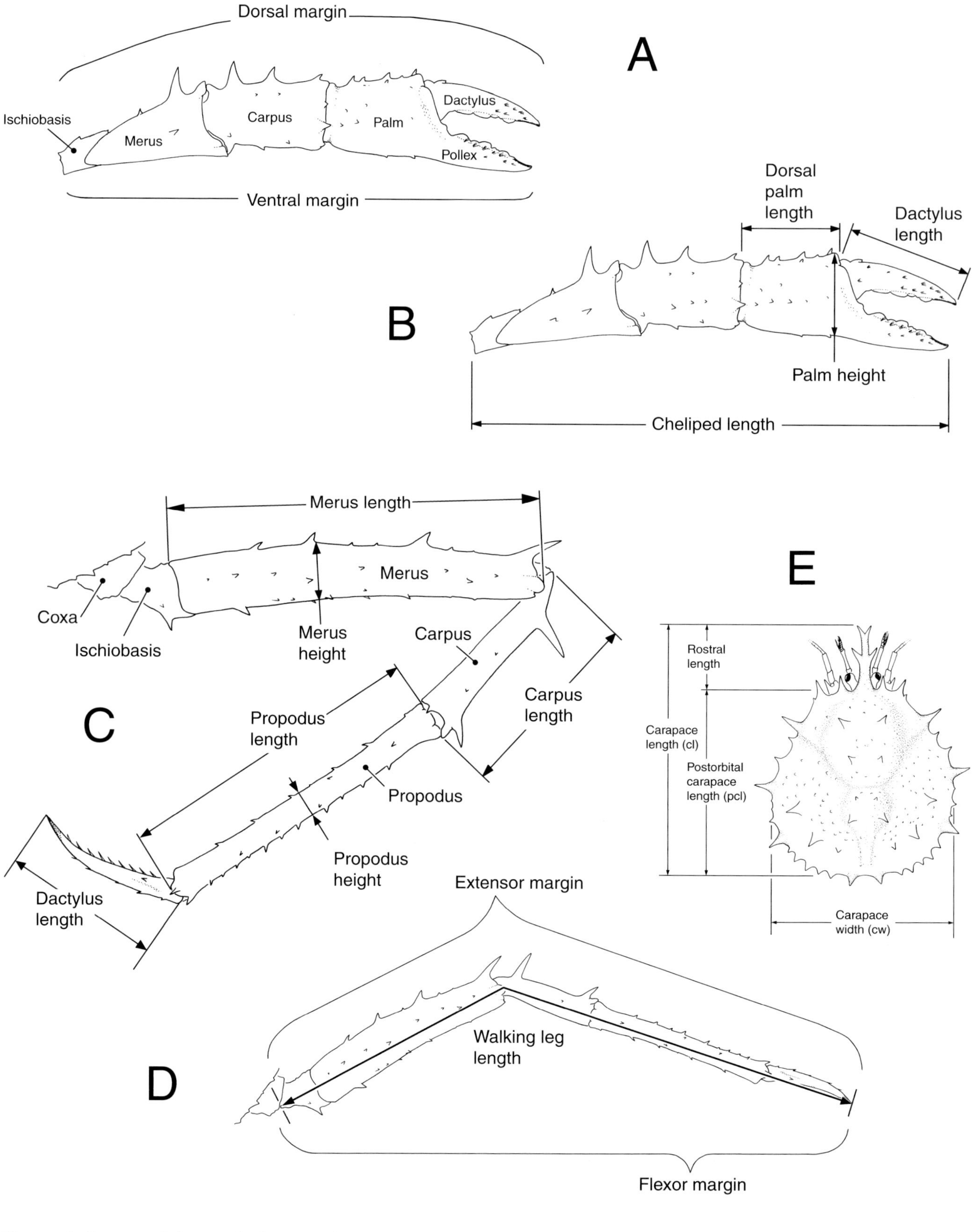

Figure 4. Pereopodal morphology and carapace measurements. A–B, cheliped (pereopod 1) morphology and measurements. C–D, walking leg (pereopods 2–4) morphology and measurements. E, carapace measurements.

gonopore is located on the pereopod 3 coxa, and male gonopore on the pereopod 5 coxa. Mature females usually have a setose surface of the pereopodal coxae. In juveniles, especially of *Lithodes* and *Neolithodes*, the dactyli of the walking legs are proportionally longer (in comparison to the propodus) in juveniles than in adults. The flexor margins of the walking leg dactyli in juveniles through adults bear a row of corneous spinules in *Paralomis*, and are unarmed in early juveniles through to adults in *Neolithodes*. Juvenile *Lithodes* have a row of corneous spinules on the flexor margins, which, in most species, is lost in late juveniles. The effects of allometry and sexual dimorphism must be carefully considered when identifying specimens.

Pereopod 5 (eighth thoracic appendage): Pereopod 5 is a very small and slender subchelate appendage, and is carried folded underneath the carapace. It is used for grooming the gills and the eggs in females, and, at present, has no taxonomic value.

ALLOMETRY AND SEXUAL DIMORPHISM

When identifying lithodids, particular attention must be given to the size and sex of specimens (see Figs 2–4 for morphological terms and measurements used herein). Lithodids exhibit considerable allometric variation in rostral and general spine length. These features nevertheless provide very useful diagnostic characters, provided body size is considered. The rostrum and general spines are proportionally longest in juveniles, becoming progressively shorter proportionally (and sometimes absolutely) with increasing body size. These allometric changes are most pronounced in species of *Lithodes* and *Neolithodes*, in which the rostrum and longest carapace spines may exceed the postorbital carapace length (pcl) in early juveniles. Allometric trends in *Paralomis* are similar to, but less pronounced than in *Lithodes* and *Neolithodes*. In addition to reduction in relative spine length, a number of allometric changes are also evident in length and ornamentation of the walking leg dactyli. The dactyl length relative to propodus length decreases with increasing body size, and in juveniles, the flexor margins are lined with a row of corneous spines. These corneous spines are retained in adult *Paralomis*, but lost in early juvenile *Neolithodes* and adults of most species of *Lithodes* (becoming diagnostic for those species retaining the spines as adults, e.g. *L. chaddertoni* sp. nov. and *L. jessica* sp. nov.). In most species of *Lithodes*, the marginal plates of abdominal somite 2 are separated from the submedian plates by a suture in juveniles that is present through to adults. In adults of some species, however, such as *L. aotearoa* sp. nov. and *L. richeri* Macpherson, 1990, the marginals are fused to the submedians. Thus, fusion or separation of the plates of abdominal somite 2 can be a useful diagnostic feature in adults.

Sexual dimorphism is evident but less pronounced than allometric changes. Primary sexual differences, such as abdominal symmetry or asymmetry and gonopore position, as discussed above, are the most obvious differences. However, secondary sexual differences are present in the form of pereopod length and shape, and these differences become increasingly pronounced with size. The chelipeds of males are usually distinctly unequal (except in *Lithodes jessica*), with the right chela larger and more robust than the left chela. In females, both chelae are similar, having the form of the left chela of males. The walking legs differ in length between sexes, being proportionally longer in males than in females. Typically, they differ by about 10–15%, sometimes more, being most obvious in mature, size-matched specimens. Another factor to consider when identifying specimens is the the effect of parasitic feminisation caused by rhizocephalan infections. The abdomen of specimens should always be checked for the presence of rhizocephalan externae. In males with advanced rhizocephalan infection, the pereopod morphometrics are similar to females.

To best facilitate identification of species, figures of both sexes, across their known size range, are provided where possible.

SYSTEMATICS

LITHODOIDEA Samouelle, 1819

Lithodiadae Samouelle, 1819: 90.
Lithodoidea. – McLaughlin *et al.*, 2007: 108.

Diagnosis. Crab-like. Carapace subtriangular to transversely ovate; rostrum present. Eyes with pigmented cornea; cornea distoventral on stalk. Maxilliped 3 with crista dentata and accessory tooth. Pereopods 1 chelate, subequal or right larger and more robust than left. Pereopods 2–4 developed as walking legs. Pereopod 5 subchelate, reduced, usually folded within branchial chamber. Abdomen short, broad, held folded against thoracic sternum; composed of calcified plates or nodules embedded in arthrodial membrane; somite 1 largely concealed by posterior carapace margin; abdominal plates more strongly developed on left side in females, producing asymmetry. Pleopods absent in males; females with paired pleopods on somite 1 and single uniramous pleopod on left side of somites 2–5. Uropods absent.

Composition. Lithodidae Samouelle, 1819, Hapalogastridae Brandt, 1850.

Key to families of Lithodoidea Samouelle, 1819

1. Rostrum well-developed, spiniform or truncate, extending anteriorly beyond eye. Abdominal somites 3–5 usually well-calcified; sometimes with membranous areas medially with nodules or small spinules .... Lithodidae
- Rostrum short, broad, not usually extending anteriorly beyond eye. Abdominal somites 3–5 ill-calcified, in form of a membranous sac; usually covered by small, thin plates .... Hapalogastridae

LITHODIDAE Samouelle, 1819

Lithodiadae Samouelle, 1819: 90 [type genus: *Lithodes* Latreille, 1806] [incorrect original spelling for Lithodidae Samouelle, 1819 (ICNZ Opinion 511)]
Lithodinae. – Macpherson, 1988b: 17.
Lithodidae. – McLaughlin *et al.*, 2007: 108. – De Grave *et al.*, 2009: 25.

Diagnosis. Rostrum well-developed, spiniform or truncate, extending anteriorly beyond eye. Abdominal somites 3–5 usually well-calcified; composed entirely of calcified plates or of nodules or spinules embedded in arthrodial membrane.

Composition. *Cryptolithodes* Brandt, 1848, *Glyptolithodes* Faxon, 1895, *Lithodes* Latreille, 1806, *Lopholithodes* Brandt, 1848, *Neolithodes* A. Milne-Edwards & Bouvier, 1894b, *Paralithodes* Brandt, 1848, *Paralomis* White, 1856, *Phyllolithodes* Brandt, 1848, *Rhinolithodes* Brandt, 1848, *Sculptolithodes* Makarov, 1934.

Remarks. Lithodidae includes 10 genera. The three largest genera, *Lithodes*, *Neolithodes*, and *Paralomis*, are represented in New Zealand, Australian and Antarctic waters. Phylogenetic relationships between lithodid genera were most recently studied by Hall & Thatje (2009a), who found *Neolithodes* and *Lithodes* to be more closely related to each other than either is to *Paralomis*. The genera of Lithodidae are distinguished in the key below, with those represented in New Zealand marked in bold.

Key to genera of Lithodidae

1. Carapace nearly smooth, without protuberances, granules, or spines; markedly wider than long, expanded laterally covering walking legs entirely when the legs are drawn in. Rostrum broad, compressed, truncated anteriorly .... *Cryptolithodes*
- Carapace with more or less numerous protuberances, granules or spines. Lateral margins of carapace without lateral expansions overhanging legs. Walking legs never completely covered by carapace when drawn in close to body. Rostrum variable in shape, but never truncated anteriorly 2
2. Sternite 5 (sternite of first pair of walking legs) with a deep, longitudinal, medial fissure 3
- Sternite 5 (sternite of first pair of walking legs) without a deep, longitudinal, medial fissure 5
3. Abdominal somite 2 composed of three plates (paired marginal, central plate formed by fused median and submedians) or fused into a single plate .... ***Lithodes***
- Abdominal somite 2 composed of five plates (median, paired submedian, and paired marginal 4
4. Abdominal somites 3–5 comprising small, more or less spiniform nodules in males, connected by arthrodial membrane, without well-defined plates. Females with a well-developed submedian plate on the left side of each segment, right side either with spiniform nodules connected by arthrodial membrane or with plates fused. Scaphocerite a

small spine or reduced to a short sclerite .... ***Neolithodes***

- Abdominal somites 3–5 bearing calcified nodules medially, with well-defined submedian and marginal plates on each side in both sexes. Scaphocerite well-developed, spinous .... *Paralithodes*

5. Submedian plates on abdominal somites 3–5 bearing concave, membranous areas. Abdominal somite 2 with a median plate and a marginal pair fused with the submedian plates. Rostrum more or less elongated, with two rounded terminal knobs........ *Phyllolithodes*

- Submedian plates on abdominal somites 3–5 lacking concave, membranous areas 6

6. Abdominal somite 2 composed of five distinct plates. Marginal plates on abdominal somites 3–5 fused with the submedian plates ......*Rhinolithodes*

- Abdominal somite 2 entire, formed by a single plate 7

7. Rostrum thick, non-spiniform, hammer-shaped. Scaphocerite small, rudimentary. Median plate on abdominal somites 3–5 subdivided *Sculptolithodes*

- Rostrum more or less spiniform. Scaphocerite well-developed. Median plates on abdominal somites 3–5 entire.. 8

8. Rostrum formed by a median spine and one dorsal spine or granule................................ *Glyptolithodes*

- Rostrum formed by a median spine and at least one pair of dorsal spines.. 9

9. Median plate on abdominal somite 3 subdivided into three. Walking leg 3 shorter than carapace width ....*Lopholithodes*

- Median plate on abdominal somite 3 undivided. Walking leg 3 equal to or longer than carapace width .... ***Paralomis***

Lithodes Latreille, 1806

Lithodes Latreille, 1806: 39 [type species: *Cancer maja* Linnaeus, 1758: 269. Gender: masculine].

Pseudolithodes Birstein & Vinogradov, 1972: 356 [type species: *Pseudolithodes zenkevitchi* Birstein & Vinogradov, 1972 = *Lithodes santolla* (Molina, 1782). Gender: masculine].

Diagnosis. Carapace pyriform, not covering bases of walking legs; regions indicated; gastric region elevated above other regions; cardiac region triangular, separated from gastric region by deep groove; cervical groove shallow, indistinct. Rostrum comprised of median spine (usually distally bifid) and one or two pairs of dorsal spines; prominent ventral rostral spine present. Abdominal somite 2 comprising 3 plates (a wide 'central' plate and a pair of narrow marginal plates) or single plate in adults of some species (all plates fused). Abdominal somites 3–5 with median regions composed of calcareous nodules joined by arthrodial membrane; submedian and marginal plates well demarcated in both sexes; submedian plates in females better developed on left side and fused with marginal plates; marginal plates usually subdivided. Sternite 5 (between pereopods 2) with deep median fissure. Scaphocerite absent or reduced to small spine. Walking legs (pereopods 2–4) similar in form, third walking leg longest, always longer than pcl; dactyli of adults with or without corneous spines along flexor margin.

Composition

L. aequispina Benedict, 1895 [northern Pacific Ocean]
L. aotearoa sp. nov. [New Zealand, southwestern Pacific Ocean]
L. australiensis sp. nov. [Australia, southwestern Pacific Ocean]
L. ceramensis Takeda & Nagai, 2004 [Indonesia, western Pacific Ocean]
L. chaddertoni sp. nov. [Western Australia, eastern Indian Ocean]
L. confundens Macpherson, 1988b [southwestern Atlantic Ocean]
L. couesi Benedict, 1895 [northern Pacific Ocean]
L. ferox Filhol, 1885b (= *L. pyriformis* Birstein & Vinogradov, 1972) [Eastern Atlantic and Brazil]
L. formosae Ahyong & Chan, 2010 [Taiwan, western Pacific Ocean]
L. galapagensis Hall & Thatje, 2009b [Galapagos Islands, eastern Pacific Ocean]
L. jessica sp. nov. [New Zealand, southwestern Pacific Ocean]
L. longispina Sakai, 1971 [Japan, Western Pacific Ocean]
L. macquariae sp. nov. [Macquarie Ridge, Southern Ocean]
L. maja Linnaeus, 1758 [North Atlantic Ocean]
L. mamillifer Macpherson, 1988c [southwestern Indian Ocean]
L. manningi Macpherson, 1988b [western Atlantic Ocean]
L. megacantha Macpherson, 1991 [French Polynesia, central Pacific Ocean]
L. murrayi Henderson, 1888 [southwestern Indian Ocean]
L. nintokuae Sakai, 1978 [northern-central Pacific Ocean]
L. panamensis Faxon, 1893 [eastern Pacific Ocean]
L. paulayi Macpherson & Chan, 2008 [Guam, western Pacific]

L. rachelae sp. nov. [Great Australian Bight, Southern Ocean]

L. richeri Macpherson, 1990 [southwestern Pacific Ocean]

L. robertsoni sp. nov. [New Zealand, southwestern Pacific Ocean]

L. santolla (Molina, 1782) (= *Pseudolithodes zenkevitchi* Birstein & Vinogradov, 1972) [southwestern Atlantic Ocean, southeastern Pacific Ocean]

L. turkayi Macpherson, 1988b [southwestern Atlantic Ocean, southeastern Pacific Ocean]

L. turritus Ortmann, 1892 [western Pacific Ocean]

L. unicornis Macpherson, 1984 [Valdivia Bank, southeastern Atlantic Ocean]

L. wiracocha Haig, 1974 [eastern Pacific Ocean].

REMARKS. Prior to the present study, *Lithodes* included 22 species. Macpherson (1988a) listed 15 species of *Lithodes* worldwide of which seven occur in the Atlantic Ocean and eight in the Indo-Pacific. Since then, seven species of *Lithodes* have been described, one from the Indian Ocean and six from the Pacific Ocean: *L. mamillifer* Macpherson, 1988b [southwestern Indian Ocean], *L. richeri* Macpherson, 1990 [New Caledonia], *L. megacantha* Macpherson, 1991 [French Polynesia], *L. ceramensis* Takeda & Nagai, 2004 [Indonesia], *L. paulayi* Macpherson & Chan, 2008 [Guam], *L. galapagensis* Hall & Thatje, 2009b [Galapagos Islands], and *L. formosae* Ahyong & Chan, 2010 [off Dasi, northeastern Taiwan]. Four species are known from New Zealand to Macquarie Island, all new, and four from around the Australian mainland, of which three are new species. The results of the present study increase the number of known species of *Lithodes* to 29. The diversity of *Lithodes* in the southwestern Pacific appears to be remarkably high compared to other regions, with six species from New Zealand and southeastern Australia alone. Seven species are known from the entire Atlantic Ocean, and a further 13 species are known from throughout the remainder of the Pacific Ocean. At present, only four species of *Lithodes* are known from the Indian Ocean, including *L. chaddertoni* sp. nov. and *L. rachelae* sp. nov. described below.

KEY TO ADULT *LITHODES* FROM NEW ZEALAND AND AUSTRALIA

1. Flexor margin of walking leg dactyli lined with small corneous spines 2
- Flexor margin of walking leg dactyli smooth 3

2. Chelae in both sexes equal, subcylindrical, nonspinose [New Zealand]................ *L. jessica* sp. nov.
- Chelae dimorphic, palm highest distally, surfaces spinose [Western Australia]..................................... ....*L. chaddertoni* sp. nov.

3. Rostrum as long as pcl in specimens exceeding 75 mm pcl [Vanuatu to southeastern Australia, possibly Indonesia]........... *L. richeri* Macpherson, 1991
- Rostrum not exceeding 0.7 pcl in specimens exceeding 75 mm pcl..4

4. Dorsal carapace spines short, stout, about 0.1 pcl or less in specimens exceeding 50 mm pcl [Macquarie Ridge to Auckland Islands, New Zealand] .... *L. macquariae* sp. nov.
- Dorsal carapace spines long, slender, exceeding 0.3 pcl in specimens exceeding 50 mm pcl (though often about 0.1 pcl or less in specimens exceeding 120 mm pcl)..5

5. Abdominal somite 2 marginal plates fused with 'central' plate. Posterior branchial margin with 8–12 spines or teeth [New Zealand] .... *L. aotearoa* sp. nov.
- Abdominal somite 2 marginal plates not fused with submedian plate, separated by suture. Posterior branchial margin with 5–8 spines or teeth............6

6. Male pereopod 4 merus length 1.3 pcl; merus length: height ratio exceeding 9 [southwestern Australia] ....*L. rachelae* sp. nov.
- Male pereopod 4 merus length less than 1.2 pcl; merus length:height ratio less than 8.....................7

7. Male pereopod 4 merus length up to 1.1 pcl in specimens exceeding 100 mm pcl. Posterior branchial margin with 7 or 8 spines or teeth [southeastern New Zealand] .... *L. robertsoni* sp. nov.
- Male pereopod 4 merus length less than 1.1 pcl in specimens exceeding 100 mm pcl. Posterior branchial margin with 5 or 6 spines or teeth [southeastern Australia] .... *L. australiensis* sp. nov.

Lithodes aotearoa sp. nov.

(Figs 5–12, Pl. 1A–B, 4C)

Lithodes murrayi. — Dell, 1963: 62. — Yaldwyn & Dawson, 1970: 279–282, figs 1–3. — Dawson & Yaldwyn, 1985: 70 [New Zealand occurrences]. — McLay, 1988: 22. — Takeda in Amaoka *et al.*, 1990: 360. — Webber, 1997: 81, fig. 4. — O'Shea *et al.*, 1999: 49, fig. 15. — Zaklan, 2002: 766 [New Zealand occurrences]. — Webber & Naylor, 2004a: 78–79. — Naylor *et al.*, 2005: 41–42. — Ahyong *et al.*, 2007: 154. [Not *L. murrayi* Henderson, 1888].

Lithodes longispina. — Dawson & Yaldwyn, 1985: 70. — McLay, 1988: 32 [Not *L. longispina* Sakai, 1971].

Lithodes spp. — O'Driscoll *et al.*, 2003: 62.

Lithodes cf. *longispinus*. — Webber & Naylor, 2004a: 79. — Naylor *et al.*, 2005: 43–44. — Ahyong *et al.*, 2007: 153.

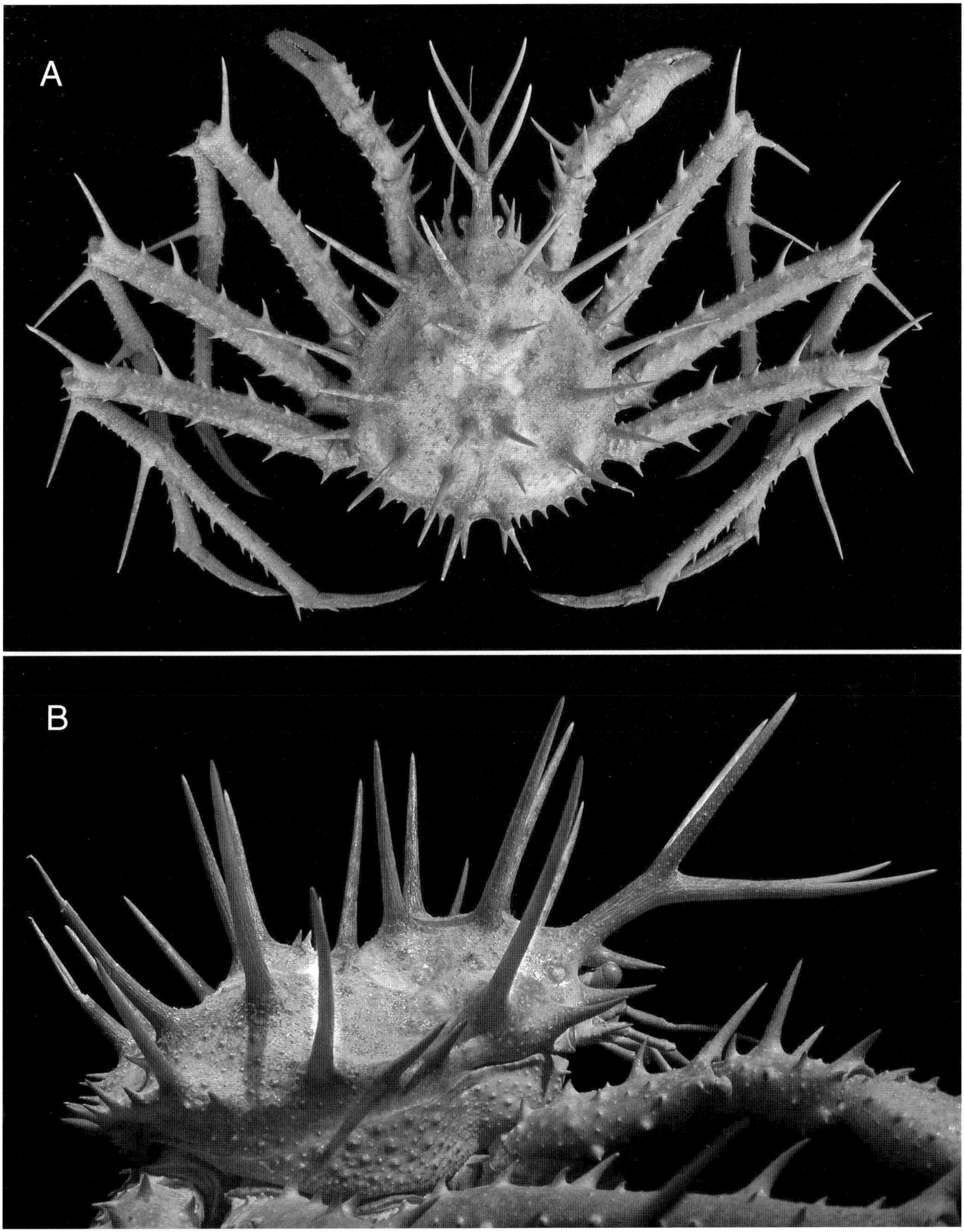

Figure 5. *Lithodes aotearoa* sp. nov., male holotype, cl 152.3 mm, pcl 88.6 mm, cw 85.3 mm, off Hawkes Bay (NIWA 34924). A, dorsal habitus. B, carapace, right lateral view.

Type material. *Holotype*: NIWA 34924, male (cl 152.3 mm, pcl 88.6 mm, cw 85.3 mm), off Hawkes Bay, 39°29.44–28.51′S, 178°25.05–25.48′E, 1000–980 m, KAH9907/37, Z10169, 3 Jun 1999.

Paratypes: NIWA 34907, 1 female (cl 96.0 mm, pcl 55.3 mm, cw 52.9 mm), off Hawkes Bay, 39°29.44–28.51′S, 178°25.05–25.48′E, 1000–980 m, KAH9907/37, Z10169, 3 Jun 1999; NMNZ Cr11738–11739, 1 ovigerous female (pcl 125.1 mm, cw 125.7 mm), 1 female (cl 175.5 mm, pcl 130.8 mm, cw 127.7 mm), Hawke Seavalley, 39°38′S, 177°55′E, 1050 m, mud bottom, coll. G. Gibbs, 21 Sep 2001; NMNZ Cr11691, 1 male (cl 198.4, pcl 135.5 mm, cw 133.3 mm), vicinity of Hawkes Bay, coll. A. Claudatus of Star Fish Supply Ltd, early Feb 1990; NMNZ Cr11692, 1 male (cl 176.7, pcl 106.7 mm, cw 102.7 mm), vicinity of Hawkes Bay, coll. A. Claudatus of Star Fish Supply Ltd, early Feb 1990.

Other material examined. *West Norfolk Ridge*: NMNZ Cr11149, 1 male (cl 197.2 mm, pcl 145.7 mm, cw 143.3 mm), 34°19.14–17.19′S, 168°24.63–21.85′E, 842–831 m, TAN0308/140, 2 Jun 2003.

Challenger Plateau: NMNZ Cr11722, 1 male (cl 190.5 mm, pcl 136.1 mm, cw 137.9 mm), 37°13.7′S, 167°49.9–51.9′E, 1109–1123 m, FV *Arrow*, A02/139/82, 9 Sep 1983; NMNZ Cr4699, 1 male (pcl 128.1 mm, cw 127.0 mm), N Challenger Plateau, 37°19.7–20.0′S, 168°32.7–43.9′E, 1078–1082 m, orange roughy trawl, bottom temperature 5.4°C, FV *Albert Sandford*, coll. R. Stewart, 1 Oct 1986; NMNZ Cr11736, 1 female (cl 189.4, pcl 143.6 mm, cw 137.7 mm), W of Westport, 37°24.3′S, 168°59.0′E, 1020 m, FV *Arrow*, A4/137/83, 17 Oct 1983; NMNZ Cr11723, 1 male (crushed, pcl ~123 mm; with *Briarosaccus callosus* Boschma, Rhizocephala), 41°41.2–39.5′S, 168°44.2–44.8′E, 1043–1056 m, FV *Arrow*, A03/48/83, 16 Oct 1983; NMNZ Cr4698, 1 ovigerous female (pcl 161.3 mm, cw 156.0 mm), Challenger Plateau, coll. R. Stewart, 30 Sep 1986.

Bay of Plenty: NMNZ Cr12036, 1 female (cl 104+ mm, pcl 69.6 mm, cw 68.5 mm), 37°22.80′S, 176°54.54′E, 1000 m, coll. G. Gibbs, 20 Nov 2001; NMNZ Cr14899, 1 male (cl 41.7 mm, pcl 20.1 mm, cw 18.3 mm), N of Lottin Point, 37°28.39–28.09′S, 178°21.00–19.80′E, 926–853 m, BS689 (NZOI R47), 18 Jan 1979; NIWA 42886, 1 male (cl 194.8 mm, pcl 154.0 mm, cw 148.1 mm), 1 female (cl 189.8 mm, pcl 127.1 mm, cw 124.0 mm), Bay of Plenty, coll. Moana Pacific Fisheries, 2008; NMNZ, 1 female (pcl 140.7 mm, cw 138.9 mm), Bay of Plenty, 37°17.5′S, 176°58.67′E, 1000 m, crab pot, FV *Savannah*, coll. G. Gibbs, 20 Nov 2001.

East Cape: NMNZ Cr11714, 1 male (cl 182.7 mm, pcl 141.6 mm, cw 148.0 mm), off East Cape, 37°24.9′S, 179°10.0′E, 1014–1109 m, RV *Wanaka*, WK01/78/85, 5 Jul 1985; NMNZ Cr4875, 1 male (cl 176.4 mm, pcl 131.5 mm, cw 129.7 mm), E of East Cape, 37°40.4–39.1′S, 179°11.4–10.1′E, 1123–1125 m, FV *Wanaka*, WK3/23/85, coll. M. Clark, 10 Dec 1985; NMNZ, 1 male (cl 69.0 mm, pcl 35.9 mm, cw 33.3 mm), 37°40.4–38.9′S, 179°25.7–27.5′E, 911–1106 m, trawl, Trip 774, tow 15, coll. A. France, 14 Jul 1994; NMNZ Cr11721, 1 male (pcl 118.7 mm, cw 114.6 mm), SE of East Cape, 37°53.9–57.6′S, 178°55.8–56.8′E, 952–1052 m, FV *Twofold Bay*, coll. N. Bagley, 6 Jun 1985.

Hawkes Bay: NMNZ Cr11740, 1 female (pcl 131.4 mm, cw 127.0 mm), Ritchie Bank, 39°29.8′S, 178°23.4′E, 900–947 m, orange roughy trawl, FV *Arrow*, A01/70/87, 7 Jul 1987.

Louisville Ridge: NMNZ Cr11729, 1 ovigerous female (cl 145.0 mm, pcl 96.2 mm, cw 84.8 mm), 40°58′S, 165°04′W, 800 m, FV *Peterson*, coll. M. Miranovich, 2 Mar 1995.

Wairarapa coast: NMNZ Cr14895, 1 juvenile female (cl 63.5+ mm, pcl 38.8 mm, cw 34.0 mm), 30.5 nautical miles [56 km] off Cape Turnagain, 1006–1189 m, FV *Twofold* Bay, A.D. Carruthers, 28 Dec 1984; NMNZ Cr11727, 1 female (cl 170.0 mm, pcl 125.5 mm, cw 121.2 mm), off Cape Palliser, 40–41°30′S, 176°00–30′E, 803–1170 m, J18/-/84; NMNZ Cr12030, 1 male (pcl 73.3 mm, cw 71.5 mm), off Wairarapa, 41°24.7–23.8′S, 176°08.4–12.1′E, 1103–1071 m, bottom trawl, RV *James Cook*, J18/4/84, 16 Oct 1984; NIWA 34920, 1 male (cl 42.9, pcl 21.0 mm, cw 19.6 mm), 41°46.95–46.83′S, 175°23.98–24.24′E, 1040–1053 m, TAN0616/79, 13 Nov 2006; NIWA 29294, 1 damaged female (pcl 21.3 mm), 41°46.96–46.87′S, 175°23.98–24.13′E, 1053–1050 m, TAN0616/83, 13 Nov 2006; NMNZ Cr11601, 1 male (cl 219.0 mm, pcl 163.7 mm, cw 165.9 mm), off Wairarapa coast, 700 m, D. Wylie, 17 Oct 1986; NMNZ Cr11584, 1 ovigerous female (pcl 165.0 mm, cw 155.5 mm), off Wairarapa coast, 700 m, coll. D. Wylie, 17 Oct 1986.

Marlborough coast: NMNZ Cr14896, 1 female (pcl 19.2 mm, cw 17.3 mm), 1 crushed male (pcl approx. 18.0 mm), 35 km E of Clarence River mouth, 42°16.30′S, 174°20.80′E, 860–790 m, BS665, NZOI R23, "pink spines, legs and chelae, white body, black eyes", 14 Jan 1979; NIWA 34921, 1 ovigerous female (cl 164.3 mm, pcl 116.5 mm, cw 114.9 mm), 42°29.10–28.87′S, 173°36.91–36.28′E, 1079–1023 m, TAN0616/99, 15 Nov 2006.

Chatham Rise: NMNZ Cr12029, 1 female (pcl 55.7 mm, cw 54.4 mm), NW of Chatham Islands, 42°42.0–41.7′S, 178°01.0–03.7′W, 1025–1055 m, COR/223/89, 13 Aug 1989; NMNZ Cr11734, 1 male (pcl 66.9 mm, cw 64.9 mm), 42°44.3–44.0′S, 177°38.4–50.1′W, 1158–1121 m, Trip 2014, tow 131, J. Yeoman, 20 Feb 2005; NMNZ Cr6560, 1 male (pcl 88.7 mm, cw 86.5 mm), NW of Chathams, 42°46.2–46.5′S, 177°41.7–44.6′W, 988–993 m, FV *Cordella*, COR/221/89, 13 Aug 1989; NIWA 34901, 1 male (cl 103.0 mm, pcl 58.1 mm, cw 57.7 mm), 42°47.17–45.66′S, 179°52.51–52.57′W, 978–1030 m, TAN9908/25, Z10189, 21 Jun 1999; NIWA 34922, 1 ovigerous female (cl 136.7 mm, pcl 99.0 mm, cw 96.3 mm), 42°47.22–47.16′E, 176°43.33–42.66′W, 996–1009

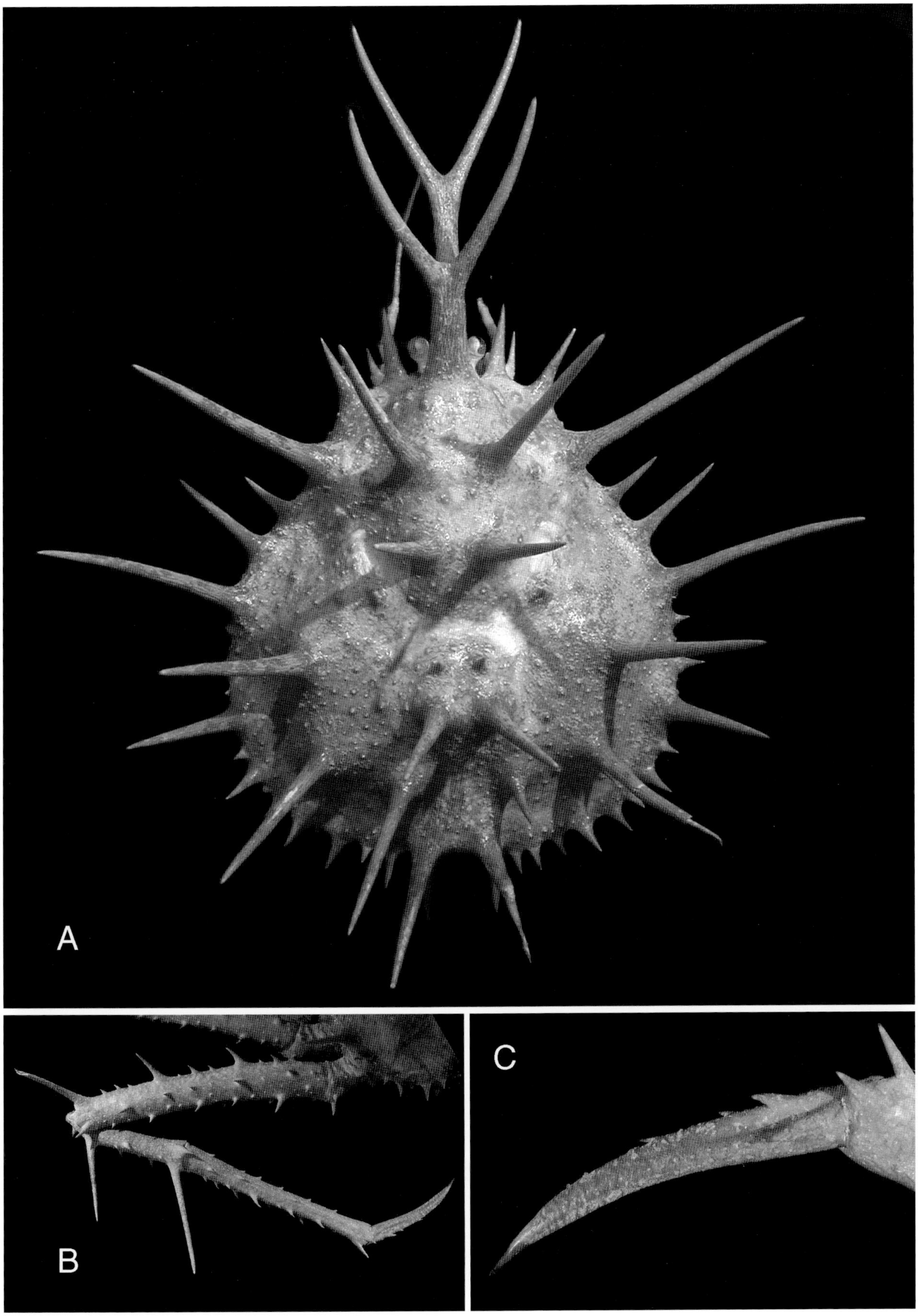

Figure 6. *Lithodes aotearoa* sp. nov., male holotype, cl 152.3 mm, pcl 88.6 mm, cw 85.3 mm, off Hawkes Bay (NIWA 34924). A, carapace. B, left pereopod 4, dorsal. C, left pereopod 4 dactylus.

m, TAN0705/163, 16 Apr 2007; NIWA 33752, 1 male (cl 212.0 mm, pcl 181.3 mm, cw 185.0 mm), 42°52.77–52.93′S, 176°27.54–25.52′E, 838–844 m, TAN0709/6, 7 Jul 2007; NIWA 42895, 1 male (pcl 148.0 mm, cw 149.9 mm), 42°58.89–58.15′S, 174°14.86–18.84′E, 988–993 m, TAN0701/124, 20 Jan 2007; NIWA 48960, 1 male (pcl 190.4 mm, cw 196.1 mm), 43°31.1′S, 174°38.8′E, 512–547 m, MFish Observer, trip 2814, tow 81, FV *San Enterprise*, coll. Ted Turton, 2009; NIWA 48961, 1 male (pcl 189.0 mm, cw 189.7 mm), 43°31.1′S, 174°38.8′E, MFish Observer, trip 2814, tow 81, FV *San Enterprise*, coll. Ted Turton, 2009; NMNZ, 1 male (cl 198.2 mm, pcl 148.1 mm, cw 152.6 mm), 44°00.5–00.8′S, 178°40.5–43.7′E, 775 m, FV *Kalinovo* K01/063/81, 3 Dec 1981; NIWA 48962, 1 male (cl 215.0 mm, pcl 178.9 mm, cw 184.5 mm), 44°02′S, 174°46′E, MFish Observer, trip 2814, tow 82, FV *San Enterprise*, coll. Ted Turton, 2009; NIWA 61189, 1 ovigerous female (pcl 117.4 mm, cw 115.3 mm), 44°03.78–03.39′S, 178°57.10–55.32′E, 757–765 m, TAN9901, Z9615, 13 Jan 1999; NIWA 34612, 1 male (cl 128.4 mm, pcl 80.9mm, cw 79.2 mm), 44°05.73–06.63′S, 174°43.55–45.23′W, 1051–1020, TAN0709/109, 21 Jul 2007; NIWA 34898, 1 male (cl 81.4 mm, pcl 45.5 mm, cw 43.6 mm), 44°06.49–07.69′S, 178°32.50–37.99′E, 906–926 m, Z10853, 7 Apr 1999; NIWA 34905, 1 juvenile female (pcl 31.9 mm, cw 29.4 mm), 44°06.58–06.81′S, 178°27.10–27.81′E, 963–955 m, TAN0011/043, 3 Nov 2000; NIWA 42894, 1 male (pcl 140.5 mm, cw 131.0 mm), 44°07.29–07.99′S, 179°02.27–06.35′W, 361–364 m, TAN0701/62, 8 Jan 2007; NMNZ Cr11761, 1 female (cl 199 mm, pcl 168.0 mm, cw 161.0 mm), 44°0.15′S, 178°31.25′E, 775–790 m, FV *Kalinovo*, K01/064/81; NMNZ Cr6561, 1 male (pcl 138.8 mm, cw 140.4 mm), W of Chathams, 44°17.2–17.4′S, 179°56.0–59.2′W, 828–840 m, FV *Cordella*, COR/204/89, 9 Aug 1989; NIWA 60582, 1 male (pcl 179.2 mm, cw 178.5 mm), 44°22.83–22.66′S, 179°54.68–56.27′E, 952–957 m, FV *San Waitaki*, SWA0901/82, 16 Nov 2009; NIWA 61159, 1 juvenile male (pcl 42.2 mm, cw 41.1 mm), 44°22.83–22.66′S, 179°54.68–56.27′E, 952–957 m, FV *San Waitaki*, SWA0901/82, 16 Nov 2009; NIWA 60566, 1 male (cl 206.2 mm, pcl 169.0 mm, cw 171.1 mm), 44°25.78′S, 178°42.07–41.67′W, 697–812 m, FV *San Waitaki*, SWA0901/74, 15 Nov 2009; NIWA 60567, 1 female (cl 230.5 mm, pcl 190.5 mm, cw 183.9.1 mm), 44°25.78′S, 178°42.07–41.67′W, 697–812 m, FV *San Waitaki*, SWA0901/74, 15 Nov 2009; NIWA 42893, 1 ovigerous female (cl 158.0 mm, pcl 118.9 mm, cw 116.0 mm), 44°26.80–26.73′S, 175°37.05–32.85′E, 760–779 m, TAN0701/86, 13 Jan 2007; NMNZ Cr11693, 1 male (pcl 170.0 mm, cw 173.3 mm), 44°27.9′S, 173°40.3′E, 727 m, FV *Shinkai Maru*, net 32, 8 Mar 1983; NIWA 60553, 1 ovigerous female (cl 191.4 mm, pcl 145.4 mm, cw 137.0 mm), 1 female (cl 141.1 mm, pcl 99.2 mm, cw 98.8 mm), 44°32.78–33.62′S, 175°23.73–23.77′S, 939–985 m, SWA0901/61, FV *San Waitaki*, 13 Nov 2009; NMNZ Cr14900, 1 female (pcl 28.7 mm, cw 26.3 mm), SW of Chatham Islands, 44°33.65′S, 177°34.07′W, 895–910 m, COR9004/152, FV *Cordella*, 21 Nov 1990; NIWA 60555, 1 juvenile female (pcl 29.5 mm, cw 27.0 mm), 44°39.75–40.42′S, 177°29.43–29.48′W, 1024–1043 m, FV *San Waitaki*, SWA0901/013, 6 Nov 2009; NIWA 34915, 1 female (pcl 87.8 mm, cw 85.8 mm), 44°48.75–48.49′S, 172°32.2–35.02′E, 921–934 m, TAN9511/001, 5 Oct 1995; NMNZ Cr9213, 2 juvenile females (pcl 29.2–32.1 mm, cw 26.6–28.9 mm), 44°50.6′S, 175°29.4′E, 1044–1050 m, RV *James Cook*, JC090015/5, 20 Sep 1990; NMNZ Cr11145, 1 juvenile female (cl 54.5 mm, pcl 25.7 mm, cw 23.1 mm), 44°50.6′S, 175°29.4′E, 1044–1050 m, RV *James Cook*, JC090015/5, 20 Sep 1990; NIWA 60530, 1 male (pcl 50.3, cw 49.3 mm), 42°48.89–48.33′S, 179°30.57–28.53′E, 1042–1049 m, TAN0509/11, 21 Jun 2005..

Stewart-Snares Shelf: NMNZ Cr1112, 1 male (cl 209.0 mm, pcl 167.8 mm, cw 171.9 mm), SE of Solander Island, Foveaux Strait, in crayfish pot, 382 fathoms [699 m], coll. C. de Rich, Nov. 1960; NIWA 60557-60561, 2 males (pcl 123.4–143.0 mm, cw 124.8–141.9 mm), 3 ovigerous females (pcl 121.1–138.4 mm, cw 115.6–137.9 mm), W of Stewart Island, 46°37.16′S, 166°37.80′E, 819–832 m, TAN0911/78, 15 Dec 2009.

Bounty Trough: NMNZ Cr11582, 1 male (cl 209.8 mm, pcl 161.8 mm, cw 156.5 mm), Bounty Plateau, 47°47.02–49.65′S, 179°25.17–27.40′W, 596–562 m, TAN9310/001, 15 Nov 1993.

Bounty Plateau: NIWA 61188, 1 male (cl 226.3 mm, pcl 183.7 mm, cw 180.0 mm), SE of Bounty Islands, 48°05.99′S, 179°56.99′W, 252 m, FIS *Endeavour*, F122, 26 Jan 1965; NMNZ Cr11583, 1 male (cl 194.0 mm, pcl 157.4 mm, cw 162.2 mm), 48°20.74–21.58′S, 179°53.71–57.42′W, 538–541 m, RV *Tangaroa*, TAN9310/002, 15 Nov 1993; NMNZ Cr11733, 1 male (cl 165.9 mm, pcl 117.0 mm, cw 113.4 mm), Bounty Plateau, 48°22.5′S, 179°56.0′E, 547–561 m, RV *Shinkai Maru*, Cruise II, prob. Stn 2/80, 17 Nov 1975; NIWA 60616, 1 juvenile female (cl 32.3 mm, pcl 13.4 mm, cw 10.8 mm), NE of Antipodes Islands, 49°18.63–18.06′S, 179°47.88–48.33′E, 1506–1476 m, TAN0307/59, 29 Apr 2003.

Campbell Plateau: NIWA 42799, 1 male (pcl 102.5 mm, cw 97.4 mm), 46°57.58–57.04′S, 170°11.11–15.43′E, 855–946 m, TAN9805/02, 7 Apr 1998; NIWA 60565, 1 male (pcl 151.0 mm, cw 150.5 mm), 1 female (pcl 128.6 mm, cw 121.8 mm; with *Briarosaccus callosus* Boschma, Rhizocephala), 48°13.47′S, 169°53.74′E, 802–831 m, TAN0911/5, 28 Nov 2009; NIWA 60551, 1 male (cl 240.0 mm, pcl 195.0 mm, cw 199.1 mm), Pukaki Rise, 48°45.42′S, 172°35.90′E, 700–717 m, TAN0911/11, 20 Nov 2009; NIWA 34891, 1 male (cl 233.8 mm, pcl 194.0 mm, cw 198.6 mm), 49°15.99′S, 168°42.99′E, 785 m, Z9312, 15 Sep 1998; NMNZ Cr4858, 1 female (cl 81+ mm, pcl 64.4 mm, cw 61.9 mm), SW of Urry Bank, 49°30.6–39.54′S, 174°08.6–01.8′E, 798 m, FV *Oyang*, #7, Tow 5, coll. C.D. Roberts, 7 May 1987; NMNZ Cr11694, 1 male (pcl 106.6 mm, cw 108.6 mm), SE of Campbell

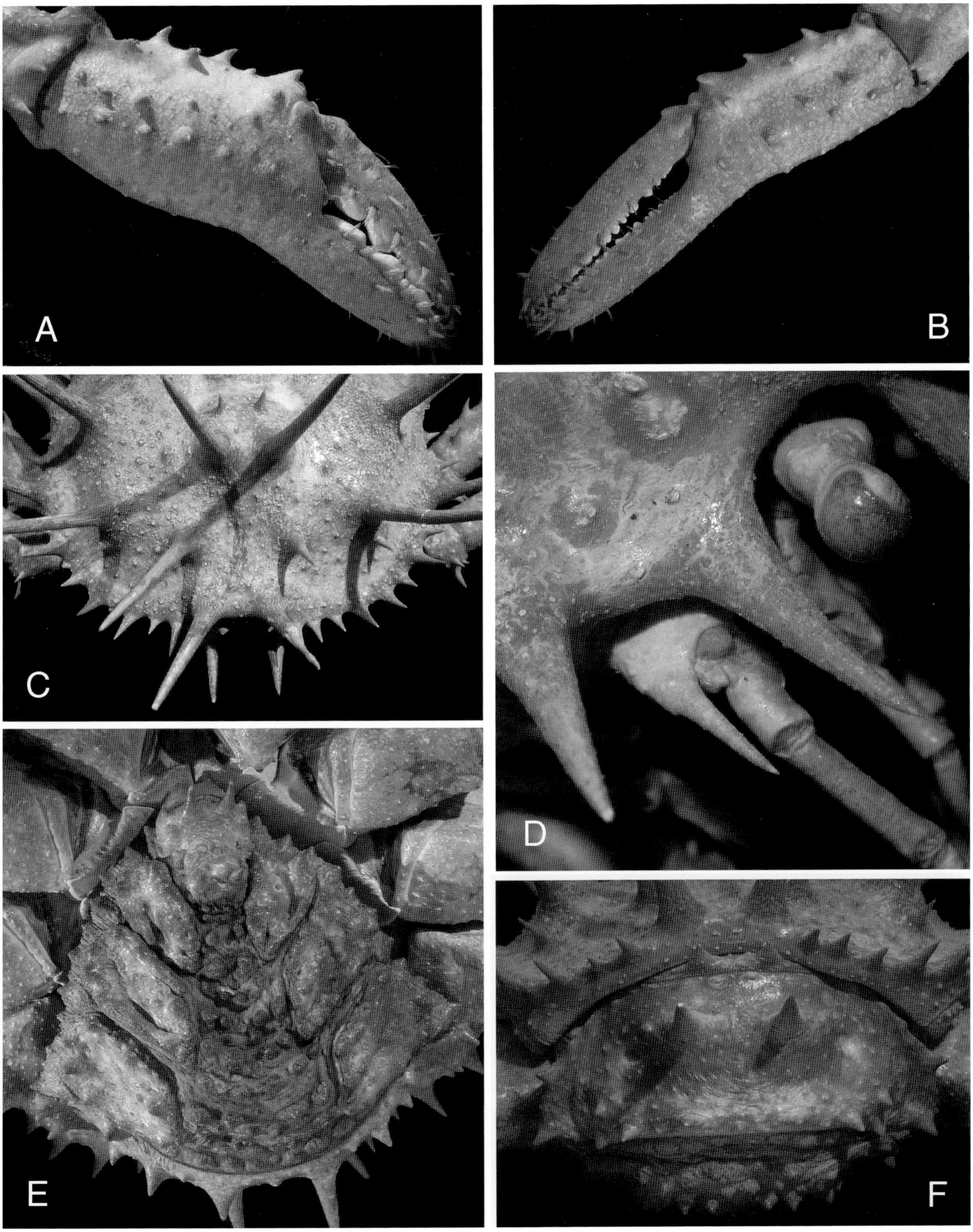

Figure 7. *Lithodes aotearoa* sp. nov., male holotype, cl 152.3 mm, pcl 88.6 mm, cw 85.3 mm, off Hawkes Bay (NIWA 34924). A, right chela. B, left chela. C, carapace, posterior surface. D, right antenna. E, abdomen. F, abdominal somite 2.

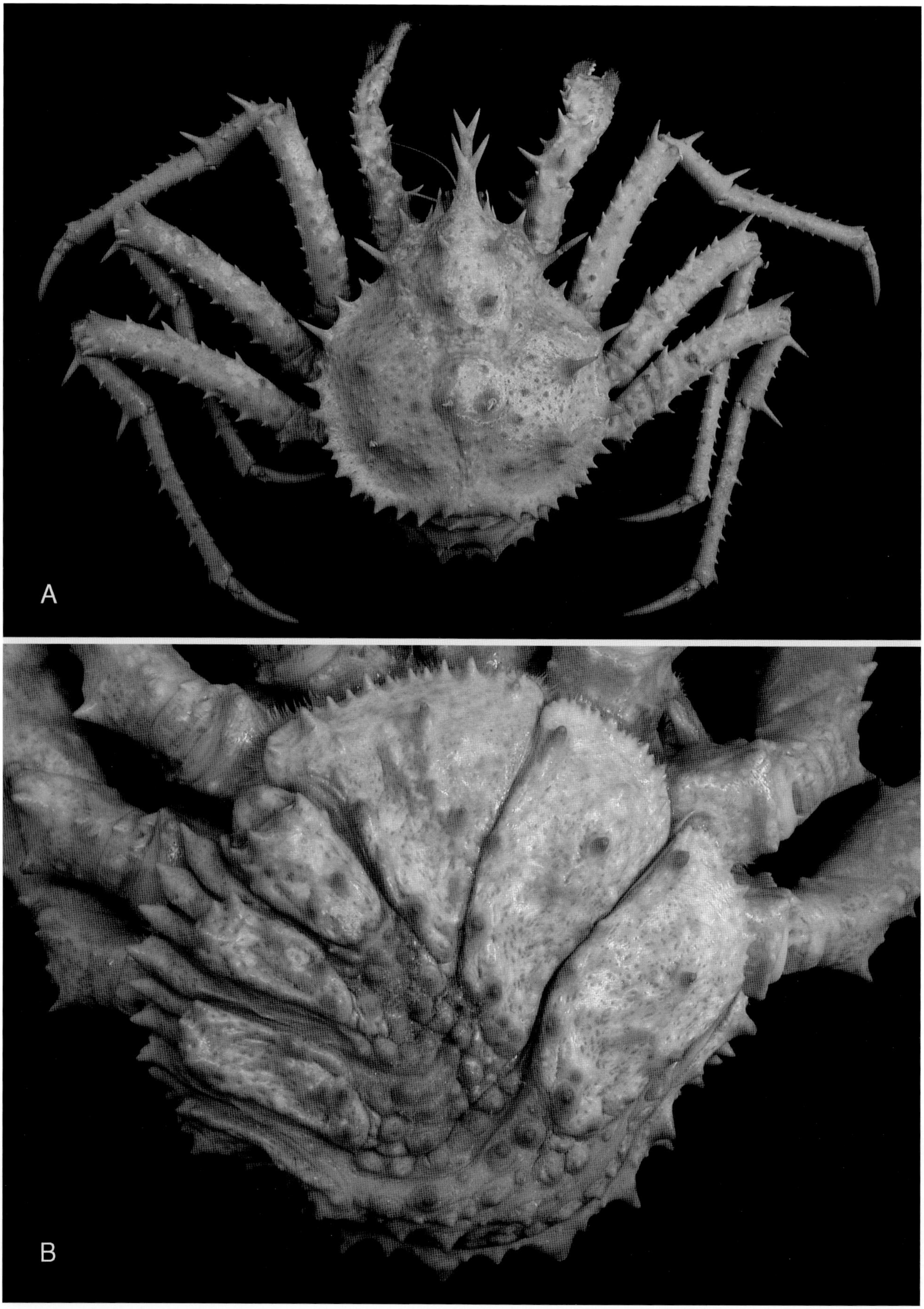

Figure 8. *Lithodes aotearoa* sp. nov., ovigerous female, pcl 125.5 mm, off Cape Palliser (Cr11727). A, dorsal habitus. B, abdomen.

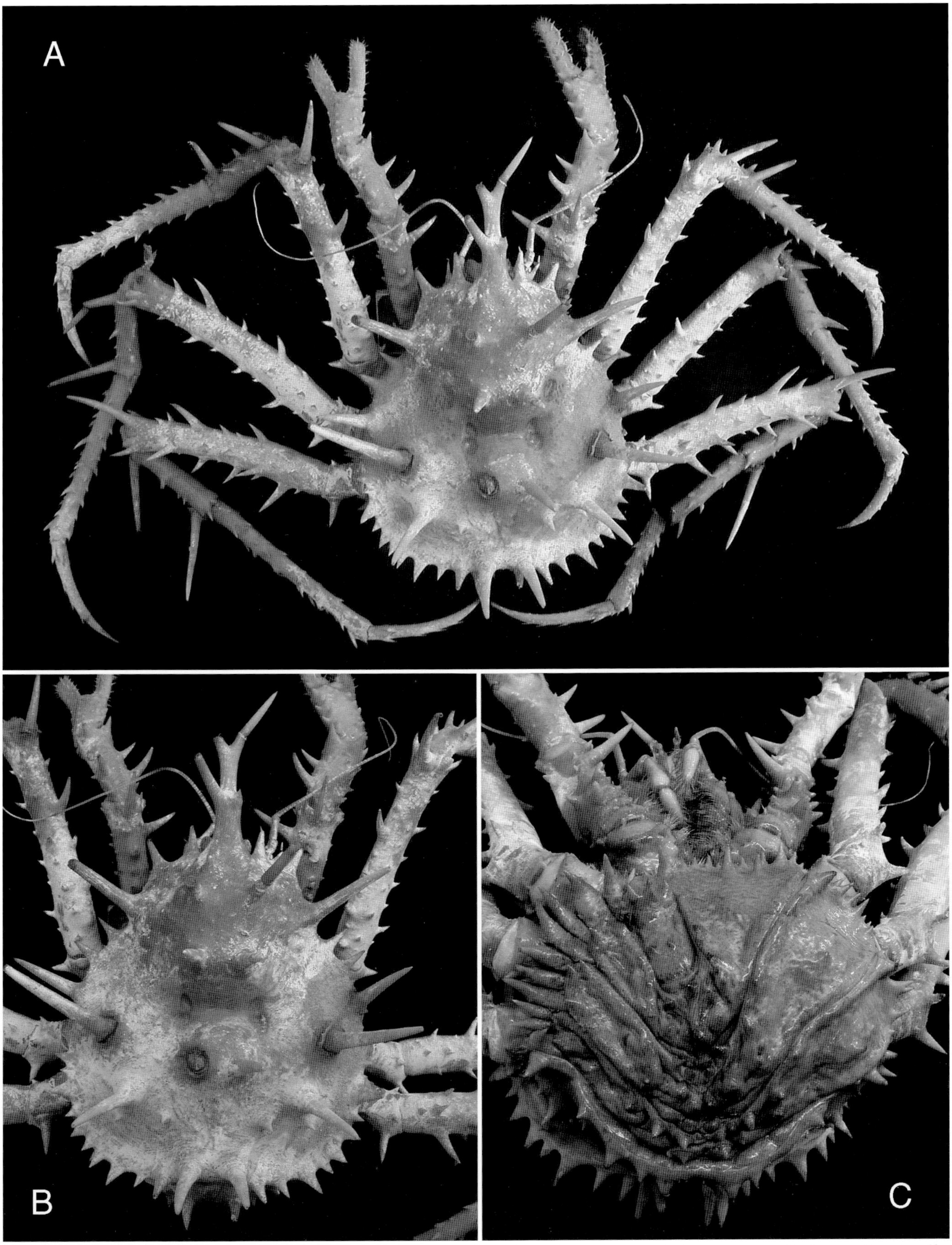

Figure 9. *Lithodes ?aotearoa* sp. nov., aberrant ovigerous female, cl 145.0 mm, pcl 96.2 mm, cw 84.8 mm, Louisville Ridge (Cr11729). A, dorsal habitus. B, carapace. C, abdomen.

Island, 53°S, 170°E, 690–875 m, stn 139 & 140, FV *Wessermunde*, 13 May 1979; NMNZ Cr11695–11696, 1 male (pcl 87.7 mm, cw 83.6 mm), 1 female (pcl 80.5 mm, cw 81.1 mm), SE of Campbell Island, 53°46.7–48.5′S, 168°27.5–35.0′E, 870–880 m, trawl, FV *Wessermunde*, stn 136, 12 May 1979; NMNZ Cr7352, 1 male (cl 206.8 mm, pcl 153.3 mm, cw 156.5 mm), SE of Campbell Island, 53°28.0′S, 171°22.2′E, 420 m, FV *Ryuyo Maru*, JQET, tow 168, coll. H. Kavale, 30 Aug 1987.

New Zealand, no specific locality: NMNZ Cr11728, 1 male (pcl 170.0 mm, cw 178.0 mm), between Ninety Mile Beach and Wairarapa, RV *Kalinovo*, K01/?/81, 1981; NMNZ Cr14898, 1 male (pcl 31.7 mm, cw 29.4 mm), SE New Zealand, from MAF, 1985; NIWA 34904, 1 female (cl 82.3 mm, pcl 40.8 mm, cw 37.2 mm), New Zealand, MAF, 16 Jan 1989; NMNZ Cr11252, 1 male (cl 178.7 mm, pcl 140.0 mm, cw 129.4 mm; with *Briarosaccus callosus* Boschma, Rhizocephala), New Zealand, no other data; NMNZ Cr11253, 1 female (pcl 154.3 mm, cw 132.8 mm; with *Briarosaccus callosus* Boschma, Rhizocephala), New Zealand, no other data; NMNZ Cr11735, 1 male (cl 185.6 mm, pcl 143.0 mm, cw 146.3 mm), floating off Wellington wharf, coll. J. Annala, Dec 1985.

Diagnosis. Carapace dorsal surface with slender spines becoming reduced to short conical spines above about 120 mm pcl (usually about 0.1 pcl); gastric region with 4 upright spines, anterior pair longer than half length of hepatic spine, posterior pair half length of hepatic spine; cardiac and intestinal regions each with 2 long upright spines; branchial surface with 2 or 3 long upright spines. Branchial margins spinose; 1st primary marginal branchial spine subequal to or shorter than hepatic spine, 2nd primary marginal branchial spine about half length of hepatic spine or less; posterior branchial margin with 8–12 spines. Rostrum not exceeding 0.7 pcl; proximal half strongly upraised by 30–45°. Abdominal somite 2 comprising single plate in adults. Antennal peduncle article 2 with outer spine not reaching midlength of article 4. Chelipeds unequal; palms with low tubercles or spines on dorsal surface. Walking legs spinose; surface between major spines smooth or with very few, scattered spines. Pereopod 4 merus about as long as or shorter than pcl, length about 7 times height or less. Dactylus 0.5–0.7 propodus length; flexor margin unarmed.

Description. *Carapace*: Pyriform, 0.97–1.14 times longer than wide; regions indicated; dorsal surface armed with long, slender spines (generally reduced to short conical spines of about 0.1 pcl above about 120 mm pcl) and scattered, widely separated granules or minute secondary spines, surface otherwise smooth. Gastric region convex, with 2 pairs of long, upright spines. Cardiac region with pair of anterior tubercles or small spines and posterior pair of long upright spines. Hepatic spine longest of carapace spines (0.6 pcl in holotype), directed anterolaterally. Intestinal region with pair of long upright spines on posterior margin. Branchial surface with 2 or 3 long upright spines as follows: 1 spine at level of pereopod 3 coxa, 1 or 2 in transverse row at level of pereopod 4 coxa, innermost varying from a spiniform tubercle to as long as outer spine, outer spine often with short spine at base. Branchial margins spinose; anterior branchial margin with 2 spines, posterior spine longer than anterior; 1st primary marginal branchial spine subequal to or shorter than hepatic spine; lateral branchial margin with 2 or 3 short spines; 2nd primary marginal branchial spine quarter to half length of hepatic spine; posterior branchial margin with 8–12 spines.

Pterygostomian region granulate or tuberculate, with small, anterior, submarginal spine.

Rostrum 0.18–0.71 pcl (becoming proportionally shorter with increasing size), comprising proximal and distal portions; proximal portion angled dorsally by at least 30°, with pair of long, divergent, dorsal spines that overreach midlength of distal portion of rostrum; distal portion subhorizontal, distally bifurcate for one-third to two-thirds length, forming pair of strongly divergent spines; ventral rostral spine elongate, extending anteriorly beyond apices of outer orbital spine. Posterior orbital margin concave; outer orbital spine directed anteriorly, about as long as anterolateral spine, reaching beyond cornea.

Ocular peduncle: Longer than cornea; unarmed or with indistinct, scattered granules.

Antennule: Peduncle unarmed, reaching anteriorly beyond antennal peduncle by about half length of distal antennular peduncle article.

Antenna: basal antennal article unarmed; outer margin of article 2 with long slender spine, not reaching beyond proximal quarter of article 5; article 3 unarmed; scaphocerite minute, blunt, shorter than article 4 (or produced to a slender spine on right side in one specimen, male, NMNZ Cr11735); article 4 unarmed, about half as long as article 5.

Abdomen: Somite 1 with triangular teeth on posterior margin. Somite 2 sparsely granulate, submedian plates fused with median by 40–60 mm pcl; median surface with pair of slender submedian spines; posterior margin with 6 stout spines or angular prominences; lateral margins irregularly dentate. Somites 3–5 with nodular median plates; surface of submedian plates irregular, nodular, with scattered acute tubercles; marginal plates of males subdivided, with angular or sharp apices; marginal plates of females with left margin crenulate to spinose (fused with submedians), right marginals subdivided, with angular to spiniform apices. Somite

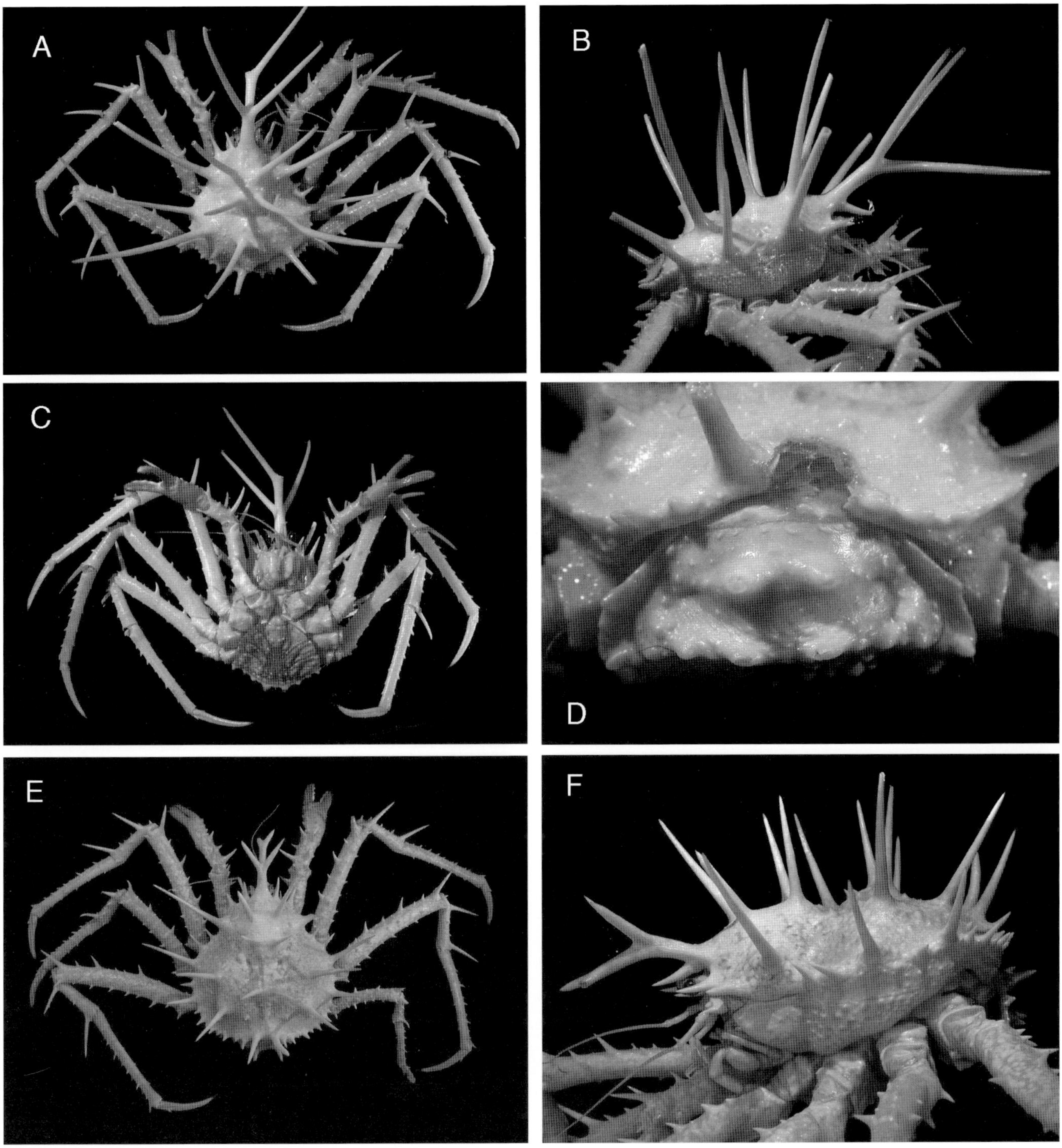

Figure 10. *Lithodes aotearoa* sp. nov. A–D, juvenile female, pcl 25.7 mm, Chatham Rise (NMNZ Cr11145). E–F, female, pcl 69.6 mm, Bay of Plenty (NMNZ Cr12036). A, E, dorsal view. B, F, lateral view. C, ventral view. D, abdominal somite 2.

6 of both sexes as long as or longer than wide, with 2 proximal and 4 distal spines or prominences. Telson rounded, unarmed, at most with 4 blunt granules.

Pereopod 1 (chelipeds): Dimorphic but spination similar. Surfaces of both chelipeds smooth or sparsely covered with granules in addition to major spines. Coxa with blunt tubercles and tufts of setae, unarmed; ischiobasis with 3 stout ventral spines and about 6 blunt tubercles. Merus inner margin, with slender subdistal spine and 2 or 3 smaller spines; ventral margin with two rows of 2 or 3 low conical spines; dorsal and lateral surface spinose, spines largest distally. Carpus with prominent spines on dorsal and lateral surfaces; dorsal margin with row of 4 stout spines; lateral margin with 4 or 5 stout spines of similar size to dorsal row; ventral margin small scattered acute tubercles. Palms of both chelipeds in both sexes with similar ornamentation; with small spines and acute tubercles on dorsal, lateral and ventral surfaces, inner surface with acute tubercles; dorsal margin with row of about 5 small conical spines; midlateral surface with 2 rows of 3 or 4 small spines, of similar size to dorsal row; ventral surface with scattered tubercles, smaller than lateral and dorsal spines.

Major cheliped 1.17–1.45 pcl (male), 1.09–1.36 pcl (female); upper palm length 0.96–1.39 times height (male), 1.14–1.48 (female); occlusal margins of fingers corneous for distal third, proximally with 3 calcareous nodules; dactylus dorsal margin broadly convex, with rows of tufts of golden setae and blunt proximal tooth, 1.21–1.42 times longer than dorsal margin of palm (male), 1.25–1.62 (female).

Minor cheliped 1.22–1.36 pcl (male), 0.96–1.35 pcl (female); upper palm length 1.21–1.53 times height (male), 1.19–1.57 (female); occlusal margin corneous for slightly less than distal half, proximally crenulate; dactylus dorsal margin broadly convex, with rows of tufts of golden setae and blunt proximal tooth, 1.61–1.72 times longer than dorsal margin of palm (male), 1.59–1.83 (female).

Pereopods 2–4 (walking legs 1–3): Similar, slender, elongate; ornamentation similar in both sexes; segments spinose, surface between major spines smooth or with scattered spines or granules. Pereopod 4 longest. Distal margins of coxae with low, blunt tubercles. Ischiobasis with 3 or 4 non-setose distal spines. Merus subcircular to ovate in cross section; extensor margin with 8–10 spines of which 2 or 3 are distinctly longer in addition to long, prominent, distal spine; flexor margin with 2 rows of 5–8 spines; upper surface with row of 6–8 spines. Carpus extensor margin with distal and second proximal spines longest, exceeding two-thirds carpus length; surface scattered with 8 or 9 small spines. Propodus ovate in cross section; with 7–9 spines on extensor margin and 6–8 spines on dorsal surface; flexor margin with 6 or 7 smaller spines. Dactylus curved, rounded in cross section distally; proximally with 2 spines on either side adjacent to articulation; extensor margin with 1 or 2 small spines ; flexor margin smooth, unarmed, with few widely spaced setae; apex corneous.

Pereopod 2 length 2.34–2.66 pcl (male), 1.78–2.54 pcl (female). Merus 0.83–0.96 pcl (male), 0.58–0.90 pcl (female); length:height ratio 5.21–7.22 (male), 3.77–6.21 (female). Carpus 0.50–0.55 merus length (male), 0.51–0.62 (female). Propodus 0.78–0.87 merus length (male), 0.84–0.96 (female); length:height ratio 8.42–12.29 (male), 7.17–10.62 (female). Dactylus 0.52–0.66 propodus length (male), 0.53–0.64 (female).

Pereopod 3 length 2.81–2.66 pcl (male), 1.89–2.72 pcl (female). Merus 0.94–1.05 pcl (male), 0.63–0.96 pcl (female); length:height ratio 5.72–7.34 (male), 4.20–6.35 (female). Carpus 0.50–0.53 merus length (male), 0.50–0.61 (female). Propodus 0.79–0.88 merus length (male), 0.86–0.92 (female); length:height ratio 8.74–11.45 (male), 7.85–10.77 (female). Dactylus 0.50–0.60 propodus length (male), 0.54–0.61 (female).

Pereopod 4 length 2.68–2.98 pcl (male), 1.89–2.77 pcl (female). Merus 0.91–1.06 pcl (male), 0.62–0.95 pcl (female); length:height ratio 5.56–7.16 (male), 4.55–6.43 (female). Carpus 0.52–0.55 merus length (male), 0.53–0.64 (female). Propodus 0.87–0.92 merus length (male), 0.89–0.96 (female); length:height ratio 9.12–13.04 (male), 8.56–10.41 (female). Dactylus 0.49–0.66 propodus length (male), 0.57–0.62 (female).

Egg size: 2.2–2.3 mm diameter.

COLOUR IN LIFE. Usually deep-purplish red overall, whitish along cervical groove, often with whitish or pinkish blotches on pereopods and carapace; occasionally pale pinkish-red or whitish overall in largest specimens (Pl. 1A, B).

ETYMOLOGY. Named *aotearoa*, for the occurrence of the species in New Zealand; used as a noun in apposition.

REMARKS. For several decades, *Lithodes murrayi* Henderson, 1888 (type locality: Prince Edward Island, southwest Indian Ocean) and *L. longispina* Sakai, 1971 (type locality: Japan) were believed to occur in New Zealand waters (Dell 1963; Yaldwyn & Dawson 1970; Dawson & Yaldwyn 1985). Examination of collections on which New Zealand records of both species were based, however, revealed that neither *L. murrayi* nor *L. longispina* occurs in the region. Moreover, the New Zealand specimens previously reported as *L. murrayi* and *L. longispina* are referable to a single undescribed species, *L. aotearoa* sp. nov. The initial New Zealand record of *L. murrayi* was based on a large specimen (pcl 167.8 mm, NMNZ Cr1112) from Foveaux Strait with relatively short dorsal spines (Dell 1963; Yaldwyn & Dawson 1970), though still considerably longer than those of the much smaller type specimens of *L. murrayi* from the southwestern Indian Ocean (pcl 64–72 mm). New Zealand records of *L. longispina* were based on specimens smaller than 120 mm pcl in which the dorsal spines are greatly elongate as in similarly sized Japanese specimens of *L. longispina*. Over the past two decades, numerous New Zealand specimens of *Lithodes* became available through several sources including commercial fishermen, fishery surveys and other deep-

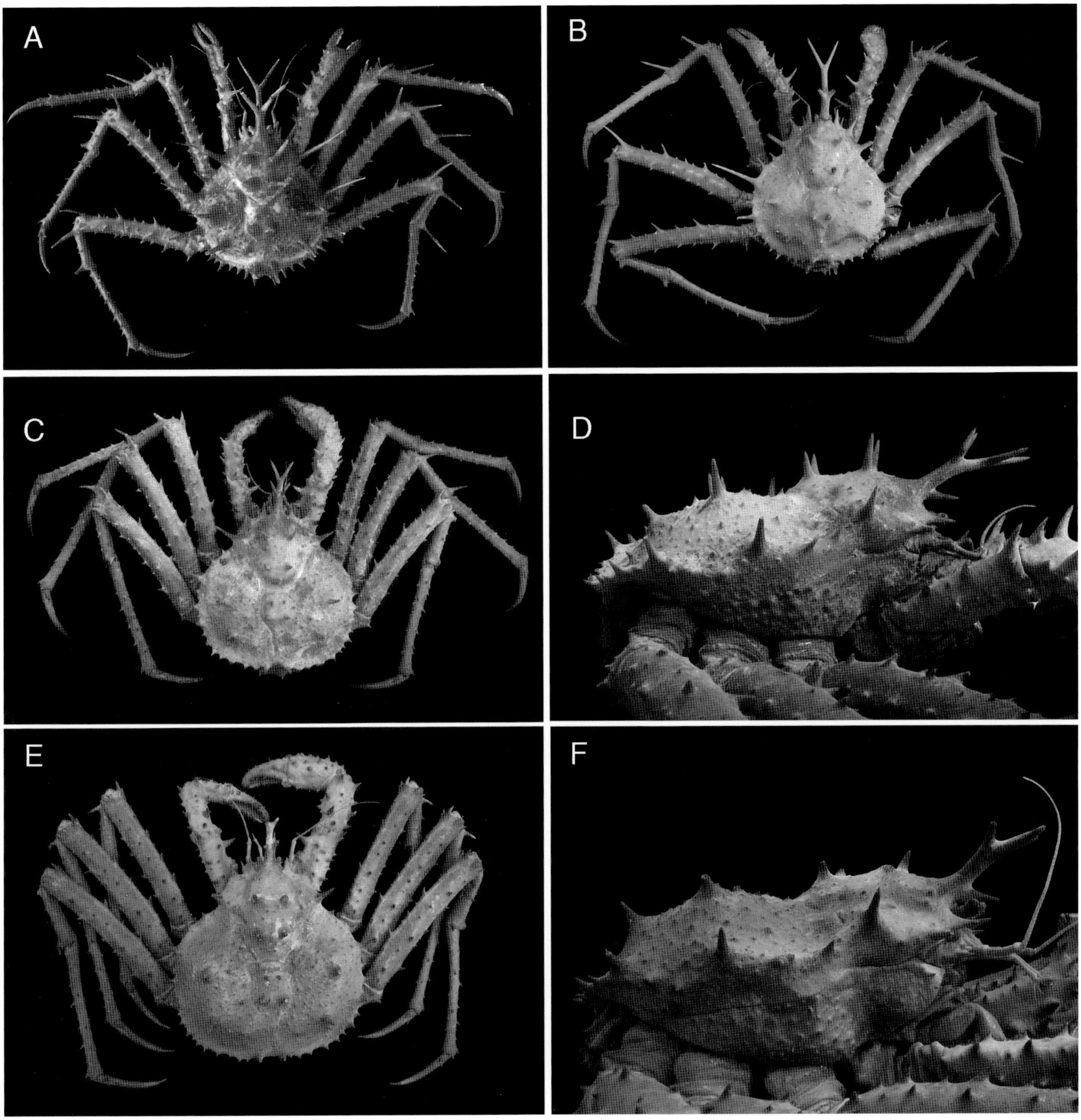

Figure 11. *Lithodes aotearoa* sp. nov. A, male, pcl 80.9 mm, Chatham Rise (NIWA 34612). B, male paratype, pcl 106.7 mm, off Hawkes Bay (NMNZ Cr11692). C–D, male, pcl 163.7 mm, off Wairarapa coast (NMNZ Cr11601). E–F, male, pcl 167.8 mm, SE of Solander Island, Foveaux Strait (NMNZ Cr1112). A–C, E, dorsal habitus. D, F, carapace, right lateral view.

water research programmes. The difficulty of assigning 'intermediate-sized' New Zealand specimens to either *L. murrayi* or *L. longispina* has been acknowledged (e.g. Webber & Naylor 2004a), but the authority afforded to the original identifications of New Zealand specimens has been such that their veracity has not been reassessed until now. As in other lithodids, particularly species of *Lithodes* and *Neolithodes*, the proportional length of the dorsal spines is inversely related to body size. Failure to account for allometry by previous workers has prevented recognition that New Zealand records of *L. longispina* and *L. murrayi* were based on phases of *L. aotearoa*. Despite significant allometric variation in lithodids, spine length at any given size varies between species. In *L. longispina* and *L. aotearoa*, long dorsal spines persist into maturity, though in the largest specimens, dorsal spines are correspondingly short. Thus, spine length remains a very useful diagnostic character providing body size is considered. *Lithodes murrayi*, a 'short-spined' species, attains little more than half the size of *L. aotearoa* (pcl 108 mm versus 194 mm), with spines always considerably shorter and stouter than those of the latter at any given size. The carapace spines of *L. aotearoa* are of similar form to adult *L. murrayi* only in specimens exceeding 170 mm pcl, well above the maximum size of the latter. Apart from dorsal spine shape and length, *L. murrayi* is readily distinguished from *L. aotearoa* in bearing 5 or 6 versus 8–12 posterior branchial marginal spines.

Lithodes aotearoa closely resembles *L. australiensis* sp. nov. and *L. rachelae* sp. nov. from Australia, *L. robertsoni* sp. nov. from New Zealand, and to a lesser extent *L. longispina* from Japan, with which the first two new species were previously misidentified (Dawson & Yaldwyn 1985; McLay 1988). The five species share the numerous, greatly elongated spines on the carapace and pereopods in specimens up to about 120 mm pcl, and the strongly upraised proximal portion of the rostrum. The new species differs from *L. longispina* in bearing more pronounced and more numerous carapace spines, especially on the branchial surface and margins: these differences are best observed in specimens of 100 mm pcl or less. *Lithodes aotearoa* bears 8–12 posterior branchial marginal spines versus versus 5 or 6 in *L. longispina*. The branchial surfaces of *L. aotearoa* bear two or three long dorsal spines of similar length to the cardiac spines: one anterior spine at the level of the second walking leg, and one or two posterior spines at the level of the third walking leg. In contrast, *L. longispina* bears only the long anterior branchial spine, with the posterior branchial spines replaced by short acute tubercles. In specimens up to 100 mm pcl, both *L. aotearoa* and *L. longispina* bear a long hepatic spine, two long primary marginal branchial spines, and a pair of long intestinal spines. In *L. longispina*, the anterior branchial margin is unarmed or bears only one or two minute spines, and the posterior branchial margin is armed with not more than six stout, well-spaced spines or tubercles of which two or three are longer than the remainder. In specimens of *L. aotearoa* exceeding 120 mm pcl, the major carapace spines are reduced in size, although the positions and numbers of spines are consistent with that of smaller specimens. Features distinguishing *L. aotearoa* from *L. rachelae*, *L. australiensis*, and *L. robertsoni* are outlined under the respective accounts of those species.

The dorsal spine length in *L. aotearoa* becomes proportionally shorter with increasing body size. The longest carapace spine (hepatic) is longer than the pcl at 17.3 mm pcl (NMNZ Cr14896), 0.9 pcl at 55.3 mm pcl (NIWA 34907), 0.6 pcl in the 88.6 mm pcl holotype, and 0.1 pcl in the 145.7 mm specimen (NMNZ Cr11149). Apart from reduction in spine size, other allometric changes include fusion of the median and marginal plates of abdominal somite 2 (fused by 40–60 mm pcl), the flexor spination of the dactyli of the walking legs (present in juveniles up to about 20 mm pcl), the relative length of the walking legs (proportionally longer with increasing size), and stoutness of the meri of the walking legs (becoming less slender with increasing size). In addition, leg length and meral stoutness is sexually dimorphic. At any given size, females have proportionally shorter walking legs and longer corresponding meri than males (by about 15–30%).

In the majority of specimens, including most adults, the rostral apex is deeply bifurcate. In several specimens, however, the apex is shallowly bifurcate as in *L. murrayi*, and in four specimens (NIWA 33752, 60557, 60567; NMNZ Cr11761), the apex is unilobate as in the holotype of *L. unicornis* Macpherson, 1984. Four specimens (NMNZ Cr11723, Cr11252–11253, NIWA 60565) were infected by the rhizocephalan parasite, *Briarosaccus callosus*. In all but one specimen, the scaphocerite is obsolete, represented by a small, blunt nodule. In a 142.8 mm pcl male (NMNZ Cr11735), the scaphocerite is produced to a sharp point on the left side, and developed as a slender spine on the right side.

An ovigerous female specimen from the Louisville Ridge (NMNZ Cr11729, Fig. 9) differs from other specimens of *L. aotearoa* in having three separate plates on abdominal somite 2 (rather than being fused into a single plate), seven posterior branchial marginal spines, more prominently spinose margins of the abdomen and proportionally longer dactyli of the walking legs (0.7 versus 0.5–0.6). In the three-segmented abdominal somite 2 and posterior branchial marginal spines, the specimen resembles *L. robertsoni* sp. nov. Apart from the long walking leg dactyli, however, limb morphometry of the Louisville Ridge specimen agrees well with *L. aotearoa*, and thus differs from *L. robertsoni*, and *L. australiensis*. The aberrant Louisville Ridge specimen could represent an undescribed species but until further material becomes available, the specimen is tentatively assigned to *L. aotearoa*.

Lithodes aotearoa is the largest species of lithodid in New Zealand waters, attaining carapace dimensions of at least cl 240.0 mm, pcl 195.0 mm and cw 199.1 mm, with a leg span of at least 1.3 m (NIWA 60551). Females are ovigerous by at least 116.5 mm pcl (NIWA 34921), but appear to be mature by about 90 mm pcl based on abdomen development and the presence of coxal setae.

Distribution. New Zealand, from the southern West Norfolk Ridge south to The Snares, including the Challenger Plateau, Bay of Plenty, Chatham Rise and

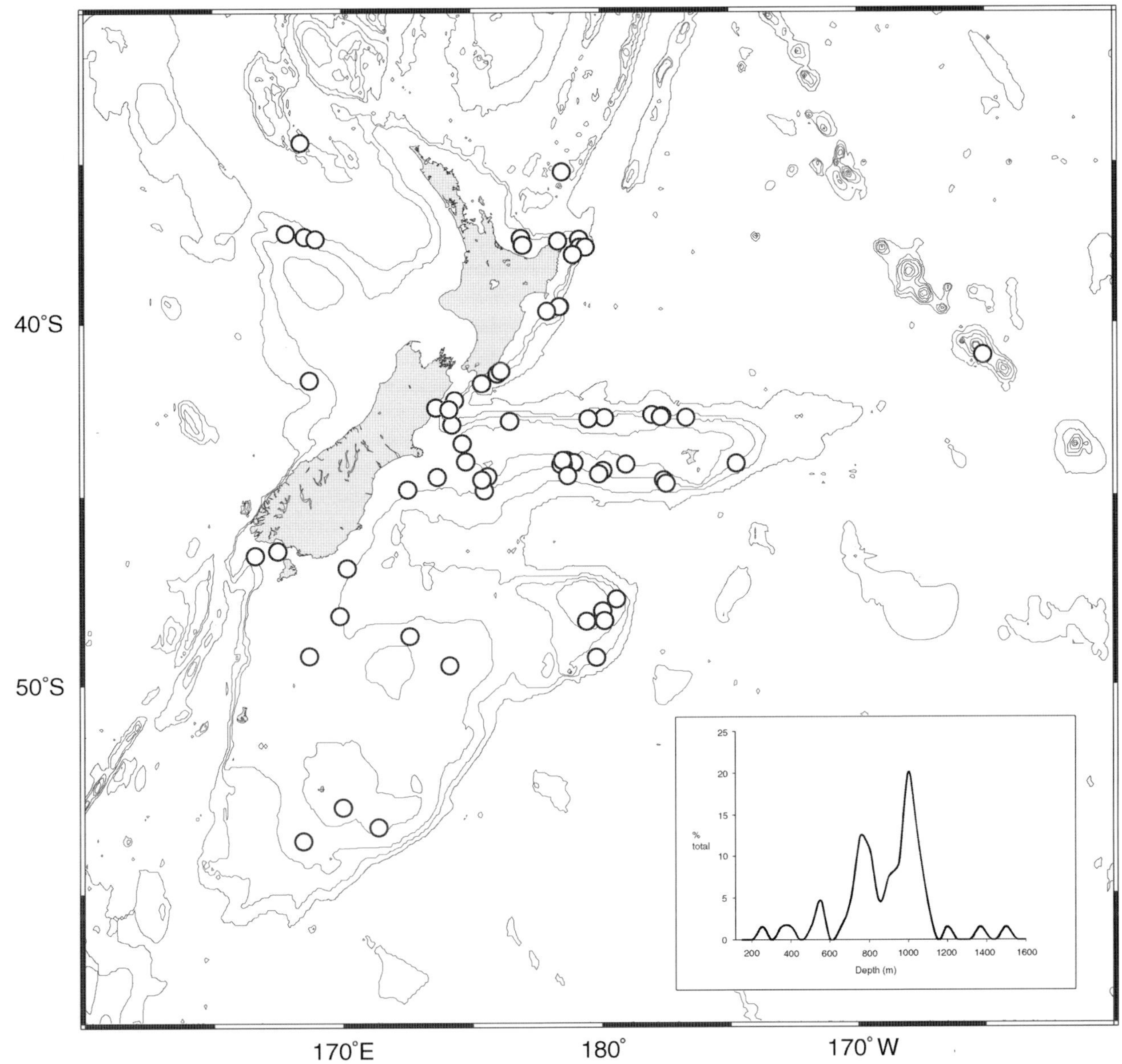

Figure 12. Geographic and bathymetric distribution of *L. aotearoa* sp. nov.

Campbell Plateau; 252–1506 m, usually 950–1100 m. An underwater image of a *Lithodes* sp. taken at 1242 m at Rumble II seamount on the southern Kermadec Ridge, appears to be *L. aotearoa* (see cover).

Lithodes australiensis sp. nov.

(Figs 13–17, Pl. 1C)

Lithodes longispina. — Dawson & Yaldwyn, 1985: 70. — Poore *et al.*, 1998: 77. — Davie, 2002: 72–73 [Australian occurrences]. — Poore, 2004: 266, 268, pl. 16c. [Not *L. longispina* Sakai, 1971]

Type material. (All New South Wales, Australia). *Holotype*: AM P35607, male (cl 161.9 mm, pcl 112.2 mm, cw 107.9 mm), E of Brush Island, 35°30′S, 150°52′E, 1065 m, FRV *Kapala*, K83-19-12, coll. K. Graham, 6 Dec 1983.

Paratypes: AM P80452, 1 male (cl 76.5 mm, pcl 39.0 mm, cw 34.8 mm), 2 females (pcl 18.8–31.6 mm, cw 15.8–28.3 mm), off Cape Hawke, 32°08′S, 153°09′E, 1007–1053 m, FRV *Kapala*, K84-10-04, 18 Jul 1984; AM P35601, 1 male (cl 166.0, pcl 97.0 mm, cw 90.7 mm), E of Newcastle, 32°57–54′S, 152°44–46′E, 988–1033 m, FRV *Kapala*, K83-13-04, coll. K. Graham, 19 Oct 1983; AM P81160, 1 male (pcl 40.0 mm, cw 35.5 mm), E of Shoalhaven Heads, 34°51–55′S, 151°15–14′E, 1042–1061 m, FRV *Kapala*, K83-14-05, coll. K. Graham, 26 Oct 1983;

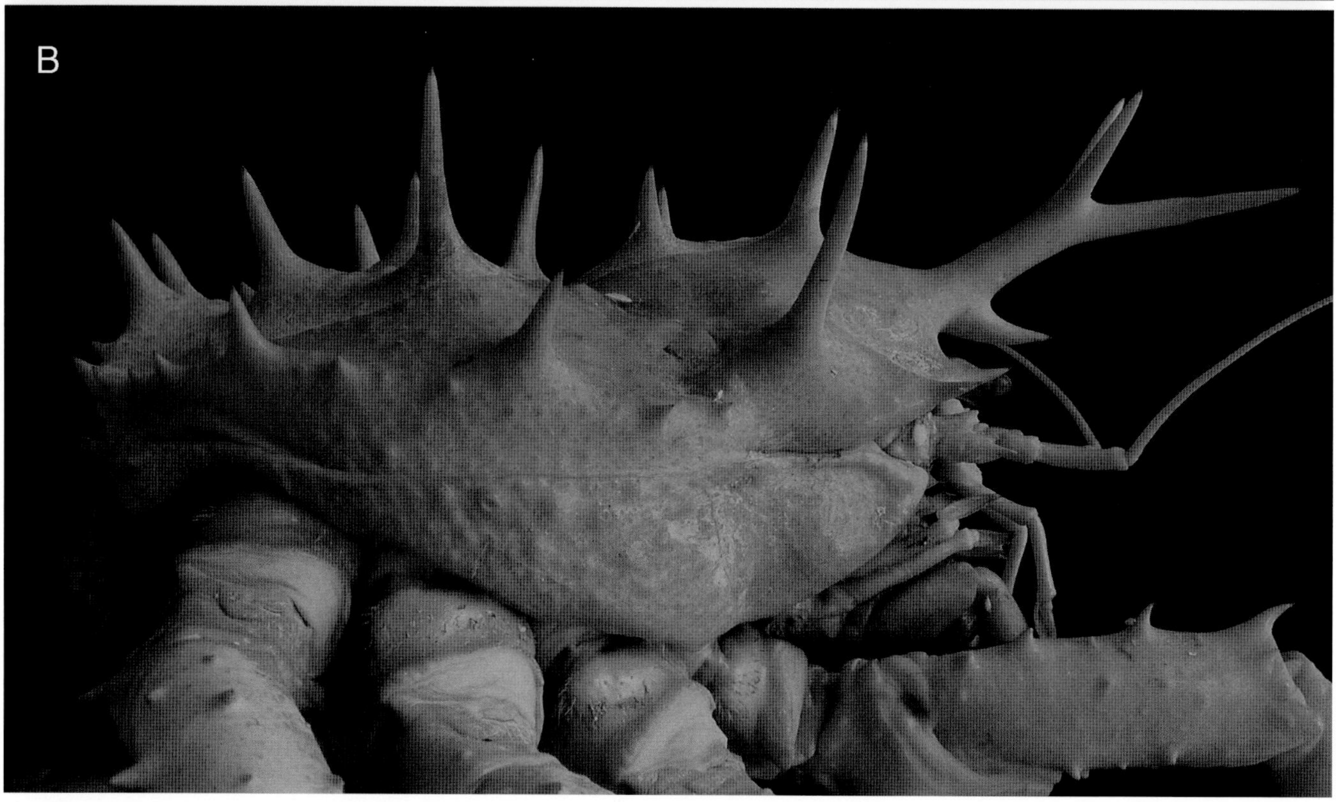

Figure 13 *(above)*. *Lithodes australiensis* sp. nov., male holotype, cl 161.9 mm, pcl 112.2 mm, cw 107.9 mm, E of Brush Island (AM P35607). A, dorsal habitus. B, carapace, right lateral view.

Figure 14 *(opposite)*. *Lithodes australiensis* sp. nov., male holotype, cl 161.9 mm, pcl 112.2 mm, cw 107.9 mm, E of Brush Island (AM P35607). A, right chela. B, left chela. C, carapace. D, right orbit and antenna. E, abdominal somite 2. F, abdomen.

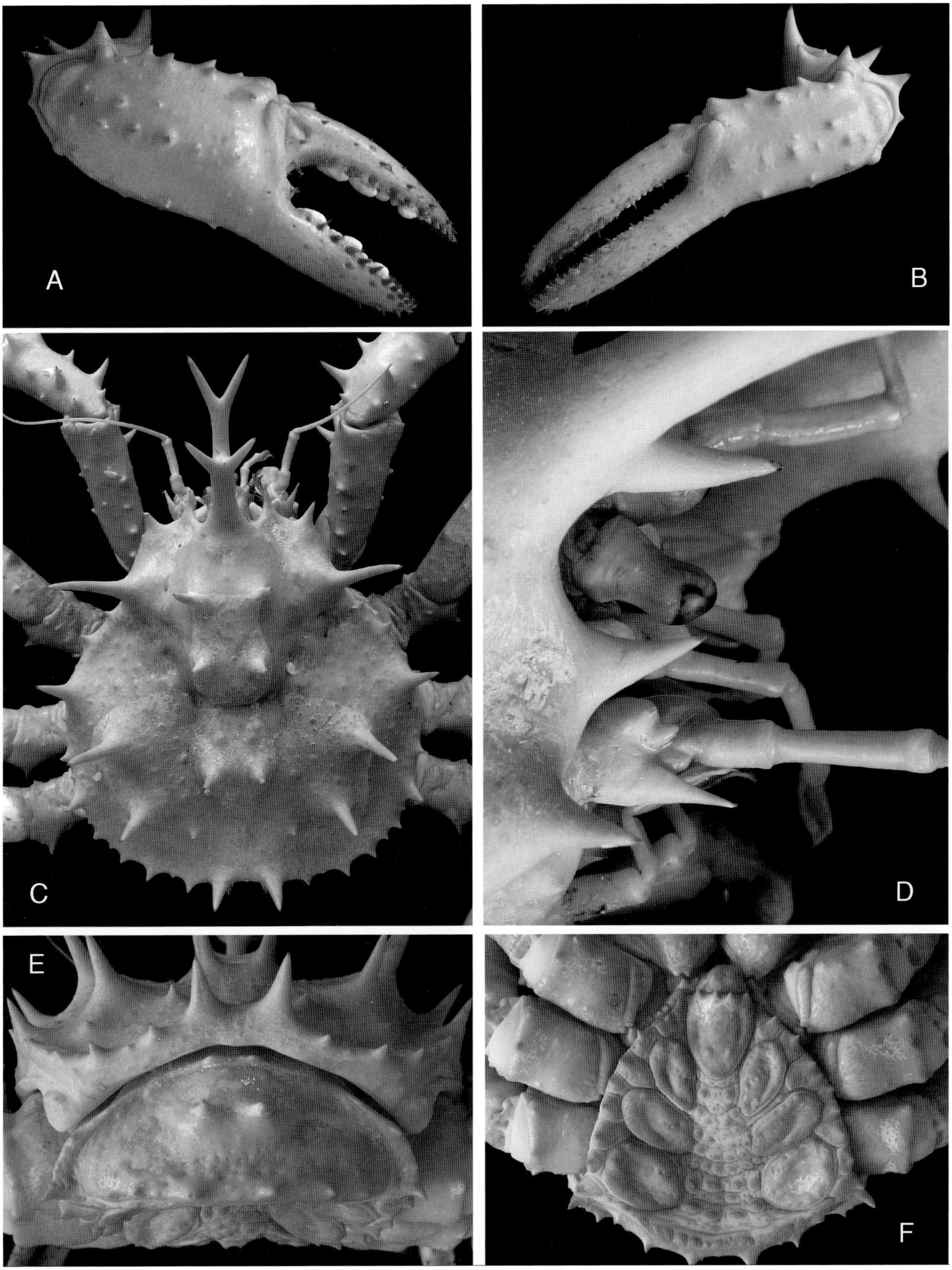
A
B
C
D
E
F

AM P35595, 1 male (pcl 50.5 mm, cw 45.5 mm), E of Shoalhaven Heads, 34°53–56′S, 151°15–13′E, 1079–1116 m, FRV *Kapala*, K83-14-06, coll. K. Graham, 26 Oct 1983; AM P80453, 1 male (cl 50.7 mm, pcl 23.9 mm, cw 20.6 mm), 2 females (cl 53.1-53.2 mm, pcl 24.4–26.0 mm, cw 21.5–21.6 mm), E of Shoalhaven Heads, 34°53–56′S, 151°15–13′E, 1079–1116 m, FRV *Kapala*, K83-14-06, coll. K. Graham, 26 Oct 1983; AM P35599, 1 male (cl 199.7 mm, pcl 132.8 mm, cw 127.5 mm), NE of Brush Island, 35°29′S, 150°52′E, 942 m, FRV *Kapala*, K84-04-03, coll. K. Graham, 3 Apr 1984; AM P35591, 1 male (pcl 39.1 mm, cw 36.3 mm), E of Gabo Island, 37°43–41′S, 150°20′E, 961–979 m, FRV *Kapala*, K83-12-01, coll. K. Graham, 26 Sep 1983.

Other material examined. *Tasmania*: TM G2970, 1 female (pcl 38.0 mm, cw 35.4 mm), off Bicheno, E Tasmania, 1000 m, coll. G. Darcey, 25 Apr 1982; TM G3162, 1 male (cl 137.0 mm, pcl 77.6 mm, cw 70.2 mm), ENE of St Patricks Head, 41°34.7–34.4′S, 148°44.6–46.6′E, 1090–1150 m, RV *Soela* stn 16, 9 May 1987; TM G3479, 1 crushed exuvium, E Tasmania, 41°39′S, 148°39′E, 915–970 m, 27 Jul 1982; TM G3490, 1 female (cl 71.3 mm, pcl 35. 6 mm, cw 34.8 mm), 15 miles [24 km] N of St Helens Hill, eastern Tasmania, 41°00′S, 148°46′E, 540–720 m, FV *Ocean Raider*, coll. D. Whennan, 13 Mar 1991; NMV J55658, 2 males (pcl 25.9–30.6 mm, cw 24.5–28.0 mm), Southern Ocean, S of Tasmania, Pedra site, 44°15.67–15.98′S, 147°05.83–05.52′E, 730–1000 m, coll. T. O'Hara & T. Costa, 2 Apr 2007; TM G3582, 1 male (cl 47.7 mm, pcl 24.1 mm, cw 22.1 mm), Main Pedra, approx. 77 km SSE of Southeast Cape, 44°16′S, 147°04′E, 1110 m, epibenthic sled, RV *Southern Surveyor*, SS01/97/06, 21 Jan 1997; TM G3659, 1 male (pcl 60.3 mm, cw 54.4 mm), Main Pedra, approx. 77 km SSE of Southeast Cape, 44°16′S, 147°08′E, 1312 m, crab trap, RV *Southern Surveyor*, SS01/97/08, 21–25 Jan 1997; NIWA 41384, 1 juvenile female (cl 95.6, pcl 53.8 mm, cw 49.3 mm), South Tasman Rise, 47°09.0–27.99′S, 148°44.6–52.99′E, 1050 m, FV *Arrow*, 1164/4, Z9395, coll. C. Rattray, 9 Oct 1998; NIWA 34906, 1 male (pcl 40.0 mm, cw 36.1 mm), South Tasman Rise, 47°27.09–27.99′S, 148°53.01–52.99′E, 1012–1058 m, 1137/12, Z9231, 13 Aug 1998; NIWA 34899, 1 male (cl 89.7 mm, pcl 52.9 mm, cw 46.2 mm), South Tasman Rise, 47°28.02–27.99′S, 148°53.05–53.05′E, 1031–1034 m, Z9232, 7 Aug 1998; NIWA 34903, 1 damaged male (pcl 78.6 mm, cw 73.3 mm), South Tasman Rise, 49°22.6′S, 150°27.0′E, 910–1200 m, Z10303.

Diagnosis. Carapace dorsal surface with slender spines becoming reduced to short conical spines above about 120 mm pcl; gastric region with 4 upright spines, anterior pair longer than half length of hepatic spine, posterior pair half length of hepatic spine; cardiac and intestinal regions each with 2 long upright spines; branchial surface with 2 long upright spines. Branchial margins spinose; 1st primary marginal branchial spine shorter than two-thirds length of hepatic spine, 2nd primary marginal branchial spine shorter than 1st primary marginal branchial spine; posterior branchial margin with 5 or 6 short spines and 1 or 2 low tubercles. Rostrum not exceeding 0.8 pcl; proximal half strongly upraised by 30–45°. Abdominal somite 2 comprising 3 plates in juveniles through adults. Antennal peduncle article 2 with outer spine not reaching beyond article 4. Chelipeds unequal; palms with low tubercles or spines on dorsal surface. Walking legs spinose; surface between major spines smooth or with very few, scattered spines. Pereopod 4 merus shorter than 1.2 pcl (1.1 to almost 1.2 in males, 0.9 to as long as pcl in females), length less than 8 times height; propodus length 12–13 times height (male), 11 (female); dactylus flexor margin unarmed.

Description. *Carapace*: Pyriform, 1.01–1.14 times longer than wide; regions indicated; dorsal surface armed with long, slender spines and scattered, widely separated granules. Gastric region convex, with 2 pairs of long, upright spines. Cardiac region with pair of anterior tubercles and posterior pair of long upright spines. Hepatic spine longest of carapace spines (0.26 pcl in holotype), directed anterolaterally. Intestinal region with pair of long upright spines on posterior margin. Branchial surface with 2 long upright spines and 2 conical tubercles as follows: 1 spine at level of pereopod 3 coxa, 1 spine at level of pereopod 4 coxa, 2 conical tubercles (or short spines in juveniles) in transverse row slightly behind posterior spine. Branchial margins spinose; anterior branchial margin with 2 small spines, posterior spine longer than anterior; 1st primary marginal branchial spine shorter than two-thirds length of hepatic spine; lateral branchial margin with 1 short spine and 1 or 2 granules; 2nd primary marginal branchial spine shorter than 1st primary marginal branchial spine (usually about half length); posterior branchial margin with 5 or 6 short spines and 1 or 2 low tubercles. Pterygostomian region granulate or tuberculate, with small, anterior, submarginal spine.

Rostrum as long as pcl in smallest specimens to half pcl in largest specimens, comprising proximal and distal portions; proximal portion angled dorsally by at least 30°, with pair of long, divergent, dorsal spines that overreach midlength of distal portion of rostrum; distal portion subhorizontal, distally bifurcate for about half length, forming pair of strongly divergent spines; ventral rostral spine elongate, extending anteriorly beyond apices of outer orbital spine. Posterior orbital margin concave; outer orbital spine directed anteriorly,

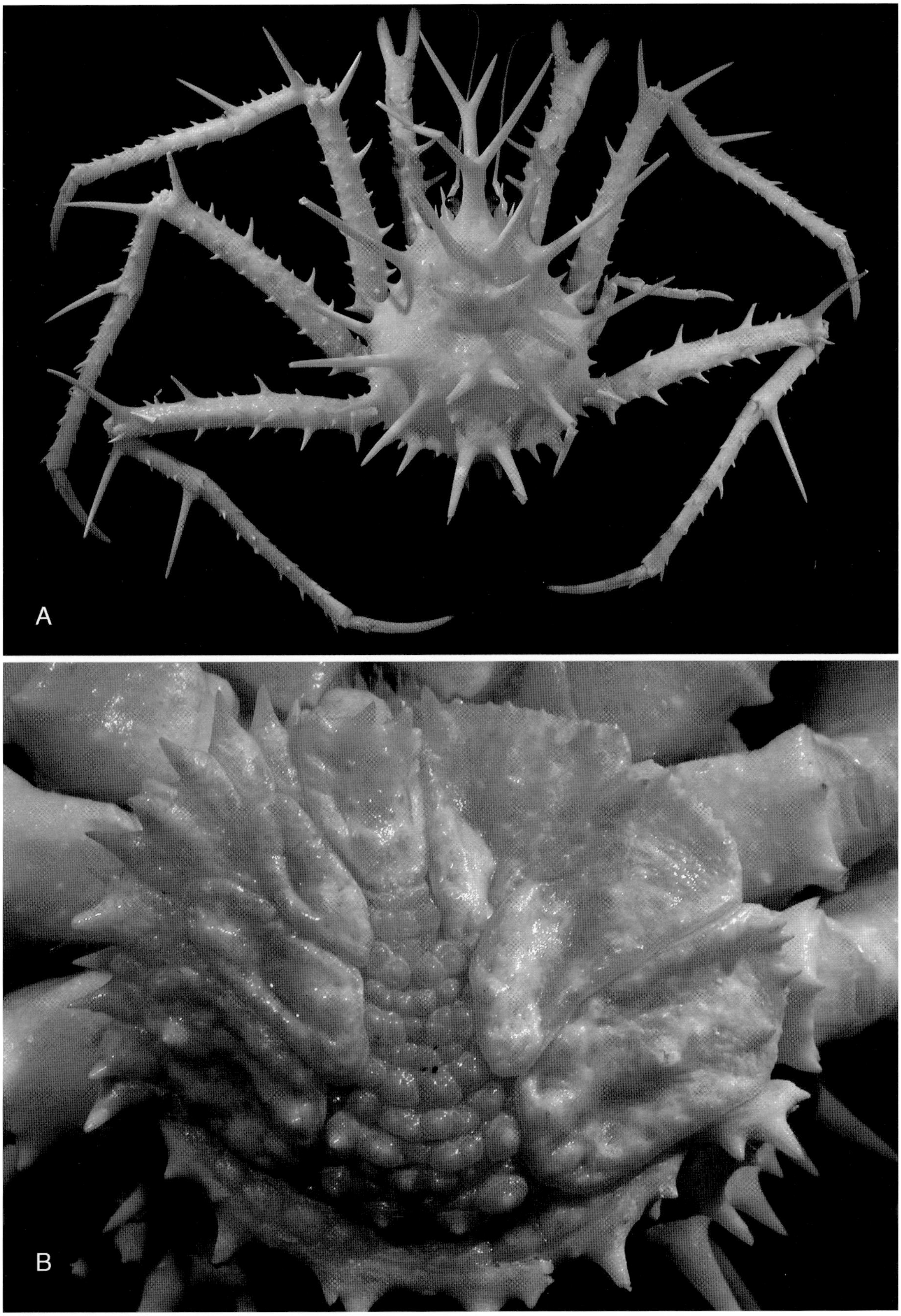

Figure 15. *Lithodes australiensis* sp. nov., female, pcl 53.8 mm, S of Tasmania (NIWA 41384). A, dorsal habitus. B, abdomen.

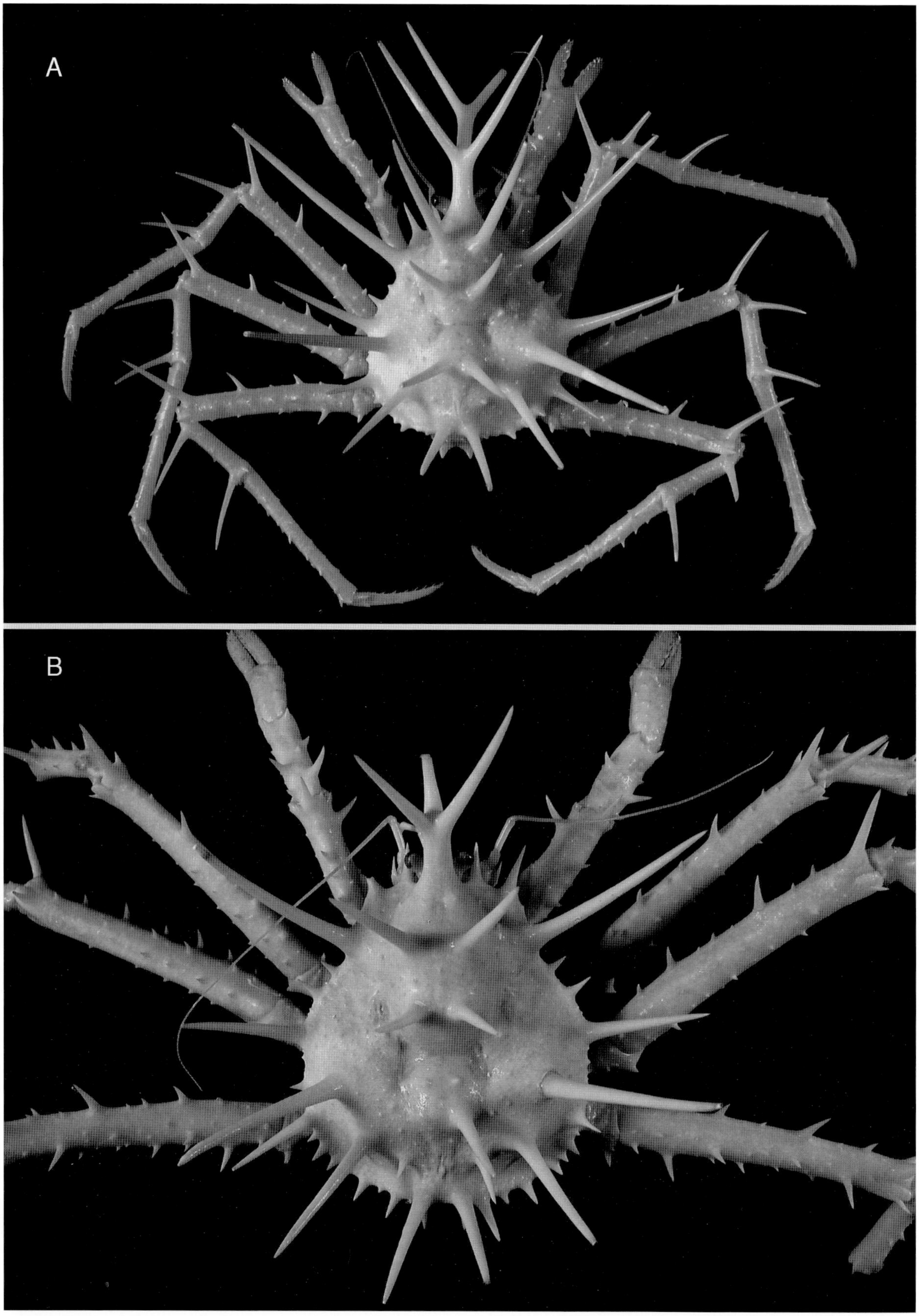

Figure 16. *Lithodes australiensis* sp. nov. A, male, pcl 24.1 mm, SSE of Southeast Cape, Tasmania (TM G3582). B, male, pcl 60.3 mm, SSE of Southeast Cape, Tasmania (TM G3659).

slightly longer than anterolateral spine, reaching to or beyond cornea.

Ocular peduncle: Longer than cornea; unarmed.

Antennule: Peduncle unarmed, reaching anteriorly beyond antennal peduncle by about three-quarters length of distal antennular peduncle article.

Antenna: Basal antennal article unarmed; outer margin of article 2 with long slender spine, not reaching beyond article 4; article 3 unarmed; scaphocerite minute, blunt, shorter than article 4; article 4 unarmed, about half as long as article 5.

Abdomen: Somite 1 with low triangular teeth on posterior margin. Somite 2 rugose; composed of 3 distinct plates separated by narrow groove; median surface with pair of slender submedian spines; posterior margin with 4–6 stout spines or angular prominences; lateral margins irregularly dentate. Somites 3–5 with nodular median plates; surface of submedian plates irregular, but unarmed; marginal plates of males subdivided, with angular or sharp apices; marginal plates of females with left margin crenulate to spinose (fused with submedians), right marginals subdivided, with angular to spiniform apices. Somite 6 of both sexes as long as or longer than wide, distal margin with 2 or 4 distal spines or prominences. Telson rounded, unarmed.

Pereopod 1 (chelipeds): Dimorphic but spination similar. Surfaces of both chelipeds smooth or with scattered granules in addition to major spines. Coxa with tufts of setae, unarmed; ischiobasis with 3 stout ventral spines and about 6 blunt tubercles. Merus inner margin with stout subdistal spine and 2 or 3 smaller spines or acute tubercles; ventral margin with two rows of 2 or 3 low conical spines; dorsal and lateral surface spinose, spines largest distally. Carpus with prominent spines on dorsal and lateral surfaces; dorsal margin with row of 3 stout spines; ventral margin with small scattered acute tubercles. Palms of both chelipeds in both sexes with similar ornamentation; with small spines and acute tubercles on dorsal, lateral and ventral surfaces, inner surface with few low tubercles; dorsal margin with row of 3 or 4 small conical spines; midlateral surface with 2 rows of 3 or 4 small spines or tubercles, of similar size to dorsal row; ventral surface with scattered tubercles, smaller than lateral and dorsal spines.

Major cheliped 1.38–1.59 pcl (male), 1.19–1.40 (female); upper palm length 1.15–1.33 times height (male), 1.42–1.55 (female); occlusal margins of fingers corneous for distal third, proximally with 3 calcareous nodules; dactylus dorsal margin broadly convex, with rows of tufts of golden setae and blunt proximal tooth, 1.22–1.31 times longer than dorsal margin of palm (male), 1.19–1.40 (female).

Minor cheliped 1.38–1.50 pcl (male), 1.27–1.40 (female); upper palm length 1.35–1.49 times height (male), 1.69–1.77 (female); occlusal margin corneous for slightly less than distal half, proximally crenulate; dactylus dorsal margin broadly convex, with rows of tufts of golden setae and blunt proximal tooth, 1.51–1.69 times longer than dorsal margin of palm (male), 1.51–1.56 (female).

Pereopods 2–4 (walking legs 1–3): Similar, slender, elongate; ornamentation similar in both sexes; segments spinose, surface between major spines smooth or with few, scattered spines or granules. Pereopod 3 and 4 subequal, longer than pereopod 2. Distal margins of coxae unarmed; surface smooth. Ischiobasis with 3 or 4 non-setose distal spines. Merus subcircular to ovate in cross section; extensor margin with 4–10 spines of which 2 or 3 are distinctly longer in addition to long, prominent, distal spine; flexor margin with 2 rows of 4 or 5 small spines; upper surface with irregular row of 5–8 small spines. Carpus about half merus length; extensor margin with distal and second proximal spines longest, exceeding one-third carpus length; surface with scattered small spines. Propodus ovate in cross section; with 7–11 small spines on extensor margin and 4–7 small spines on dorsal surface; flexor margin with 4–7 small spines. Dactylus curved, rounded in cross section distally; proximally with 2 spines on either side adjacent to articulation; with 3 or 4 small corneous spines along extensor margin; flexor margin smooth or with about 4 soft, minute, widely spaced setae; apex corneous.

Pereopod 2 length 2.91–3.21 pcl (male), 2.64–2.77 pcl (female). Merus 1.10–1.16 pcl (male), 0.96 pcl (female);length:height ratio 6.56–7.42 (male), 6.43–8.64 (female). Carpus 0.48–0.51 merus length (male), 0.51–0.52 (female). Propodus 0.81–0.86 merus length (male), 0.84–0.88 (female); length:height ratio 11.16–12.8 (male), 10.09–13.35 (female). Dactylus 0.52–0.58 propodus length (male), 0.62–0.63 (female).

Pereopod 3 length 3.12–3.48 pcl (male), 2.69–3.01 pcl (female). Merus 1.10–1.23 pcl (male), 0.95–1.04 pcl (female); length:height ratio 6.00–7.44 (male), 6.46–6.94 (female). Carpus 0.51–0.53 merus length (male), 0.52 (female). Propodus 0.85–0.88 merus length (male), 0.85–0.87 (female); length:height ratio 11.70–13.58 (male), 10.67–11.06 (female). Dactylus 0.51–0.68 propodus length (male), 0.57–0.61 (female).

Pereopod 4 length 3.12–3.48 pcl (male), 2.57–2.95 pcl (female). Merus 1.10–1.17 pcl (male), 0.92–1.02 pcl (female); length:height ratio 6.88–7.84 (male), 6.59–7.33 (female). Carpus 0.54–0.56 merus length (male), 0.54 (female). Propodus 0.89–0.93 merus length (male), 0.88–0.90 (female); length:height ratio 12.32–13.57 (male), 10.96–11.04 (female). Dactylus 0.50–0.56 propodus length (male), 0.61–0.62 (female).

Colour in life. Overall deep-red (Pl. 1C).

Etymology. Named *australiensis*, for the Australian distribution of the new species.

Remarks. *Lithodes longispina* Sakai, 1971 was first reported from Australia based on "identification of large numbers of *L. longispina* in south-east Australian waters" from collections in the Australian Museum (Dawson & Yaldwyn 1985: 72). All records of *L. longispina* from Australia and New Zealand, however, are referable to other species. As speculated by Macpherson (2001) and Poore (2004), many records of *L. longispina* from New South Wales are based on *L. richeri* Macpherson, 1990, reported below. The second *L. longispina*-like species from southeastern Australia is herein named *L. australiensis* sp. nov. *Lithodes australiensis* is a southern species, ranging from central New South Wales (~ 32°S) to the South Tasman Rise, south of Tasmania (~49°S). *Lithodes richeri* has a more northerly distribution than *L. australiensis*, ranging from Vanuatu and New Caledonia (possibly also Indonesia; see account of *L. richeri*) to the vicinity of Brush Island, southern New South Wales (~35°30′S). The ranges of the two species overlap, and are sometimes sympatric off central and southern New South Wales. Differences between *L. australiensis* and *L. richeri* are outlined under the account of the latter.

Lithodes australiensis sp. nov. is most similar to *L. robertsoni* sp. nov. and *L. aotearoa* sp. nov. from New Zealand, and *L. rachelae* sp. nov. from southwestern Australia, in having the combination of unarmed flexor margins of the dactyli of the walking legs in adults, and long, slender, dorsal spines that persist in early adults. Although *L. australiensis* has in the past been misidentified as *L. longispina*, the new species differs by its two long dorsal branchial spines: one anteriorly and one posteriorly. *Lithodes longispina* lacks the long dorsal posterior branchial spine, at most bearing a short conical tooth or tubercle. *Lithodes australiensis* is distinguished from *L. rachelae* by its proportionally shorter walking legs (see account of *L. rachelae*), and from *L. robertsoni* also by walking leg proportions (see account of *L. robertsoni*). *Lithodes australiensis* differs from *L.*

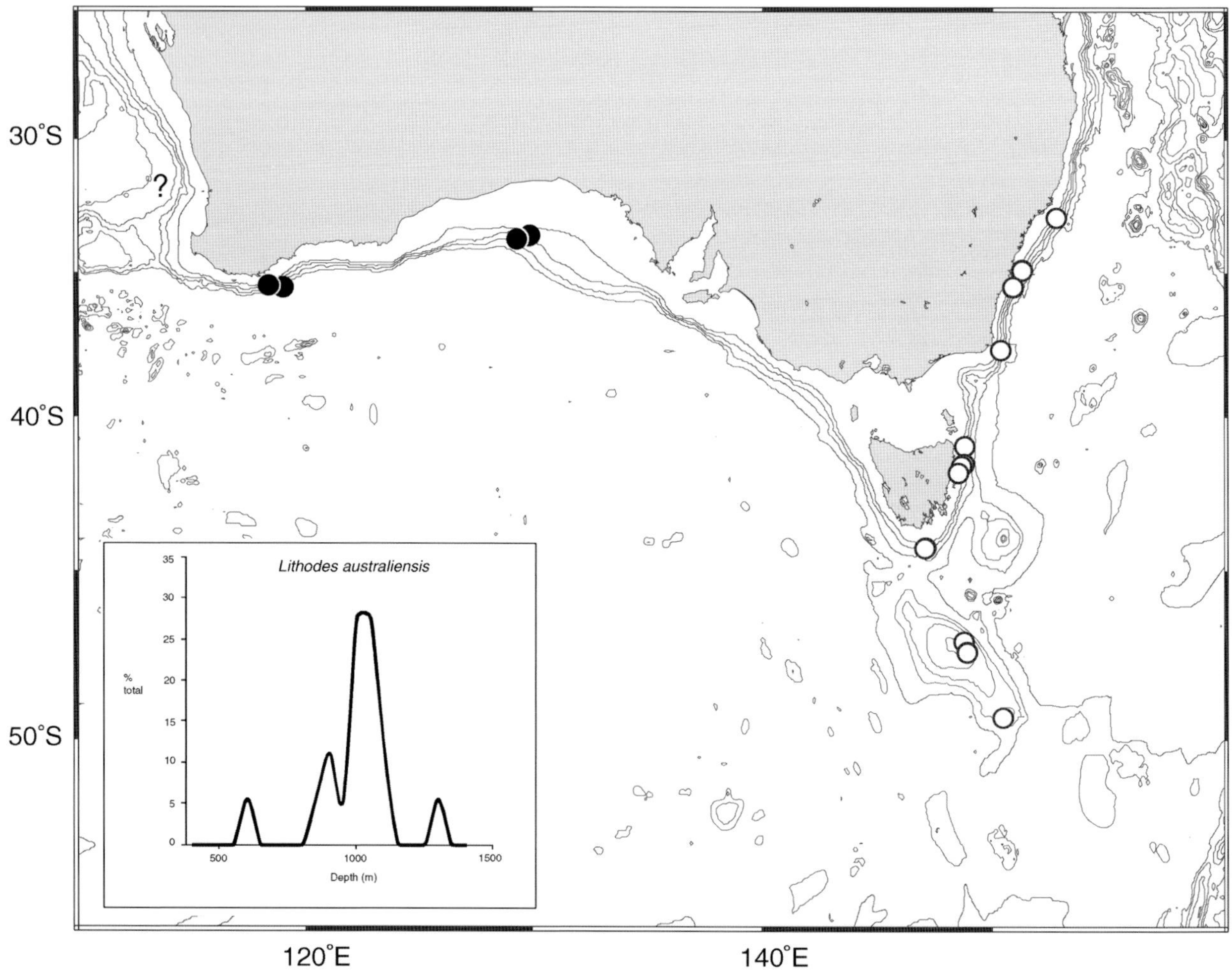

Figure 17. Geographic distribution of *Lithodes australiensis* sp. nov. (○), *L. rachelae* sp. nov. (•), and *L. chaddertoni* sp. nov. (?) Inset: bathymetric distribution of *L. australiensis*.

aotearoa in having 5 or 6 rather than 8–12 posterior branchial marginal spines and in having abdominal somite 2 composed of three plates in adults, rather than being fused into a single plate as in adult *L. aotearoa*. Additionally, adult *L. australiensis* have more elongate walking legs than adult *L. aotearoa*; in the former, the pereopod 4 merus length always exceeds the pcl (1.09–1.17 pcl), whereas in the latter, it only slightly exceeding pcl in the largest males (0.91–1.06) (see Table 1). Female *L. australiensis* also have more elongate walking legs than female *L. aotearoa* but the differences are less pronounced (Table 2). *Lithodes australiensis* possibly matures at a larger size than does *L. aotearoa* because the flexor spines on the dactyli of the walking legs are present up to about 60 mm pcl in *L. australiensis*, but are lost by about 20 mm pcl in *L. aotearoa*.

All known female *L. australiensis* are juveniles, so the size at which females mature is not presently known, although it exceeds about 53.8 mm pcl (the largest known female). The abdomen of the 53.8 mm female is already asymmetrical (with spinular and subdivided right marginal plates, and fused left marginal and submedian plates), although it has not yet achieved full asymmetry.

As in other lithodids, dorsal spine length in *L. australiensis* becomes proportionally shorter with increasing body size. The longest carapace spine (hepatic) is as long as the pcl at 24.1 mm pcl (TM G3582), 0.7 pcl at 60.3 mm pcl (TM G3659), and 0.3 pcl in the 112.2 mm pcl holotype. The row of corneous spines on the flexor margin of the walking leg dactyli, a juvenile feature, is present up to about 60 mm pcl, but the marginal plates of abdominal somite 2 remain unfused to the median plate in adult males (adult females are presently unknown, but this feature is not known to be sexually dimorphic in other species of *Lithodes*).

Distribution. Southeastern Australia, from the vicinity of Cape Hawke, New South Wales, south to Victoria, Tasmania and the South Tasman Rise; 540–1312 m, usually 1000–1100 m.

Lithodes chaddertoni sp. nov.

(Figs 17–20, Pl. 4A)

Type material. *Holotype*: NMNZ Cr11193, male (cl 190.1 mm, pcl 100.9 mm, cw 89.9 mm), Naturaliste Plateau or Broken Ridge, Indian Ocean off SW Western Australia, crab pot, coll. J. Chadderton, 2002.

Diagnosis. Carapace dorsal surface with long slender spines; gastric region with 4 upright spines, anterior pair longer than half length of hepatic spine, posterior pair half length of hepatic spine; cardiac and intestinal regions each with 2 long upright spines; branchial surface with 2 or 3 long upright spines. Rostrum exceeding 0.8 pcl. First and second primary marginal branchial spines greatly elongated, of similar length. Posterior branchial margin with 6 or 7 short spines, 4 prominent. Chelipeds unequal; palms with low tubercles or spines on dorsal surface. Walking leg dactyli with spinous flexor margin; half propodi length.

Table 1. Selected morphometric measurements of adult male *L. aotearoa* sp. nov., *L. australiensis* sp. nov., *L. rachelae* sp. nov., *L. robertsoni* sp., and *L. paulayi* Macpherson & Chan, 2008. Measurements of *L. paulayi* based on male holotype, pcl 112.0 mm, cw 102.7 mm (FLMNH 2283).

	Male pcl	P4 L/pcl (Male)	P4 merus L/pcl (Male)	P4 merus L/H (Male)	P4 propodus L/H (Male)
L. aotearoa	72.3–194.0	2.68–2.98	0.91–1.06	5.56–7.16	9.12–13.04
L. australiensis	78.0–133.8	3.12–3.48	1.10–1.17	6.88–7.84	12.32–13.57
L. robertsoni	81.7–128.1	3.02–3.20	1.05–1.10	6.70–8.16	11.43–14.08
L. rachelae	75.0–113.1	3.62–3.71	1.29–1.30	9.17–9.36	15.24–15.61
L. paulayi	112.0	3.17	1.14	9.59	17.37

Table 2. Selected morphometric measurements of adult female *L. aotearoa* sp. nov., *L. australiensis* sp. nov., *L. rachelae* sp. nov., and *L. robertsoni* sp. nov. Females of *L. paulayi* not presently known.

	Female pcl	P4 L/pcl (Female)	P4 merusL/pcl (Female)	P4 merus L/H (Female)	P4 propodus L/H (Female)
L. aotearoa	69.6–161.3	1.89–2.77	0.62–0.95	4.55–6.43	8.56–10.41
L. australiensis	53.8–84.5	2.57–2.95	0.92–1.02	6.59–7.33	10.96–11.04
L. robertsoni	70.2–71.2	2.91–2.92	0.96–0.98	6.11–6.76	11.55–11.87
L. rachelae	43.8	3.10	1.06	13.21	15.24–15.61
L. paulayi	?	?	?	?	?

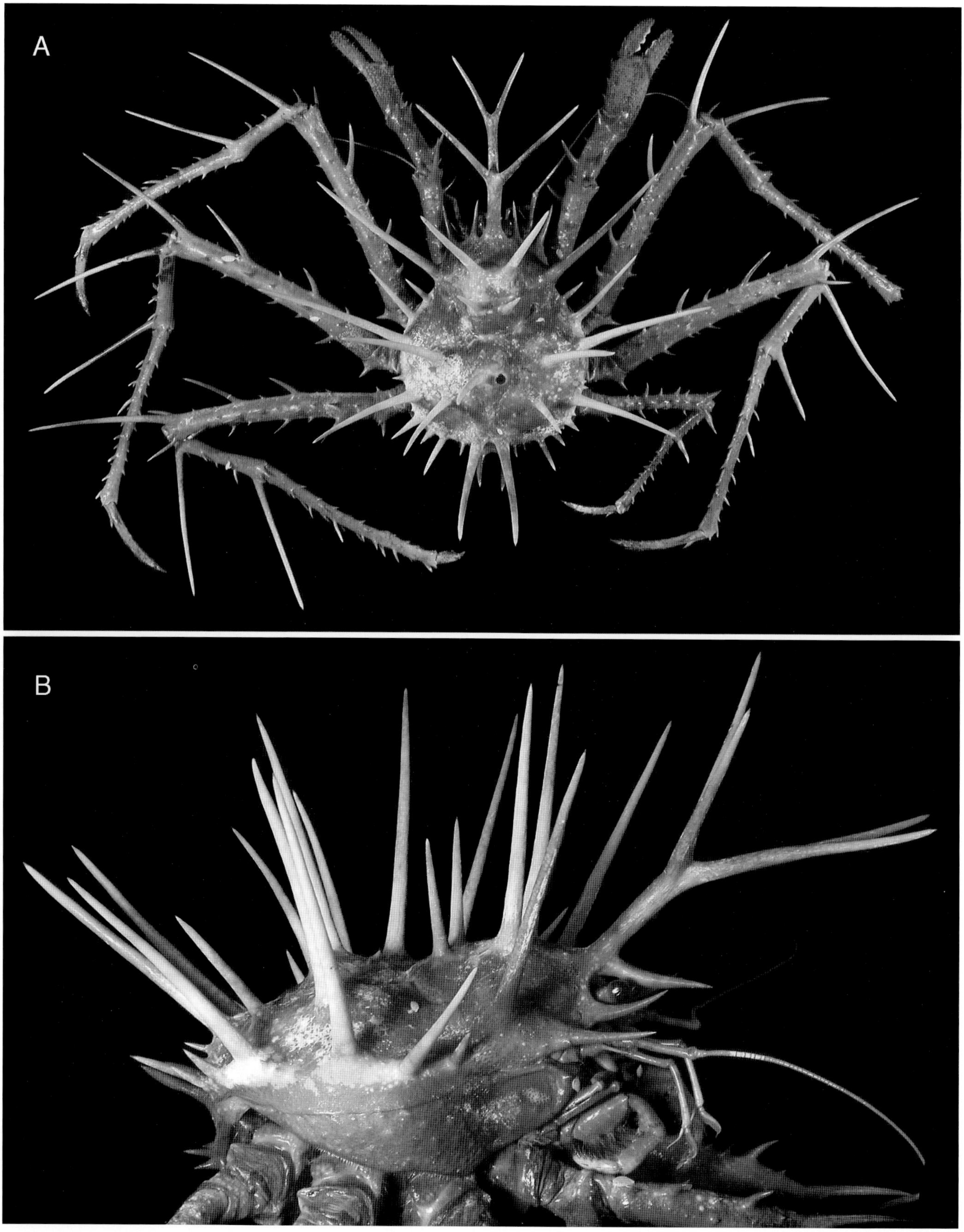

Figure 18. *Lithodes chaddertoni* sp. nov., male holotype, cl 190.1 mm, pcl 100.9 mm, cw 89.9 mm, Western Australia (NMNZ Cr11193). A, dorsal habitus. B, carapace, right lateral view.

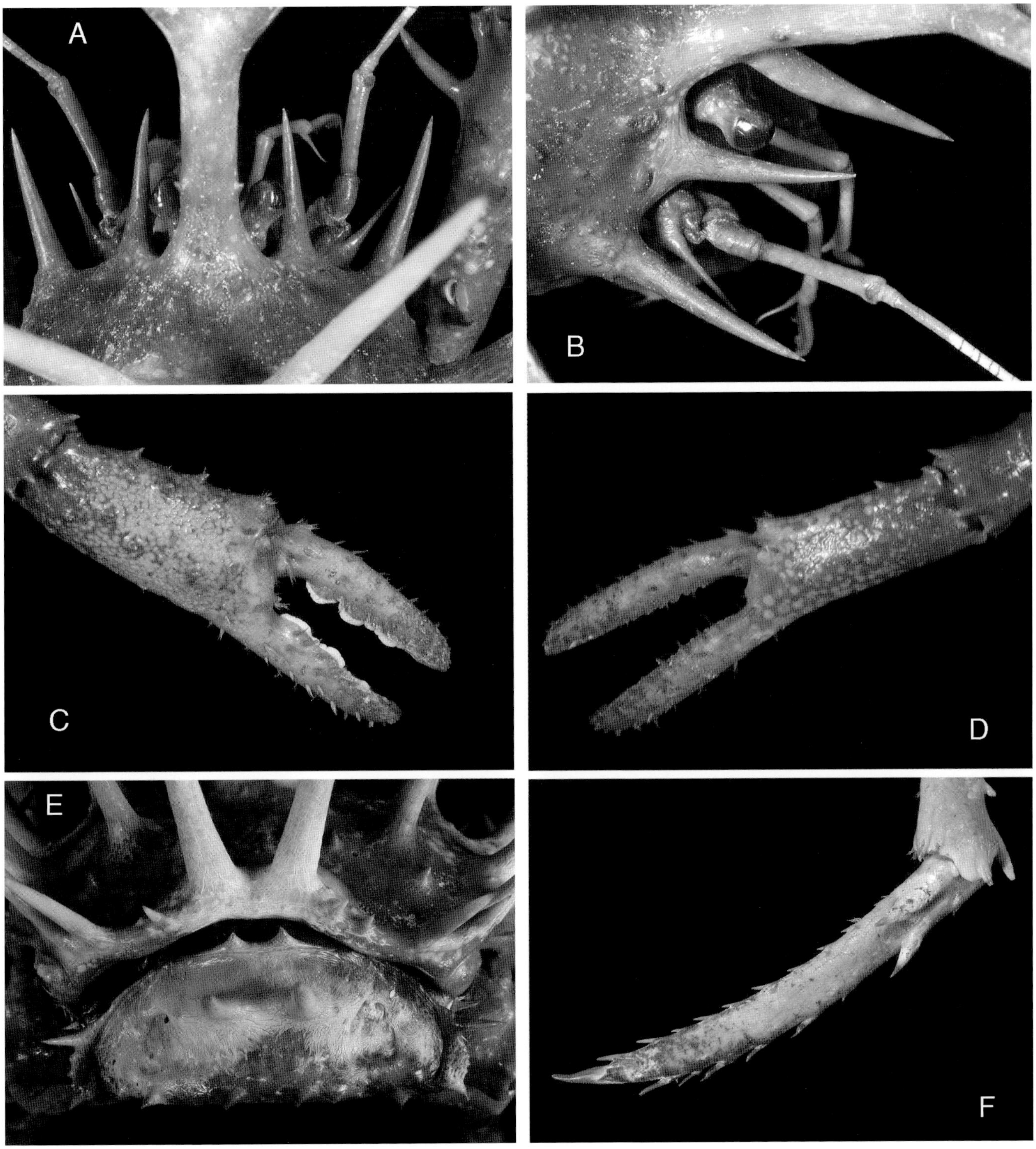

Figure 19. *Lithodes chaddertoni* sp. nov., male holotype, cl 190.1 mm, pcl 100.9 mm, cw 89.9 mm, Western Australia (NMNZ Cr11193). A, anterior carapace, dorsal view. B, right orbit and antenna. C, right chela. D, left chela. E, abdominal somite 2. F, right pereopod 3 dactylus.

Description. *Carapace*: Pyriform, about 1.13 times longer than wide; regions indicated; dorsal surface armed with long, slender spines and very few, scattered, widely separated granules or small conical tubercles, surface otherwise smooth; longest spine (hepatic spine) 0.9 pcl. Gastric region convex, with 2 pairs of long, upright spines. Cardiac region with pair of minute anterior spines and posterior pair of long upright spines, posterior pair almost as long as hepatic spine. Hepatic spine directed anterolaterally. Posterior margin of intestinal region with pair of long, upright spines. Branchial surface with 2 long upright spines: first at level of pereopod 3 coxa, second at level of pereopod 4 coxa. Branchial margins spinose; anterior branchial margin with 2 slender spines, posterior spine twice or more length of anterior, half or more length of hepatic spine; 1st primary marginal branchial spine as long as hepatic spine; lateral branchial margin with 1 or 2 short spines; 2nd primary marginal branchial spine slightly shorter than 1st primary marginal branchial spine; posterior branchial margin with 6 or 7 short spines, 4 prominent. Pterygostomian region granulate, irregular, with short subdistal tooth.

Rostrum 0.88 pcl, comprising proximal and distal portions; proximal portion angled dorsally by about 45°, with pair of long, divergent dorsal spines that overreach midlength of distal portion of rostrum; distal portion subhorizontal, distally bifurcate for half length, forming pair of strongly divergent spines; ventral rostral spine elongate, extending anteriorly beyond apices of outer orbital spine. Posterior orbital margin concave. Outer orbital spine directed anteriorly, subequal to anterolateral spines, reaching beyond cornea.

Ocular peduncle: Longer than cornea; unarmed.

Antennule: Peduncle unarmed, reaching anteriorly beyond antennal peduncle by three-quarters length of distal antennular peduncle article.

Antenna: Basal antennal article unarmed; outer margin of article 2 with slender spine that reaches to or slightly beyond article 4; articles 3–5 unarmed; scaphocerite short, blunt; article 4 about half as long as article 5. Peduncle extending beyond apex of ventral rostral spine by two-thirds length of distal peduncular article.

Abdomen: Somite 1 with pair of tubercles or triangular teeth on posterior margin. Somite 2 composed of three plates; surface sparsely granulate; median surface with pair of slender submedian spines; posterior margin with 8 spines of varying length; lateral margins irregularly dentate. Somites 3–5 with median nodules; submedian plates pitted, unarmed; marginal plates with irregularly angular margins. Somite 6 elongate, unarmed, distal margin with 4 low, blunt prominences. Telson unarmed, rounded.

Pereopod 1 (chelipeds): Dimorphic but spination similar. Surfaces of both chelipeds smooth or with scattered granules in addition to several small spines. Coxa unarmed; with tufts of setae mesially. Ischiobasis with 4 stout ventral spines and about 3 or 4 blunt tubercles. Merus inner margin, with curved subdistal spine; ventral margin with 3 low conical spines and scattered tubercles; dorsal margin with 5–7 small, scattered spines and 3 prominent spines, increasing in length distally, distalmost exceeding half carpus length; lateral surface spinose, spines largest distally. Carpus with 6 spines and 3 or 4 scattered, small spines or acute tubercles. Palm with 7–11 small scattered spines or acute tubercles on dorsal and lateral surfaces, inner and ventral surfaces unarmed.

Major cheliped 1.74 pcl; upper palm length 1.57 times height; occlusal margins of fingers corneous for distal quarter, proximally with 3 calcareous nodules; dactylus dorsal margin broadly convex, with rows of tufts of golden setae and blunt proximal tooth, 1.28 times longer than dorsal margin of palm.

Minor cheliped 1.69 pcl; upper palm length 1.76 times height; occlusal margin corneous for slightly less than distal half, proximally crenulate; dactylus dorsal margin broadly convex, with rows of tufts of golden setae and blunt proximal tooth, 1.56 times longer than dorsal margin of palm.

Pereopods 2–4 (walking legs 1–3): Similar, slender, spinose; surface between major spines smooth or with few, minute, sparsely distributed spines. Pereopod 4 longest. Coxa of pereopods 2–3 unarmed; pereopod 4 coxa with anterodistal spine. Ischiobasis with 4 distal spines, posteriormost spine of pereopod 4 longest. Merus subcylindrical; flexor margins with 5–8 spines; extensor margins with 2 or 3 prominent spines and 3–5 smaller spines, in addition to long, prominent, distal spine, exceeding half merus length; dorsal surface with two irregular rows of 4–8 spines of which 3 or 4 prominent. Carpus surface with scattered, short, slender, strongly elongated distal and second proximal spines, each exceeding carpus length. Propodus with 8–11 spines on extensor and flexor margins. Dactylus curved, laterally compressed; proximally with 2 spines on either side adjacent to articulation; with 4–6 slender corneous spines along extensor margin; flexor margin lined with 7–10 slender corneous spines; apex corneous.

Pereopod 2 length 3.12 pcl. Merus 1.12 pcl; length: height ratio 8.79. Carpus 0.51 merus length. Propodus 0.87 merus length; length:height ratio 14.80. Dactylus 0.52 propodus length.

Pereopod 3 length 3.39 pcl. Merus 1.16 pcl; length: height ratio 8.74. Carpus 0.54 merus length. Propodus 0.91 merus length; length:height ratio 15.17. Dactylus 0.52 propodus length.

Pereopod 4 length 3.5 pcl (estimated; dactylus broken). Merus 1.15 pcl; length:height ratio 8.80. Carpus 0.56 merus length. Propodus 0.92 merus length; length: height ratio 14.84.

Colour in life. Overall deep-red (Pl. 4A).

Etymology. Named in honour of John Chadderton, who collected the holotype.

Remarks. *Lithodes chaddertoni* sp. nov. is the most strongly spinose of known species of the genus, with spines measuring up to 0.87 pcl, and rostral length

Figure 20. *Lithodes chaddertoni* sp. nov., male holotype, cl 190.1 mm, pcl 100.9 mm, cw 89.9 mm, Western Australia (NMNZ Cr11193). A, carapace. B, abdomen.

0.88 pcl in adults. *Lithodes chaddertoni* resembles *L. jessica* sp. nov. from New Zealand in the combination of long dorsal spines in adults and spinose flexor margins of the walking leg dactyli. The two species are readily distinguished by cheliped form: equal with smooth palms in *L. jessica* versus distinctly dimorphic with spinose palms in *L. chaddertoni*. Additionally, *L. chaddertoni* has very few carapace tubercles and granules compared to *L. jessica*, and proportionally longer major spines, despite the large size difference (maximum 100.9 mm in *L. chaddertoni* versus 64.1 mm pcl in *L. jessica*) (compare Figs 18 and 21).

Many species of *Lithodes* have long slender dorsal spines in juveniles, but spine length is usually much reduced in adults. Thus, adult *L. chaddertoni* and *L. jessica* are apparently paedomorphic, closely resembling juvenile *L. australiensis* sp. nov. and *L. aotearoa* sp. nov. in the long dorsal spines and armed flexor margins of the dactyli of the walking legs.

Adults of other 'long-spined' species of *Lithodes* from southern Australia, namely *L. australiensis* and *L. rachelae*, lack a spinose flexor margin on the walking leg dactyli and have proportionally shorter dorsal spines. Note that flexor dactylar spines are also present in juvenile *L. australiensis* up to about 60 mm pcl; juveniles of *L. rachelae* are presently unknown.

Distribution. Presently known only from off southwestern Western Australia.

Lithodes jessica sp. nov. (Figs 21–26)

Type material. *Holotype*: NIWA 61195, ovigerous female (cl 97.8 mm, pcl 58.6 mm, cw 52.1 mm), southern Lord Howe Rise, 35°56.1'S, 166°09.7'E, 1050 m, trip 1204/27, Z9720, coll. A. Knox, 6 Mar 1999.

Paratypes: NIWA 34902, 1 ovigerous female (cl 110.5 mm, pcl 64.1 mm, cw 58.1 mm), Lord Howe Rise, no other data; NMNZ Cr11138, 1 male (cl 70+ mm, pcl 48.5 mm, cw 43.7 mm), off Napier, Hawkes Bay, 39°59'S, 178°13'E, 680–1100 m, FV *Peterson*, coll. S. O'Shea, Apr 1994.

Diagnosis. Carapace dorsal surface with long slender spines; gastric region with 4 spines, anterior pair greatly elongate, subequal to length of hepatic spine, posterior pair short, less than one-quarter length of anterior pair; cardiac and intestinal regions each with 2 long upright spines; branchial surface with 2 or 3 long upright spines. Rostrum about 0.7 pcl. First and second primary marginal branchial spines short, less than one-third length of hepatic spine. Posterior branchial margin with 7 or 8 spines. Chelipeds equal, slender;

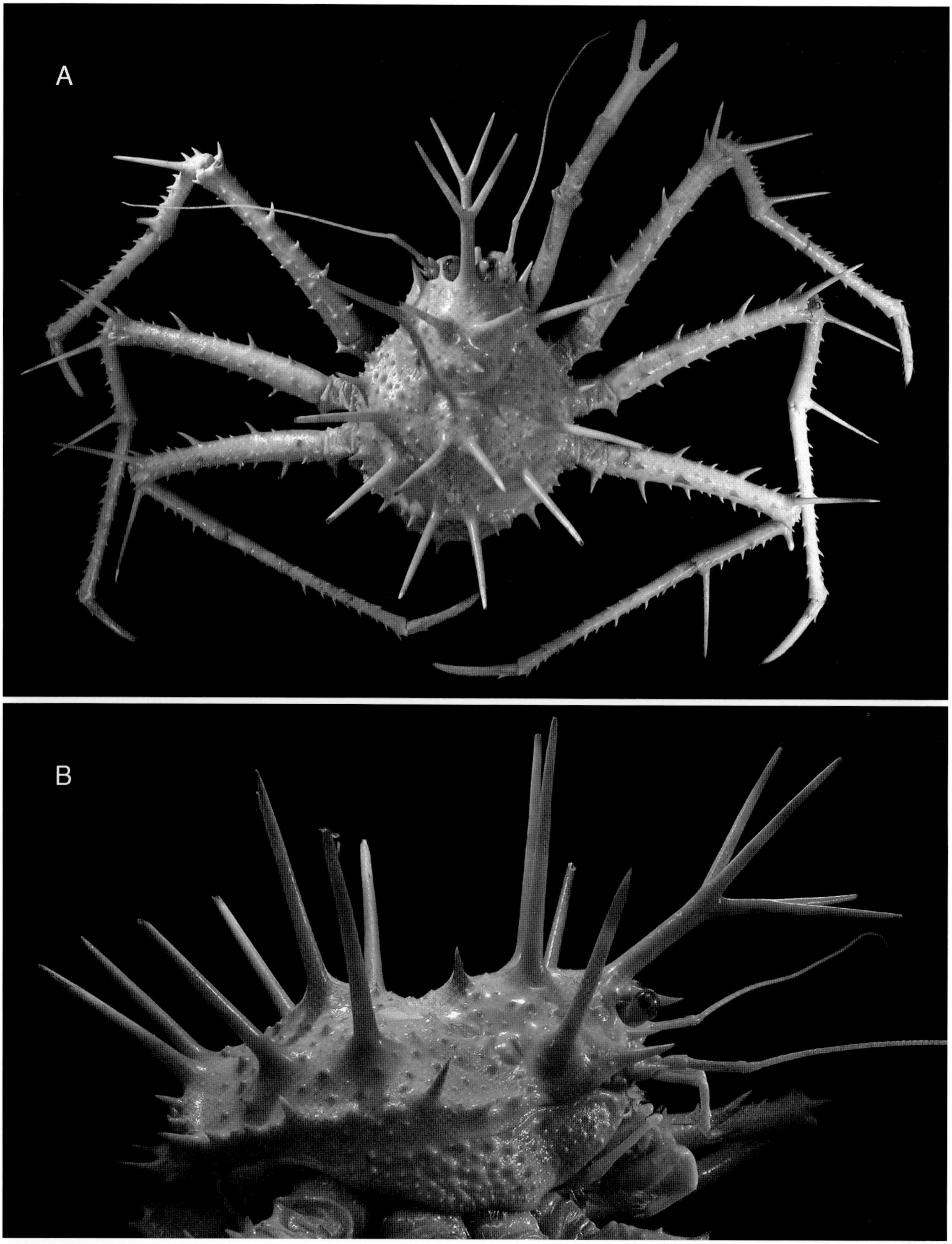

Figure 21. *Lithodes jessica* sp. nov., ovigerous female holotype, cl 97.8 mm, pcl 58.6 mm, cw 52.1 mm, Lord Howe Rise (NIWA 61195). A, dorsal habitus. B, carapace, right lateral view.

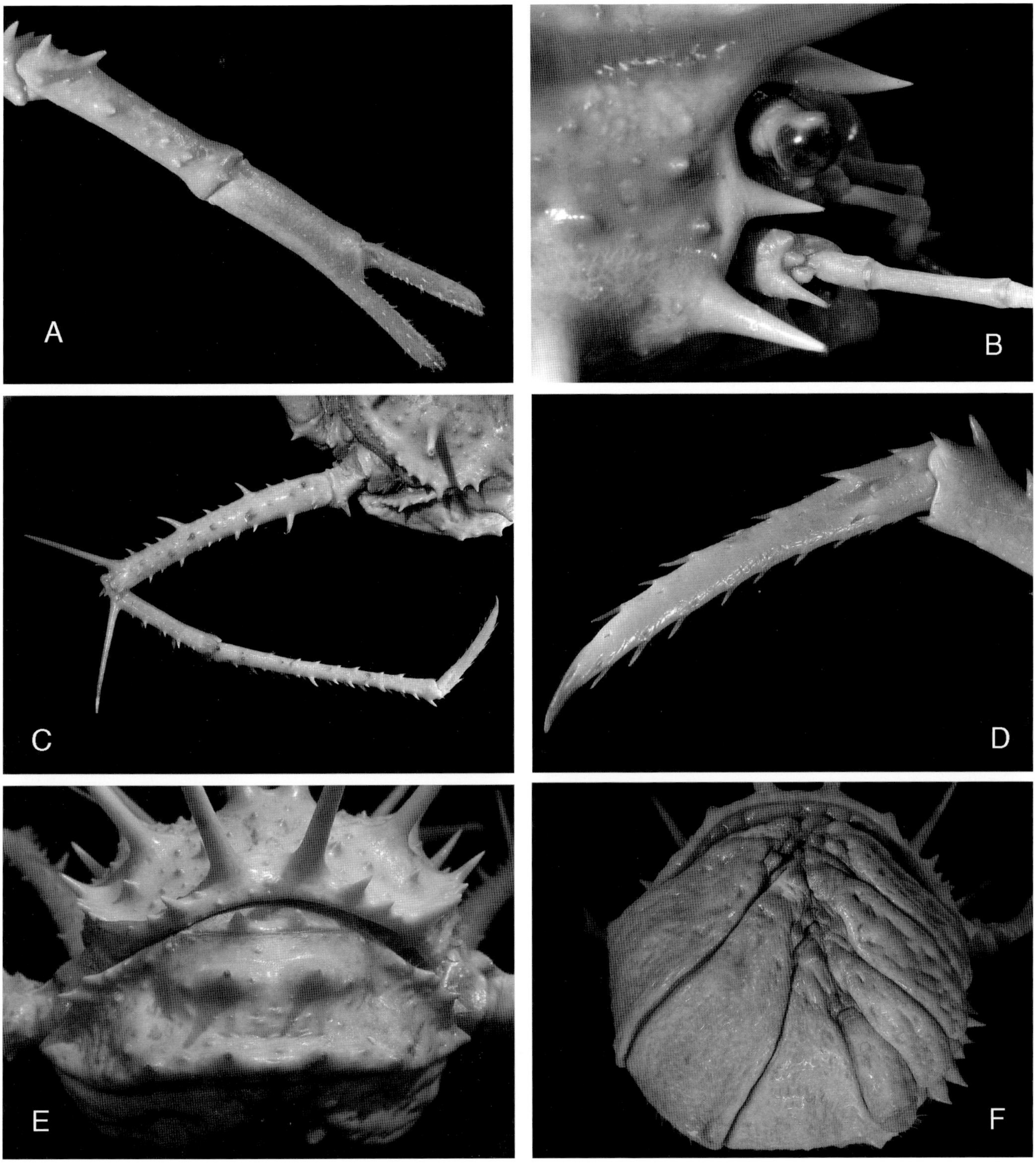

Figure 22. *Lithodes jessica* sp. nov., ovigerous female holotype, cl 97.8 mm, pcl 58.6 mm, cw 52.1 mm, Lord Howe Rise (NIWA 61195). A, right cheliped. B, right orbit and antenna. C, left pereopod 4. D, left pereopod 4 dactylus. E, posterior carapace and abdominal somite 2. F, abdomen.

palms cylindrical, smooth, with very few low tubercles or short teeth. Walking leg dactyli half propodi length; dactyli with spinous flexor margin.

Description. *Carapace*: Pyriform, about 1.10–1.12 times longer than wide; regions indicated; dorsal surface armed with long, slender spines and numerous, well-spaced granules or small conical tubercles, surface

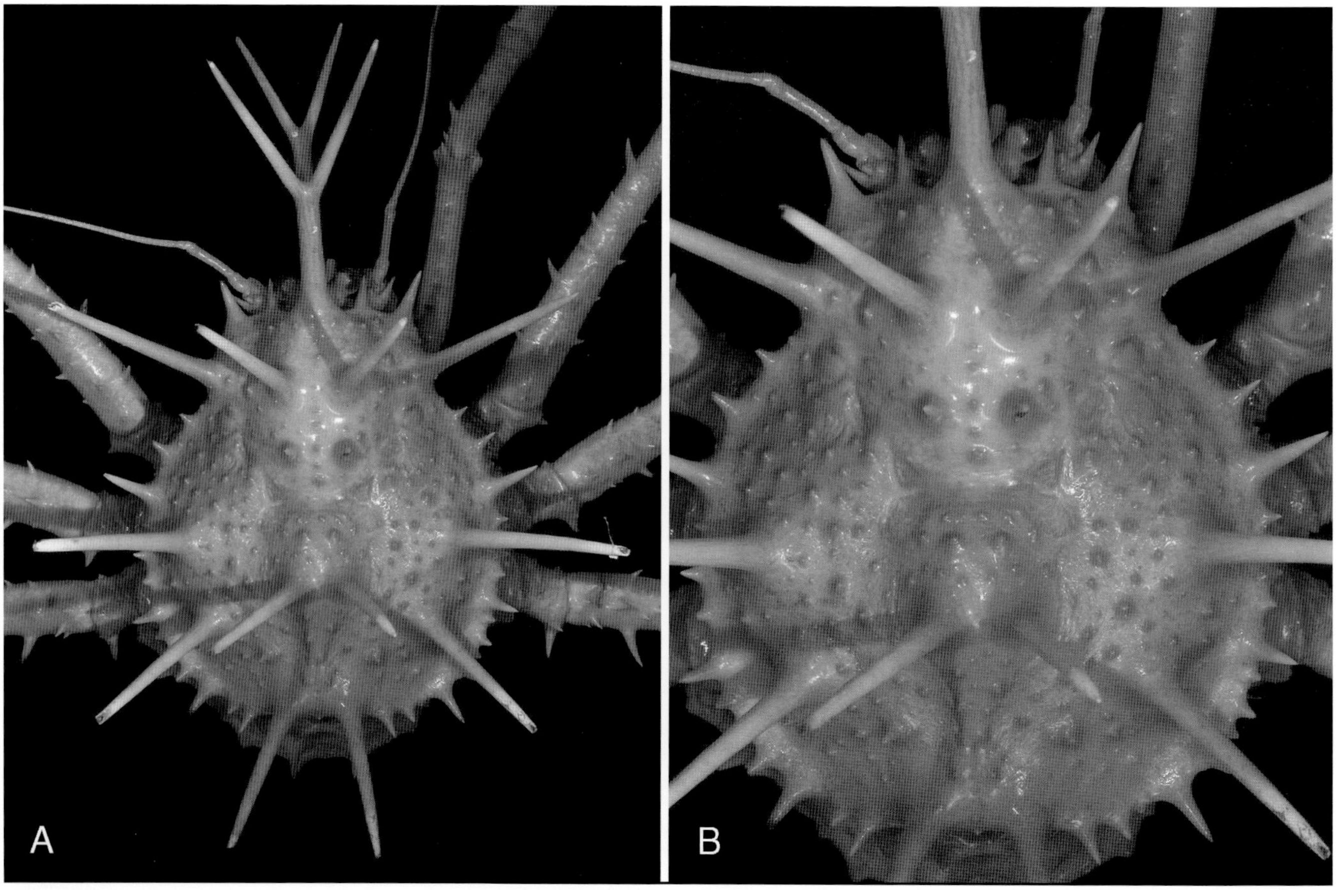

Figure 23. *Lithodes jessica* sp. nov., ovigerous female holotype, cl 97.8 mm, pcl 58.6 mm, cw 52.1 mm, Lord Howe Rise (NIWA 61195). A–B, carapace.

otherwise smooth. Gastric region convex, with 1 pair of long, upright spines anteriorly, posteriorly with pair of short spines or acute tubercles; anterior gastric spines longest of carapace spines (0.6–1.0 pcl). Cardiac region with pair of anterior tubercles or small spines and posterior pair of long upright spines. Hepatic spine directed anterolaterally. Posterior margin of intestinal region with pair of long, upright spines. Branchial surface with 2 long upright spines: first at level of anterior margin of pereopod 4 coxa, second at level of posterior margin of pereopod 4 coxa, often with short spine at base. Branchial margins spinose; anterior branchial margin with 2 short spines, posterior spine the longer; 1st primary marginal branchial spine one-quarter length hepatic spine; lateral branchial margin with 3 short spines; 2nd primary marginal branchial spine about one-eighth length of hepatic spine; posterior branchial margin with 7 or 8 spines. Pterygostomian region granulate or tuberculate, anterior margin angular.

Rostrum 0.67–0.72 pcl, comprising proximal and distal portions; proximal portion angled dorsally by about 45°, with pair of long, divergent dorsal spines that overreach midlength of distal portion of rostrum; distal portion subhorizontal, distally bifurcate for about two-thirds length, forming pair of strongly divergent spines; ventral rostral spine elongate, extending anteriorly beyond apices of outer orbital spine. Posterior orbital margin rounded; outer orbital spine directed anteriorly, shorter than first hepatic spine, not reaching beyond cornea.

Ocular peduncle: Longer than cornea; unarmed.

Antennule: Peduncle unarmed, reaching anteriorly beyond antennal peduncle by half to three-quarters length of distal antennular peduncle article.

Antenna: Basal antennal article unarmed; outer margin of article 2 with slender spine that reaches to or slightly beyond midlength of article 4; articles 3–5 unarmed; scaphocerite absent; article 4 about half as long as article 5. Peduncle extending beyond apex of ventral rostral spine by half to three-quarters length of distal peduncular article.

Abdomen: Somite 1 with pair of tubercles or triangular teeth on posterior margin. Somite 2 undivided, surface sparsely granulate, submedian plates fused with median in adults; median surface with pair of stout submedian spines; posterior margin with 4 stout spines or angular prominences; lateral margins irregularly dentate. Somites 3–5 with nodular median plates; submedian plates pitted, unarmed; marginal plates with irregularly angular margins in males; marginal

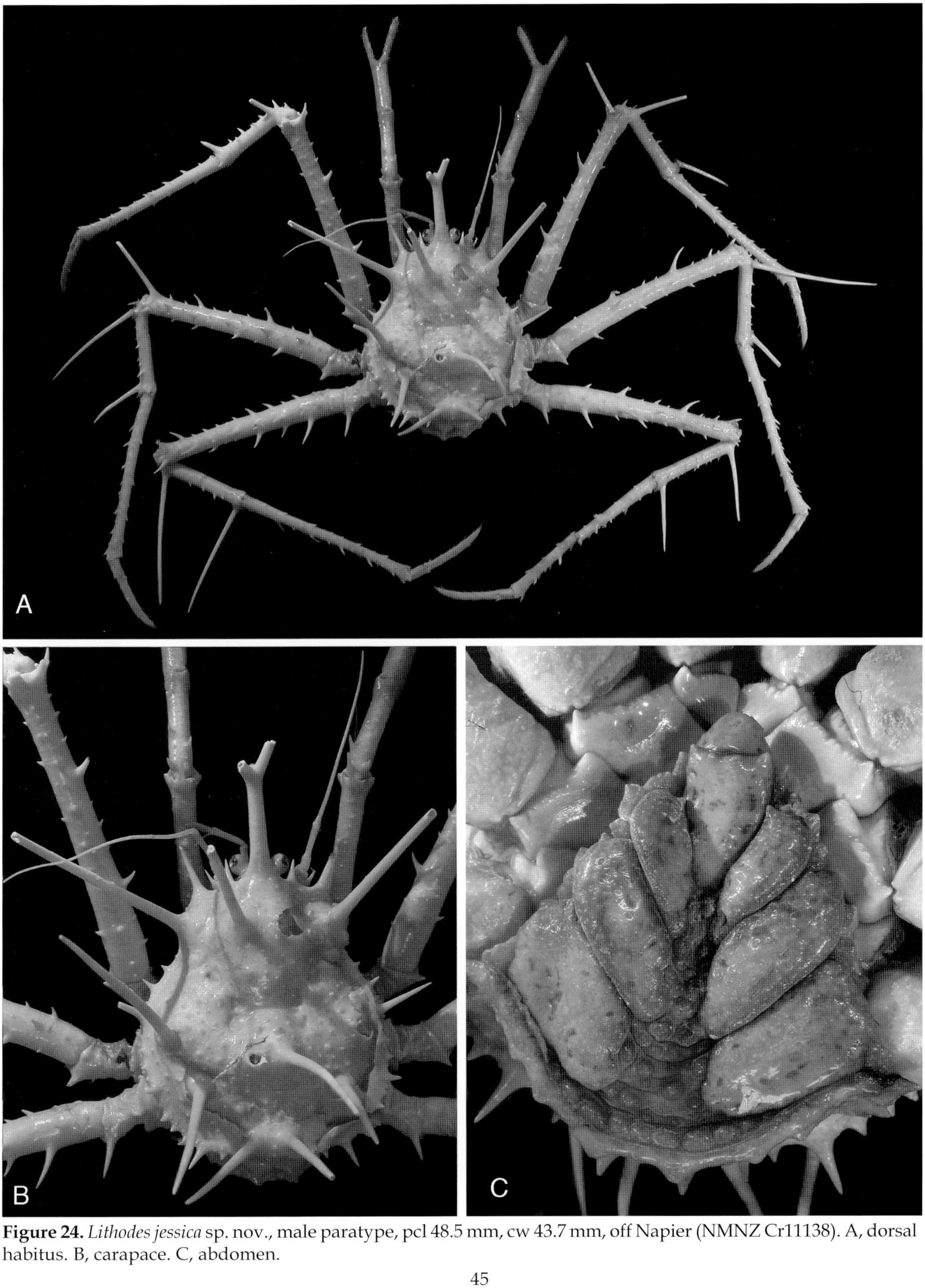

Figure 24. *Lithodes jessica* sp. nov., male paratype, pcl 48.5 mm, cw 43.7 mm, off Napier (NMNZ Cr11138). A, dorsal habitus. B, carapace. C, abdomen.

Figure 25. *Lithodes jessica* sp. nov., male paratype, pcl 48.5 mm, cw 43.7 mm, off Napier (NMNZ Cr11138). A, right cheliped. B, right cheliped. C, left pereopod 4. D, left pereopod 4 dactylus.

plates of females with left margin smooth to crenulate (fused with submedians), right marginals subdivided into 3 or 4 plates on right side, with spiniform apices. Somite 6 of both sexes elongate, unarmed, distal margin with 4 low, blunt prominences. Telson unarmed, rounded.

Pereopod 1 (chelipeds): Subcylindrical; slender, subequal, dimorphism not marked. Chelipeds 1.83 pcl (male), 1.55–1.60 pcl (female). Coxa unarmed; mesial surfaces with setal tufts; ischiobasis with 3 stout ventral spines. Merus inner margin smooth or granular; extensor margin with slender distal spine and 7–9 small spines or spinules; flexor margin with 1 or 2 spinules. Carpus longer than palm; smooth except for 4 prominent spines on proximal half and 2–4 scattered spinules distally. Palm smooth, unarmed; upper palm length 2.95–3.00 times height (male), 2.72 (female). Fingers unarmed; with rows of tufts of golden setae; occlusal margins corneous for distal third to half, remainder calcareous, crenulate; dactylus dorsal margin straight for proximal two-thirds, 1.06–1.08 times longer than dorsal margin of palm (male), 1.12 (female).

Pereopods 2–4 (walking legs 1–3): Similar, slender, spinose; surface between major spines smooth or with few, minute, sparsely distributed spines. Pereopod 4 longest. Coxae unarmed. Ischiobasis with 2 distal ventral spines, posteriormost spine of pereopod 4 longest. Merus subcylindrical; extensor and flexor margins with 2–4 longer spines amongst small, scattered spinules, respectively in addition to long, prominent, distal spine; dorsal surface with 5–7 spines and scattered spinules. Carpus surface with small scattered spines; extensor margin with distal and second proximal spines longest, exceeding three-quarters carpus length. Propodus with 7–9 spines on extensor and flexor margins. Dactylus half propodus length; curved, laterally compressed; proximally with 2 spines on either side adjacent to articulation; extensor margin with 3–5 slender corneous

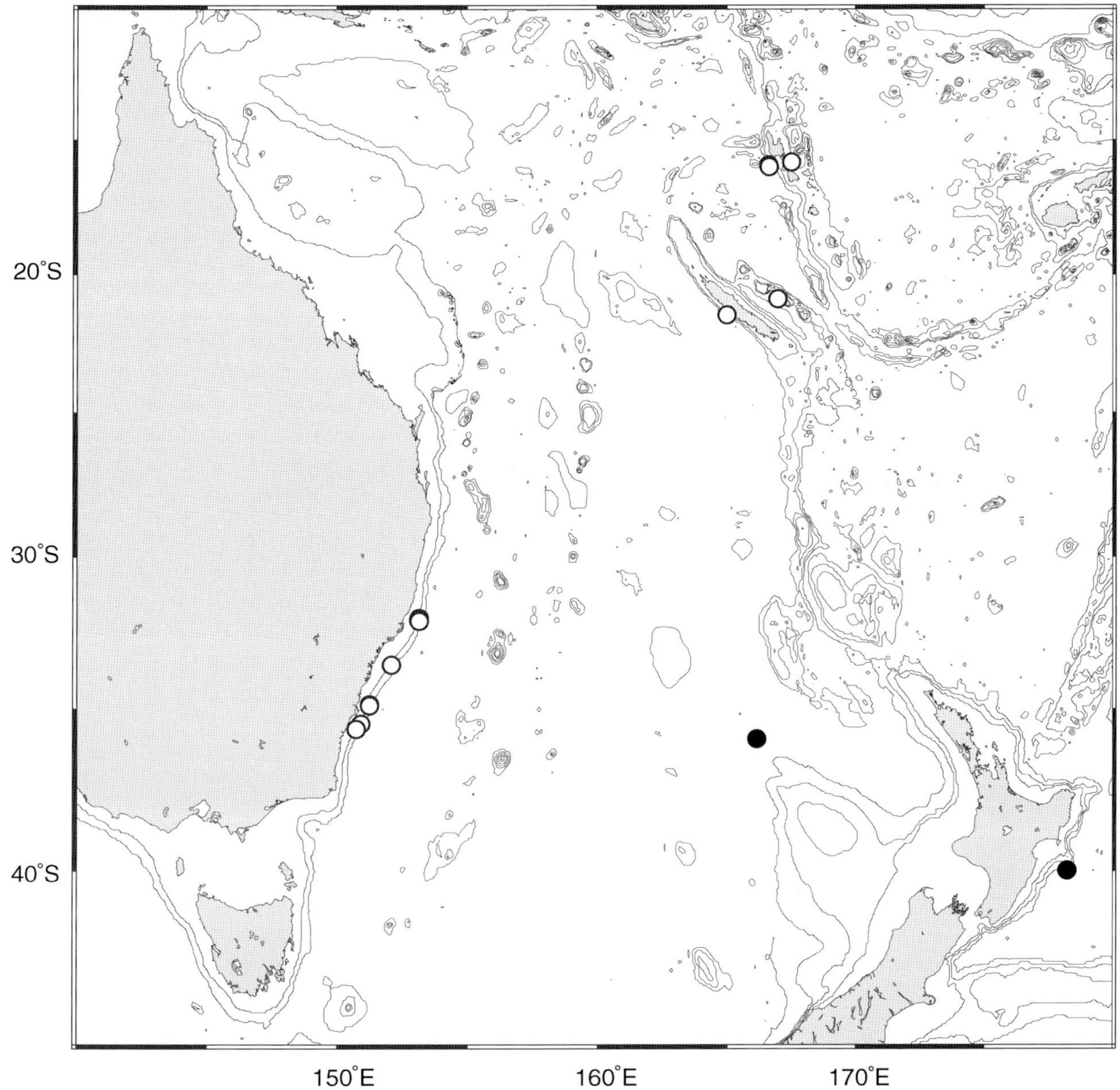

Figure 26. Geographic distributions of *Lithodes jessica* sp. nov. (•) and *L. richeri* Macpherson, 1991 (○).

spines along proximal three-quarters; length of flexor margin lined with 8 or 9 slender corneous spines; apex corneous.

Pereopod 2 length 3.45 pcl (male), 2.81–2.95 pcl (female). Merus 1.25 pcl (male), 0.94–1.02 pcl (female); length:height ratio 9.01 (male), 7.71–8.48 (female). Carpus 0.50 merus length (male), 0.57–0.60 (female). Propodus 0.84 merus length (male), 0.90–0.93 (female); length:height ratio 15.55 (male), 14.02–14.69 (female). Dactylus 0.50 propodus length (male), 0.51–0.53 (female).

Pereopod 3 length 3.42 pcl (male), 2.98–3.11 pcl (female). Merus 1.26 pcl (male), 0.98–1.04 pcl (female); length:height ratio 9.28 (male), 7.14–8.21 (female). Carpus 0.51 merus length (male), 0.59–0.61 (female). Propodus 0.83 merus length (male), 0.96 pcl (female); length:height ratio 15.50 (male), 14.02–14.14 (female). Dactylus 0.52 propodus (male), 0.49–0.51 (female).

Pereopod 4 length 3.54 pcl (male), 2.99–3.12 pcl (female). Merus 1.34 pcl (male), 0.91–1.01 pcl (female); length:height ratio 9.69 (male), 6.44–7.84 pcl (female). Carpus 0.49 merus length (male), 0.62–0.67 (female). Propodus 0.83 merus length (male), 1.02–1.04 (female); length:height ratio 16.64 (male), 14.11–15.08 (female). Dactylus 0.52 propodus (male), 0.49–0.54 (female).

Colour in life. Not known.

Etymology. Named for Jessica Ahyong; used as a noun in apposition.

Remarks. Owing to the long dorsal spines, including two long dorsal branchial spines, *Lithodes jessica* sp. nov. resembles early adults of *L. aotearoa* sp. nov. (pcl ~100 mm or less) and *L. robertsoni* sp. nov. from New Zealand, *L. chaddertoni* sp. nov., *L. rachelae* sp. nov., and *L. australiensis* sp. nov. from Australia, *L. megacantha* Macpherson, 1991 from French Polynesia, and *L. paulayi* Macpherson & Chan, 2008, from Guam. *Lithodes jessica* is most obviously distinguished from each of these species by the subequal, slender, cylindrical chelipeds, which bear only a few short spines on the merus and carpus, and none on the palm. The chelipeds of the aforementioned species are instead multispinose and distinctly dimorphic, in which the major palm is highest near the dactylar articulation. *Lithodes jessica* also differs from *L. australiensis*, *L. aotearoa*, *L. rachelae*, *L. megacantha*, *L. paulayi*, and *L. robertsoni* by the spinose versus unarmed flexor margin on the walking leg dactyli in adults, and by the longer walking legs. In this respect, *L. jessica* resembles *L. chaddertoni*, which also has a spinular flexor margin of the walking leg dactyli. Note, however, that the flexor spination of the walking leg dactyli, a juvenile lithodid feature retained by adults of *L. jessica* and *L. chaddertoni*, is also present in juvenile *L. australiensis* and *L. aotearoa* up to about 60 mm pcl and 20 mm pcl, respectively. Therefore, walking leg dactyl armature should not be used to separate specimens of *L. jessica* from small *L. australiensis* and *L. aotearoa*. Instead, the cheliped dimorphism, barely evident in *L. jessica*, is obvious even in the smallest specimens of *L. aotearoa* and *L. australiensis* examined.

Distribution. Presently known only from northern New Zealand waters between the southern Lord Howe Rise and Hawkes Bay; 680–1100 m.

Lithodes macquariae sp. nov.

(Figs 27–31, Pl. 2E)

Lithodes murrayi. — Hale, 1941: 272, pl. 3: figs. 3, 4. — Healy & Yaldwyn, 1970: 74, fig. 37. — Dawson, 1988: 225. — Davie, 2002: 73 [Macquarie Island occurrences]. — Zaklan, 2002: 766 [Macquarie Island occurrences] [Not *L. murrayi* Henderson, 1888].

Type material. *Holotype*: AM P42530, male (cl 123.2 mm, pcl 102.0 mm, cw 108.3 mm), off Garden Cove, Macquarie Island, 54°30′S, 158°57′E, 16 m, MA-7, rocky bottom, coll. R. Williams, 20 Nov 1977.

Paratypes: AM P42531, 1 male (cl 107.5 mm, pcl 89.3 mm, cw 88.3 mm), off Garden Cove, Macquarie Island, 54°30′S, 158°57′E, 16 m, 20 Nov 1977; NIWA 61160, 1 juvenile female (cl 31.3 mm, pcl 20.9 mm, cw 18.5 mm), Macquarie Ridge, S of Macquarie Island, 55°00′S, 158°46.99′E, 110 m, C730, 24 Nov 1961; NIWA 61187, 1 male (cl 136.5 mm, pcl 111.1 mm, cw 106.0 mm), Macquarie Island, 54°29.5′S, 158°58.50′E, 22 m, C732, 25 Nov 1961; NIWA 61186, 1 male (cl 90.9 mm, pcl 73.1 mm, cw 71.3 mm), Macquarie Ridge, N of Macquarie Island, 54°25.00′S, 159°01.99′E, 104 m, C733, 25 Nov 1961; NIWA 40798, 1 male (pcl 99.3 mm, cw 94.1 mm), Macquarie Ridge, 55°21.19–21.36′S, 158°26.20–26.46′E, 605–709 m, TAN0803/093 SEL 1483, 15 Apr 2008; NMV J14503, 1 female (pcl 68.3 mm, cw 62.6 mm), Garden Cove, Macquarie Island, coll. K. Bennett, 2 Feb 1976; NMV J59364, 2 males (cl 121.5–133.0 mm, pcl 100.6–113.2 mm, cw 98.3–113.5 mm), 54°35.42′S, 158°56.60–58.42′E, 250–500 m, SS0199/93, 26 Jan 1999; NMV J14550, 1 male (cl 145.1 mm, pcl 120.0 mm, cw 116.8 mm), Camp Beach, Macquarie Island, coll. P. Puckert, 3 June 1957; SAM C6852, 1 juvenile male (cl 27.8 mm, pcl 16.6 mm, cw 14.8 mm), W side of Macquarie Island, 54°42.7–41.9′S, 158°46.1–45.9′E, 100–300 m, dredge, RV *Southern Surveyor*, SS0199/60, 22 Jan 1999.

Other material examined. *Campbell Plateau*: NIWA 34897, 1 damaged juvenile male (pcl 44.4 mm, cw 40.1 mm), W of Auckland Islands, 50°11′S, 165°49.3′E, 1090 m, MFISH trip 1292/5, FV *Atlantic Elizabeth*, Z10222, coll. S. Beatson, 18 Nov 1999.

Solander Trough: NIWA 46529, 1 juvenile (cl 16.1 mm, pcl 7.9 mm cw 6.6 mm), 50°56.31′S, 164°33.16′E, 1053–998 m, TAN0306/4, 14 Apr 2003; NIWA 46524, 1 juvenile female (cl 24.9 mm, pcl 12.6 mm, cw 10.9 mm), 50°56.56′S, 164°36.55′E, 1140–1105 m, TAN0306/6, 14 Apr 2003.

Material studied from photographs. *Macquarie Ridge*: NIWA 40735, 1 juvenile male (pcl 35.9 mm), 55°21.69–21.73′S, 158°25.66–26.11′E, 501–630 m, TAN0803/91, 15 Apr 2008; NIWA 40902, 1 juvenile female (pcl 43.8 mm), 56°14.77–14.49′S, 158°30.33–30.90′E, 676–750 m, TAN0803/98, 16 Apr 2008.

Diagnosis. Carapace dorsal spines short and conical, with 4 gastrics, 2 cardiacs, 2 branchials and 2 intestinals. Hepatic margin short tubercle or spine behind anterolateral spine. Posterior branchial margin with 6–9 triangular spines. Rostrum about 0.2–0.3 pcl, with pair of short divergent dorsal spines; distal portion distally bifurcate for one-third length. Abdominal somite 2 composed of three segments, with marginal plates clearly demarcated. Chelipeds dimorphic, segments spinose. Walking legs with surface between major spines smooth or with scattered spines or granules;

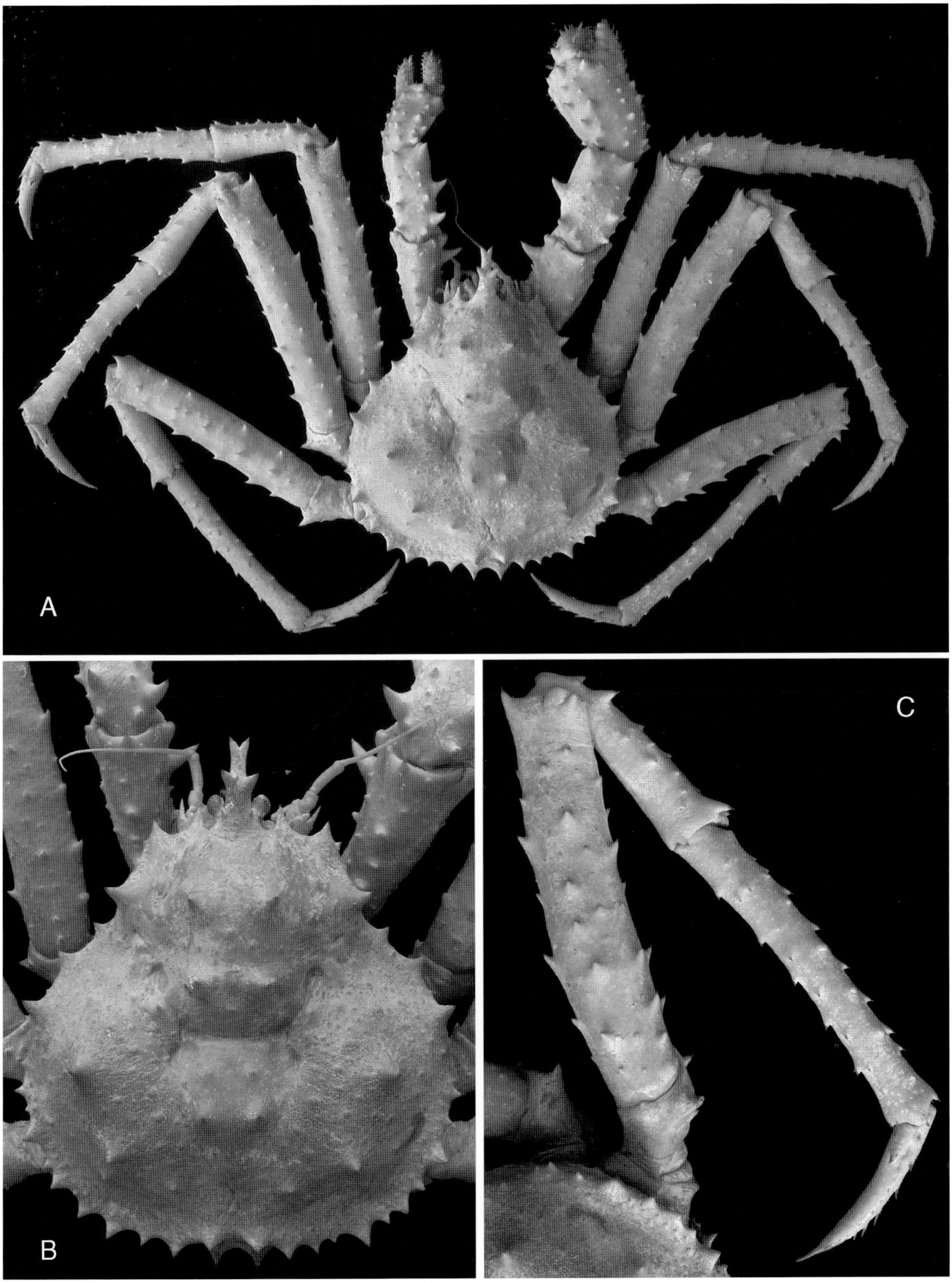

Figure 27. *Lithodes macquariae* sp. nov., male holotype, cl 123.2 mm, pcl 102.0 mm, cw 108.3 mm, off Garden Cove, Macquarie Island (AM P42530). A, dorsal habitus. B, carapace. C, right pereopod 4.

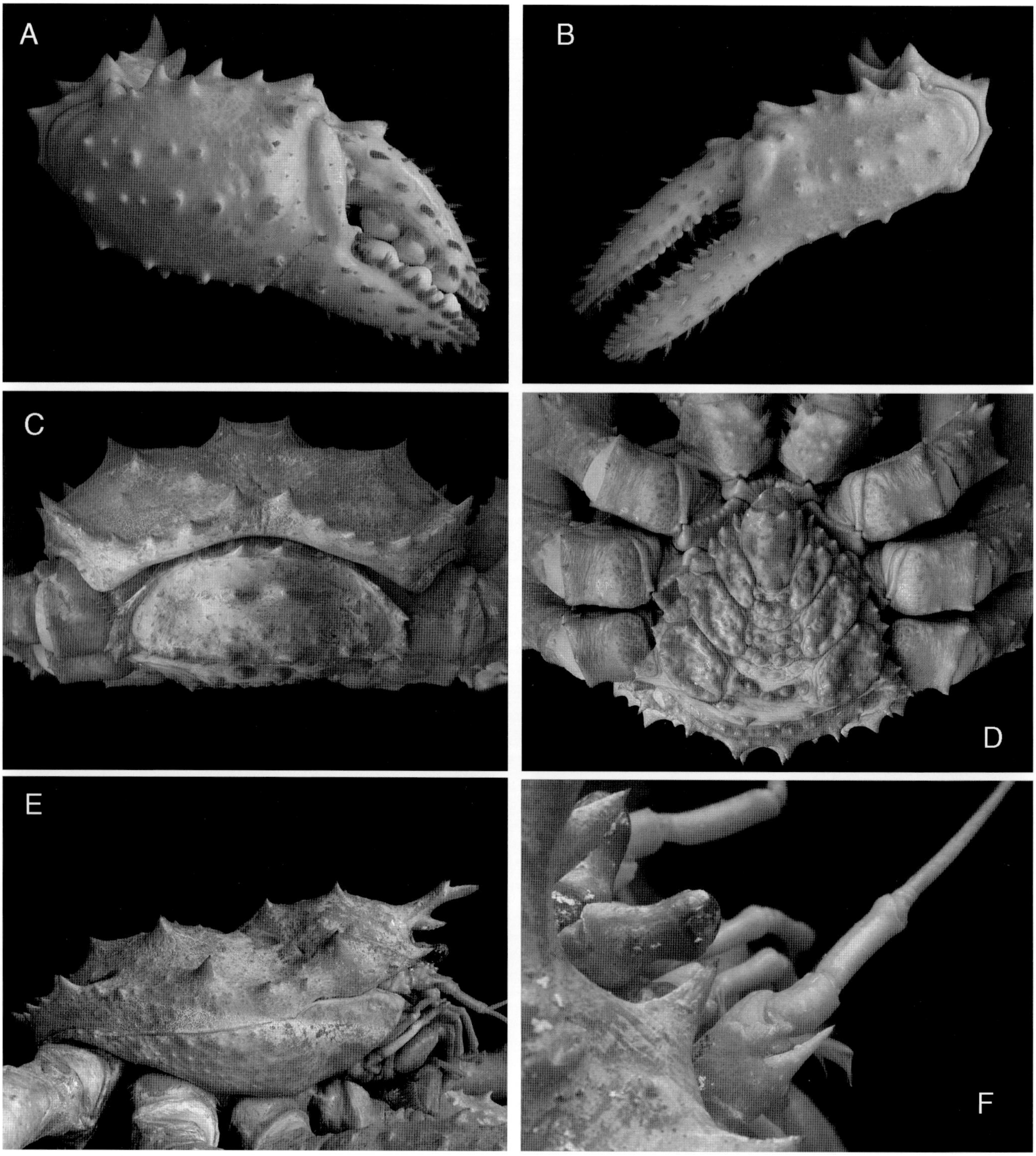

Figure 28. *Lithodes macquariae* sp. nov., male holotype, cl 123.2 mm, pcl 102.0 mm, cw 108.3 mm, off Garden Cove, Macquarie Island (AM P42530). A, right chela. B, left chela. C, posterior carapace and abdominal somite 2. D, ventral surface and abdomen. E, carapace, right lateral view. F, right orbit and antenna.

longest dorsal spines not exceeding half merus height. Walking leg 2 longest, about 3 times pcl in males; merus as long as pcl in males, shorter than pcl in females, less than 6 times longer than high; propodus shorter than merus; dactylus about 0.6 propodus length, flexor margin unarmed.

Description. *Carapace*: Pyriform, 0.94–1.09 times longer than wide; regions indicated; dorsal surface armed with short, conical spines and scattered, widely separated granules. Gastric region convex, with 2 pairs of short, conical spines. Cardiac region with pair of low blunt anterior tubercles and posterior pair of short coni-

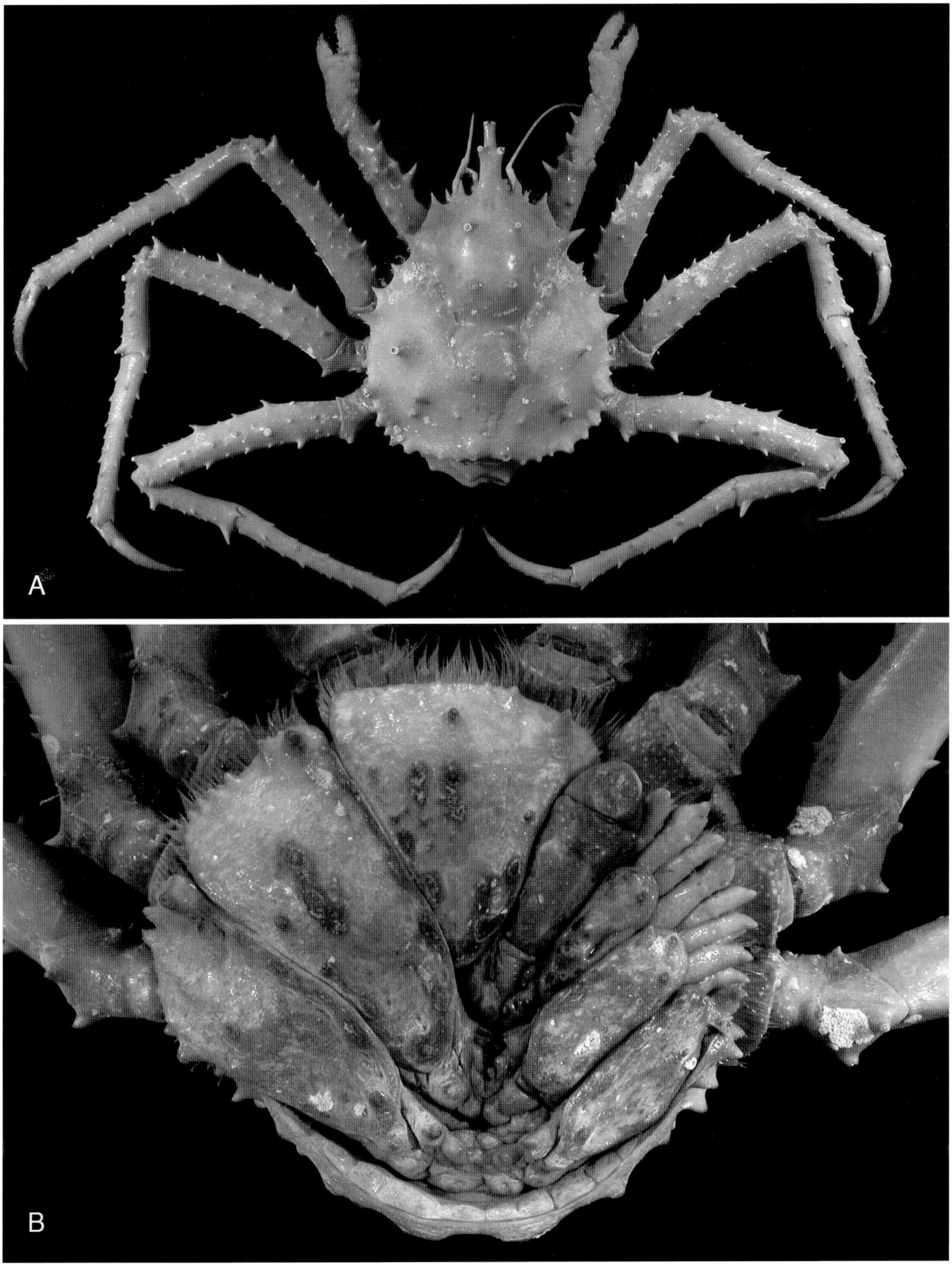

Figure 29. *Lithodes macquariae* sp. nov., female paratype, pcl 68.3 mm, cw 62.6 mm, Garden Cove, Macquarie Island (NMV J14503). A, dorsal habitus. B, abdomen.

cal spines. Hepatic margin with 1 or 2 short tubercles or small spines near midlength and anterolaterally directed conical spine near junction with branchial region. Intestinal region with pair of short, upright, conical spines on posterior margin. Branchial surface with 1 short spine at level of pereopod 3 coxa, 1 short spine at level of pereopod 4 coxa, and conical tubercle behind base. Branchial margins spinose; anterior branchial margin with 2 short spines, posterior spine the longer; 1st primary marginal branchial spine subequal to shorter than hepatic spine; lateral branchial margin with 2 or 3 short spines; 2nd primary marginal branchial spine shorter than 1st primary marginal branchial spine; posterior branchial margin with 6–9 spines. Pterygostomian region granulate or tuberculate, with small, anterior, submarginal spine.

Rostrum short, 0.17–0.24 pcl, comprising upraised proximal portion and horizontal distal portion; proximal portion angled dorsally by at least 30°, with pair of short, divergent dorsal spines; distal portion distally bifurcate for one-third length; ventral rostral spine stout, extending anteriorly beyond apices of outer orbital spine but not beyond distal antennal peduncle segement 4. Posterior orbital margin concave; outer orbital spine directed anteriorly, not reaching beyond cornea; longer than anterolateral spine.

Ocular peduncle: Longer than cornea, unarmed.

Antennule: Peduncle unarmed, reaching anteriorly beyond antennal peduncle by about three-quarters length of distal antennular peduncle article.

Antenna: Basal antennal article unarmed; outer margin of article 2 with slender spine, not reaching beyond midlength of article 4; articles 3–5 unarmed; scaphocerite minute, shorter than article 4; apex blunt, or rarely acute; article 4 unarmed, about half as long as article 5.

Abdomen: Somite 1 with pair of triangular teeth on posterior margin. Somite 2 composed of three plates, with marginal plates clearly demarcated; surface sparsely granulate; median surface with pair of stout submedian spines; posterior margin with 4 posterior spines and several small acute tubercles; lateral margins irregularly dentate. Somites 3–5 with nodular median plates; surface of submedian plates irregular, with scattered, low, acute tubercles; marginal plates of males subdivided, with angular or sharp apices; marginal plates of females with left margin crenulate to spinose (fused with submedians), right marginals subdivided, with angular to spiniform apices. Somite 6 of both sexes as long as or longer than wide, with 2 proximal 4 or distal prominences. Telson rounded, unarmed, at most with 4 blunt granules.

Pereopod 1 (chelipeds): Dimorphic but spination similar. Surfaces of both chelipeds sparsely covered with granules in addition to major spines. Coxa with blunt tubercles and tufts of setae, unarmed; ischiobasis with 3 stout ventral spines and about 6 small tubercles. Merus inner margin, with stout subdistal spine and 2 or 3 smaller spines; ventral surface with two rows of 2 or 3 low tubercles; dorsal and lateral surface spinose, spines largest distally. Carpus with prominent spines on dorsal and lateral surfaces; dorsal margin with row of 4 stout spines; lateral surface with 3 stout spines of similar size to dorsal row in addition to scattered smaller spines. Palms of both chelipeds in both sexes with similar ornamentation; with small spines and acute tubercles on dorsal, lateral and ventral surfaces, inner surface with acute tubercles; dorsal margin with row of about 4 conical spines; midlateral surface with 2 rows of 4–7 small spines; ventral surface with 2 rows of short spines or tubercles (outer with 5 or 6, inner with 2).

Major cheliped 1.51–1.60 pcl (male), 1.22 (female); upper palm length 1.07–1.22 times height (male), 1.28 (female); occlusal margins of fingers corneous for distal third, proximally with 3 calcareous nodules; dactylus dorsal margin broadly convex, with rows of tufts of golden setae and blunt proximal tooth, 1.14–1.25 times longer than dorsal margin of palm (male), 1.30 (female).

Minor cheliped 1.40–1.43 pcl (male), 1.22 (female); upper palm length 1.17–1.35 times height (male), 1.29 (female); occlusal margin corneous for slightly less than distal half, proximally crenulate; dactylus dorsal margin broadly convex, with rows of tufts of golden setae and blunt proximal tooth, 1.43–1.58 times longer than dorsal margin of palm (male), 1.45 (female).

Pereopods 2–4 (walking legs 1–3): Similar, slender, elongate; segments spinose, surface between major spines smooth or with scattered spines or granules. Pereopod 4 longest, about 3 times pcl in males. Distal margins of coxae with 2 short spines. Ischiobasis with 3 non-setose distal spines. Merus subcircular to ovate in cross section; extensor margin with 5–8 spines of which 3 or 4 are distinctly longer in addition to paired distal spines; flexor margin with 2 rows of 4–7 spines; upper surface with row of 5–7 spines in addition to scattered smaller spines. Carpus slightly longer than half merus length; extensor margin with distal and second proximal spines longest; surface with scattered small spines. Propodus ovate in cross section, shorter than merus; with 6–9 spines on extensor margin and scattered spines on dorsal surface; flexor margin with 5 or 6 smaller spines. Dactylus curved, rounded in cross section distally; proximally with 2 spines on either side adjacent to articulation; with 2–4 corneous spines along extensor margin; flexor margin unarmed; apex corneous.

Pereopod 2 length 2.58–2.73 pcl (male), 2.17 (female). Merus 0.92–0.95 pcl (male), 0.75 (female); length: height ratio 4.48–5.33 (male), 4.73 (female). Carpus 0.53–0.55 merus length (male), 0.58 (female). Propodus

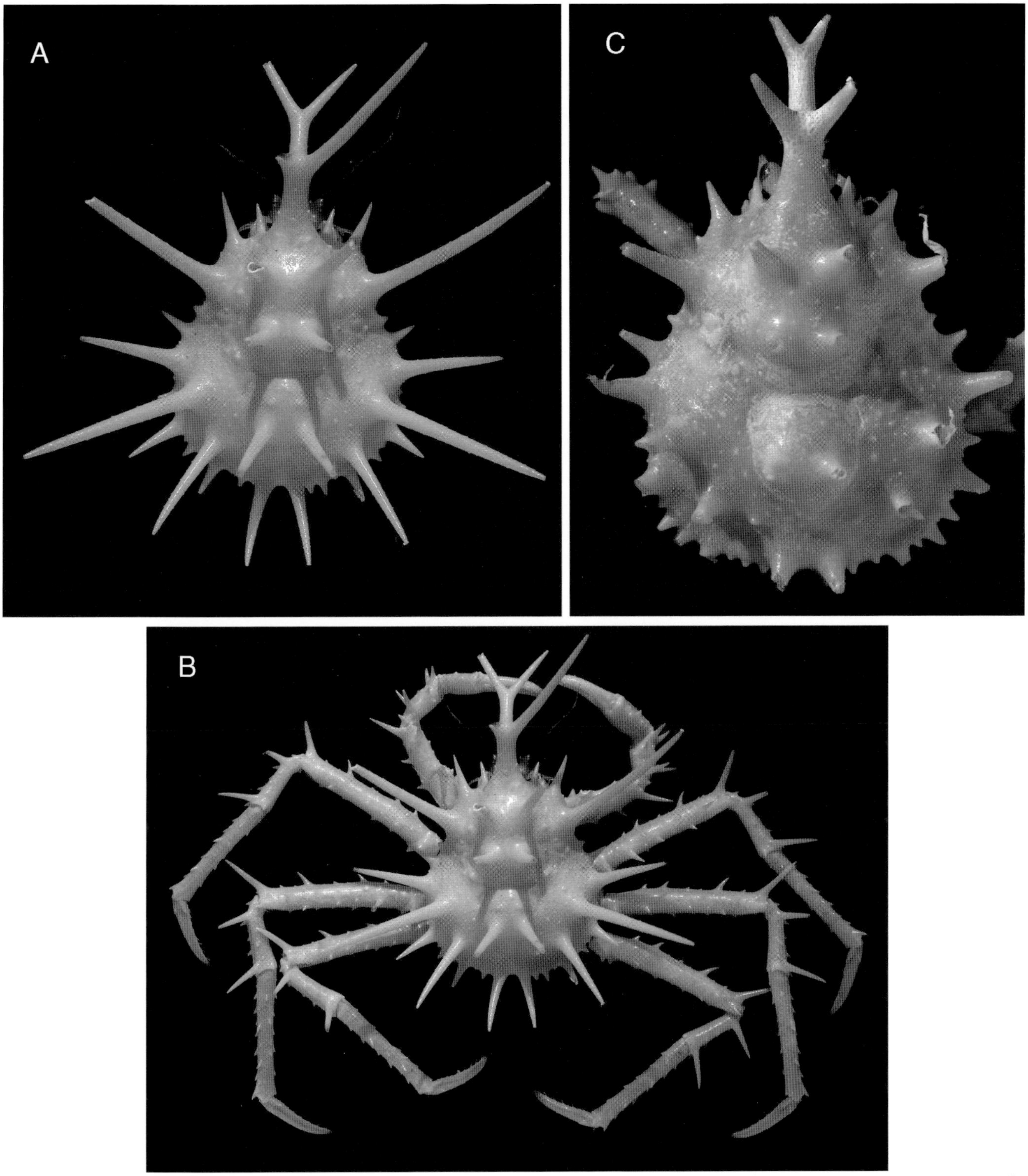

Figure 30. *Lithodes macquariae* sp. nov., A–B, juvenile male paratype, cl 27.8 mm, pcl 16.6 mm, cw 14.8 mm, W side of Macquarie Island (SAM C6852). C, female paratype, cl 31.3 mm, pcl 20.9 mm, cw 18.5 mm, Macquarie Ridge, S of Macquarie Island (NIWA 61160).

0.85–0.86 merus length (male), 0.88 (female); length: height ratio 7.55–8.80 (male), 8.41 (female). Dactylus 0.59–0.67 propodus length (male), 0.59 (female).

Pereopod 3 length 2.72–2.99 pcl (male), 2.37 (female). Merus 0.96–1.05 pcl (male), 0.82 (female); length: height ratio 5.20–5.92 (male), 5.05 (female). Carpus 0.52–0.55 merus length (male), 0.57 (female). Propodus 0.84–0.90 merus length (male), 0.87 (female); length: height ratio 8.23–9.05 (male), 8.35 (female). Dactylus 0.57–0.63 propodus length (male), 0.57 (female).

Pereopod 4 length 2.82–2.98 pcl (male), 2.48 (female). Merus 0.95–1.00 pcl (male), 0.81 (female); length: height ratio 4.76–5.90 (male), 5.21 (female). Carpus 0.52–0.56 merus length (male), 0.59 (female). Propodus 0.90–0.92 merus length (male), 0.93 (female); length: height ratio 7.91–9.76 (male), 8.98 (female). Dactylus 0.58–0.64 propodus length (male), 0.57 (female).

Colour in life. Deep-red overall. Juveniles usually paler, being pinkish overall with red spines, rostrum and cephalic appendages (Pl. 1F).

Etymology. The specific name alludes to the distribution of the species, the Macquarie Ridge; used as a noun in the genitive singular.

Remarks. *Lithodes macquariae* sp. nov. most closely resembles *L. murrayi* Henderson, 1888, with which it has been previously misidentified (Hale 1941; Healy & Yaldwyn 1970). The Macquarie Island and southwest Indian Ocean forms are no doubt closely related, but differ consistently, justifying specific separation. *Lithodes macquariae* differs from *L. murrayi* chiefly in the proportional lengths of the walking leg dactyli, measuring about 0.6 rather than less than 0.5 propodus length, and in bearing 6–9 triangular posterior branchial marginal spines (of which 6 or 7 are prominent) rather than 4–6 (of which 3 or 4 are prominent). Adults might also have a proportionally shorter rostrum, measuring about 0.2 pcl versus about 0.3 pcl in adults of 89 mm pcl or greater (Macpherson 1988c). Macpherson (1988c)

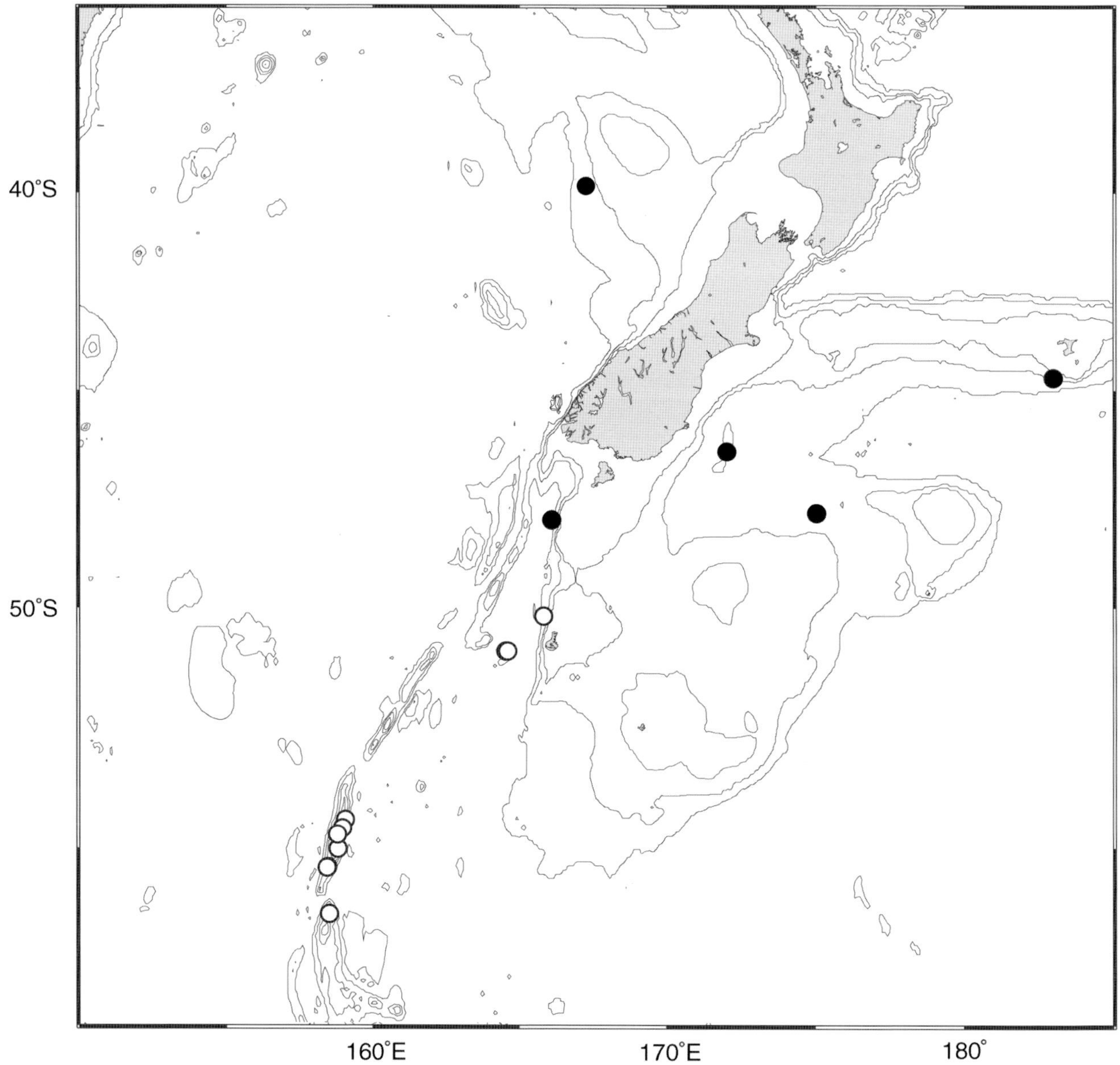

Figure 31. Geographic distributions of *Lithodes macquariae* sp. nov. (○) and *L. robertsoni* sp. nov. (●).

noted that rostral length should be used with caution in view of allometric variation and anomalies caused by damage and regeneration. However, morphometric data based on extensive series of *L. murrayi* from the Crozet Islands indicates that the rostral length follows a consistent pattern (Arnaud & Do-Chi 1977), so the differences observed in proportional rostral length might be real rather than apparent. Data from a larger series of *L. macquariae* are required to better evaluate rostral length variation. Note that Henderson's (1888) original figure of *L. murrayi* incorrectly depicts the walking leg dactyli as about 0.6 propodus length. Macpherson's (1988b) redescription and figures of *L. murrayi* show that the dactyli measure half or slightly less than half propodus length.

The smallest juveniles examined (male pcl 7.9 mm, female pcl 12.6 mm, male pcl 16.6 mm, female pcl 20.9 mm), differ from adults in expected features: a proportionally longer rostrum (1.0 pcl, 1.0 pcl, 0.7 pcl, and 0.5 pcl, respectively) with longer apical spines, a more elongate carapace (pcl 1.2, 1.2, 1.1, 1.1 times cw, respectively), proportionally longer carapace spines, and flexor spinules on the walking leg dactyli. The length of the hepatic spine is about 1.0 pcl at 7.9 and 12.6 mm pcl, and 0.8 pcl at 16.6 mm pcl. At 20.9 mm pcl, the hepatic spine is 0.3 pcl, which is considerably shorter than that of size-matched juveniles, or even juveniles of twice the size of *L. australiensis* in which the hepatic spine length exceeds the pcl. In the damaged male juvenile (pcl 44.4 mm, NIWA 34897), the flexor spinules on the walking leg dactyli are minute and barely evident. Two juvenile *L. macquariae*, a male (pcl 35.9 mm) and female (pcl 43.8 mm), were captured and photographed in 2008 by NIWA expedition TAN0803 to the Macquarie Ridge (Pl. 1F). Unfortunately, both specimens were lost at sea, but the relatively short carapace and pereopod spines are evident in the photographs.

Sexual dimorphism in *L. macquariae* is typical of other lithodids. Adult females have an asymmetrical abdomen and proportionally shorter walking legs. The examined female specimen is mature, having a fully developed abdomen and setose ventral surfaces of the pereopodal coxae.

Prior to revision by Macpherson (1988b, c), *Lithodes murrayi* was attributed a southern circumpolar distribution, having been reported from the southwestern Indian Ocean (Prince Edward Island, Crozet Islands, Reunion Islands, South Africa, Macquarie Island and New Zealand) (Yaldwyn & Dawson 1970). As shown by Macpherson (1988b, c), however, records of *L. murrayi* from South Africa are referable to *L. mamillifer* Macpherson, 1988c, and those from Chile are based on *L. turkayi* Macpherson, 1988b. Similarly, records of *L. murrayi* from the Australian and New Zealand regions are referable to *L. macquariae* and *L. aotearoa* sp. nov. *Lithodes murrayi* is thus reliably known only from the southwestern Indian Ocean; recent reports of *L. murrayi* from Peter I Island and the Bellingshausen Sea (Klages *et al.* 1995; García Raso *et al.* 2005) require verification.

Of all lithodids known from the region, *L. macquariae* is the only species to range into shallow sublittoral depths. *Lithodes macquariae* is known from 16–709 m depth around Macquarie Island, and 998–1050 m at 5° further north, near the Auckland Islands. This is consistent with the polar emergence pattern observed in other lithodids at high latitudes, particularly in the southern hemisphere (Lovrich *et al.* 2002). *Lithodes macquariae* is one of two lithodid species known from the Macquarie Ridge. The second species, *Paralomis birsteini*, can be distinguished by its undivided abdominal somite 2, which, in *L. macquariae*, comprises three plates.

Distribution. Known only from the Macquarie Ridge near Macquarie Island, north to the Solander Trough in the vicinity of the Auckland Islands; 16–1140 m.

Lithodes rachelae sp. nov.

(Figs 17, 32–34, Pl. 1D)

?*Lithodes* aff. *longispina*. — Poore *et al.*, 2008: 27.

Type material. *Holotype*: SAM C6357, male (cl 142.0 mm, pcl 82.9 mm, cw 75.0 mm), Great Australian Bight, approx. 120 nautical miles [222 km] SSE of Eucla, South Australia, 33°37.2'S, 129°52.0'E, 930–1087 m, FV *Longva III*, shot 31, coll. K. Gowlett-Holmes, 5 Apr 1989.

Paratypes: SAM 6853, 1 female (cl 80.0 mm, pcl 43.8 mm, cw 40.0 mm), SE of Albany, Western Australia, 1500 m, FV *Comet*, coll. D. Turner, 10 Nov 1990; SAM C6367, 1 male (pcl 117.3 mm, cw113.1 mm), Great Australian Bight, approx. 120 nautical miles SSE [222 km] of Eucla, South Australia, 33°45'S, 129°17'E, 999–1101 m, FV *Adelaide Pearl*, coll. K. Gowlett-Holmes & H. Cameron, 1 Aug 1988.

Other material examined. *Western Australia:* NMV J54985, 1 juvenile female (pcl 6.7 mm, cw 5.4 mm), SE of Albany, 35°26.05'S, 118°21.07'E, SS1005/028, RV *Southern Surveyor*, 23 Nov 2005.

Diagnosis. Carapace dorsal surface with slender spines; gastric region with 4 upright spines, anterior pair longer than half length of hepatic spine, posterior pair half length of hepatic spine; cardiac and intestinal regions each with 2 long upright spines; branchial surface with 2 long upright spines. First primary marginal branchial spine about two-thirds length of hepatic spine. Branchial margins spinose; posterior branchial margin with 7 or 8 short spines. Rostrum exceeding 0.6 pcl; proximal half strongly upraised by

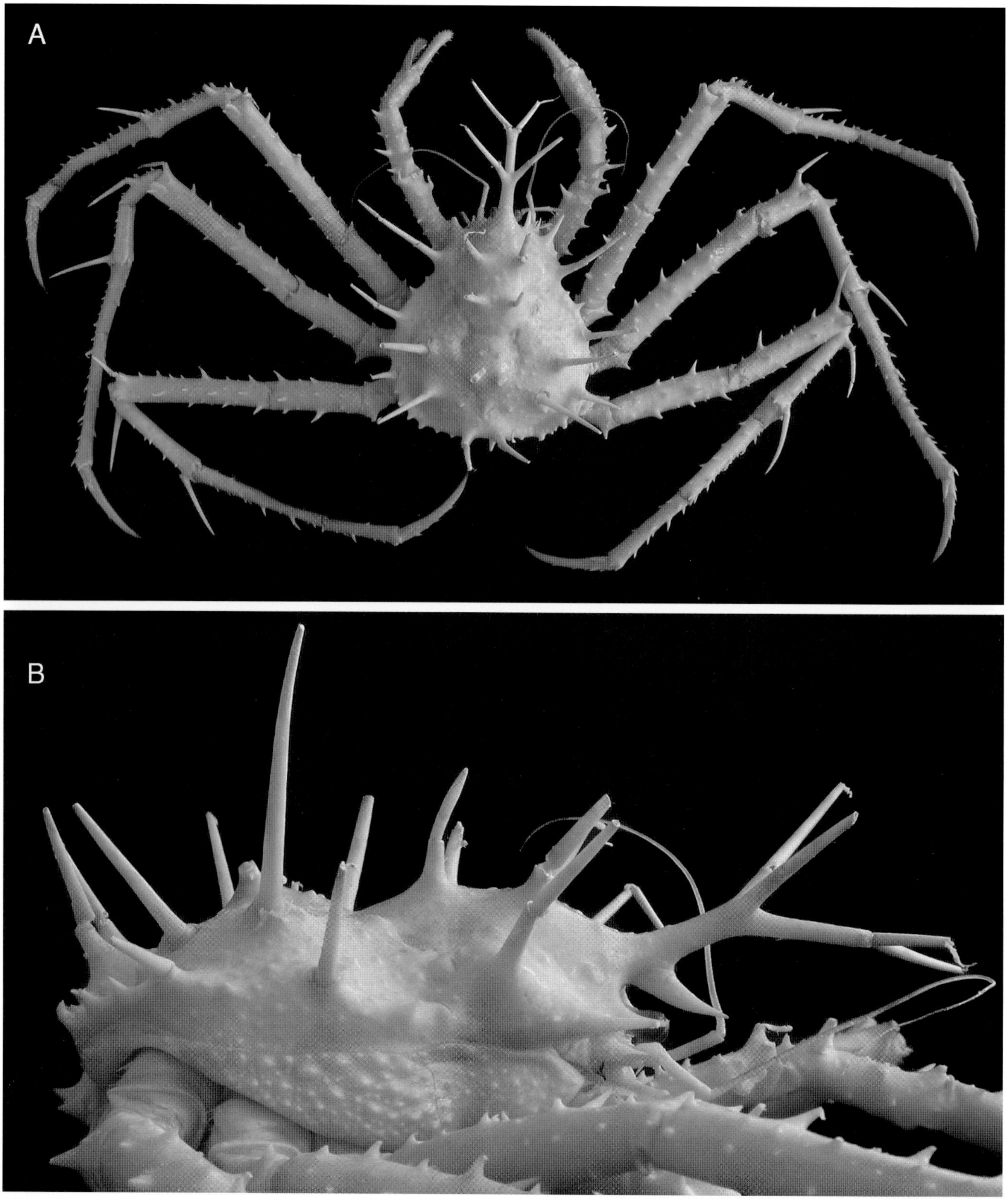

Figure 32. *Lithodes rachelae* sp. nov., male holotype, cl 142.0 mm, pcl 82.9 mm, cw 75.0 mm, Great Australian Bight, SSE of Eucla, South Australia (SAM C6357). A, dorsal habitus. B, carapace, right lateral view.

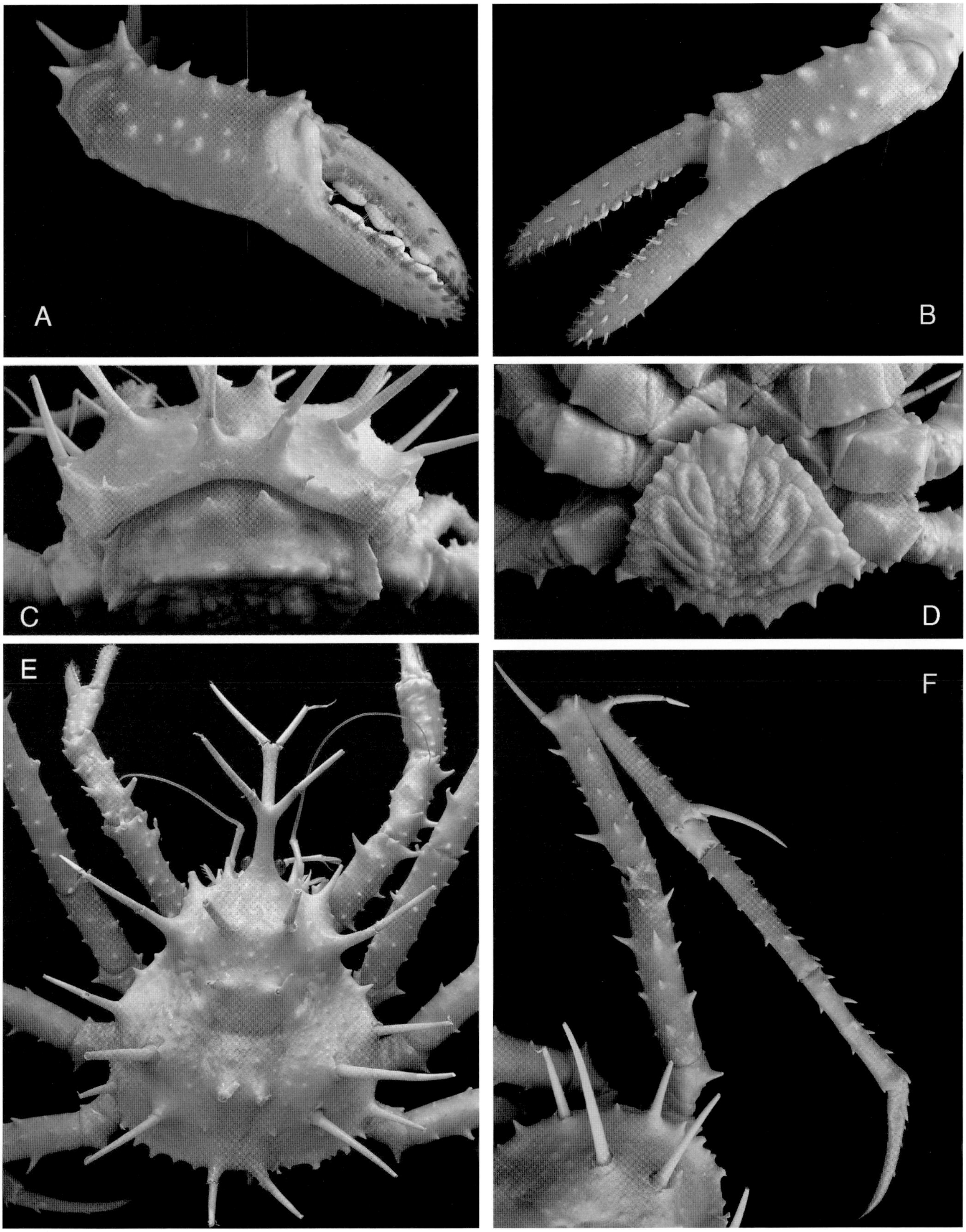

Figure 33. *Lithodes rachelae* sp. nov., male holotype, cl 142.0 mm, pcl 82.9 mm, cw 75.0 mm, Great Australian Bight, SSE of Eucla, South Australia (SAM C6357). A, right chela. B, left chela. C, posterior carapace and abdominal somite 2. D, ventral surface and abdomen. E, carapace. F, right pereopod 4.

about 30°. Abdominal somite 2 comprising 3 plates in juveniles through adults. Antennal peduncle article 2 with slender outer spine not reaching beyond article 4. Chelipeds unequal; palms with low tubercles or spines on dorsal surface. Walking legs spinose; surface between major spines smooth or with very few, scattered spines. Pereopod 4 merus about 1.3 pcl in males, as long as pcl in females; merus length about 9 times height (male), less than 8 times height (female); propodus length 15–16 times height (male), 13 times height (female); dactylus flexor margin unarmed.

Description. *Carapace*: Pyriform, 1.04–1.10 times longer than wide; regions indicated; dorsal surface armed with long, slender spines and scattered, widely separated granules. Gastric region convex, with 2 pairs of long, upright spines. Cardiac region with pair of anterior tubercles and posterior pair of long upright spines. Hepatic margin with low tubercle at midlength and spine longest of carapace spines (0.58 pcl in holotype), directed anterolaterally. Intestinal region with pair of long upright spines on posterior margin. Branchial surface with 2 long upright spines and 2 conical tubercles as follows: 1 spine at level of pereopod 3 coxa, 1 spine at level of pereopod 4 coxa, 2 conical tubercles (or short spines in juveniles) in transverse row slightly behind posterior spine. Branchial margins spinose; anterior branchial margin with 2 short spines, posterior spine the longer; 1st primary marginal branchial spine about two-thirds length of hepatic spine; lateral branchial margin with 1 or 2 short spines; 2nd primary marginal branchial spine shorter than 1st primary marginal branchial spine; posterior branchial margin with 7 or 8 short spines. Pterygostomian region granulate or tuberculate, with small, anterior, submarginal spine.

Rostrum 0.71–0.82 pcl, comprising proximal and distal portions; proximal portion angled dorsally by about 30°, with pair of long, divergent, dorsal spines that overreach midlength of distal portion of rostrum; distal portion subhorizontal, distally bifurcate for half length, forming pair of strongly divergent spines; ventral rostral spine elongate, extending anteriorly beyond apices of outer orbital spine. Posterior orbital margin concave; outer orbital spine directed anteriorly, slightly longer than anterolateral spine, reaching slightly beyond cornea.

Ocular peduncle: Longer than cornea; unarmed.

Antennule: Peduncle unarmed, reaching anteriorly beyond antennal peduncle by full length of distal antennular peduncle article.

Antenna: Basal antennal article unarmed; outer margin of article 2 with long slender spine, not reaching midlength of article 4; article 3 unarmed; scaphocerite minute, blunt, shorter than article 4; article 4 unarmed, about half as long as article 5.

Abdomen: Somite 1 with low triangular teeth on posterior margin. Somite 2 rugose; composed of 3 distinct plates separated by narrow groove; median surface with pair of slender submedian spines; posterior margin with 6 stout spines; lateral margins irregularly dentate. Somites 3–5 with nodular median plates; surface of submedian plates irregular, unarmed; marginal plates of males subdivided, with angular apices; marginal plates of females with left margin crenulate (fused with submedians), right marginals subdivided, with angular apices. Somite 6 of both sexes longer than wide, distal margin with 2 or 4 blunt prominences. Telson rounded, unarmed.

Pereopod 1 (chelipeds): Dimorphic but spination similar. Surfaces of both chelipeds smooth or with scattered granules in addition to several small spines. Coxa unarmed; with tufts of setae mesially. Ischiobasis with 2 or 3 blunt ventral spines and scattered tubercles. Merus inner margin, with straight subdistal spine; ventral margin with scattered tubercles; dorsal margin with 5–7 small, scattered spines and 3 prominent spines, increasing in length distally, distalmost shorter than half carpus length; lateral surface spinose, spines largest distally. Carpus with 6 spines and 3 or 4 small, scattered spines or acute tubercles. Palm with 13–15 small scattered spines or tubercles on dorsal and lateral surfaces, inner and ventral surfaces unarmed.

Major cheliped 1.48–1.65 pcl (male), 1.39 (female); upper palm length 1.17–1.42 times height (male), 1.42 (female); occlusal margins of fingers corneous for distal fifth, proximally with 3 or 4 calcareous nodules; dactylus dorsal margin broadly convex, with rows of tufts of golden setae and blunt proximal tooth, 1.22–1.24 times longer than dorsal margin of palm (male), 1.40 (female).

Minor cheliped 1.54–1.60 pcl (male), 1.68 (female); upper palm length 1.37–1.64 times height (male), 1.55 (female); occlusal margin corneous for slightly less than distal half, proximally crenulate; dactylus dorsal margin broadly convex, with rows of tufts of golden setae and blunt proximal tooth, 1.54–1.60 times longer than dorsal margin of palm (male), 1.68 (female).

Pereopods 2–4 (walking legs 1–3): Similar, slender, spinose; surface between major spines with small, scattered spines. Pereopod 4 longest. Coxa of pereopods 2–4 unarmed. Ischiobasis dorsodistal margin with two small spines or blunt tubercles; ventrodistal margin with two spines, posteriormost spine of pereopod 4 longest. Merus subcylindrical; extensor and flexor margins with small scattered spines and 3 or 4 major spines in addition to long, prominent, distal spine, not exceeding half merus length; dorsal surface with irregular row of 6–10 spines of which 3 or 4 prominent. Carpus surface with scattered, short spines and strongly elongated distal and second proximal spines, each exceeding half carpus length. Propodus extensor margins with 12–15 spines; flexor margins with 6–10

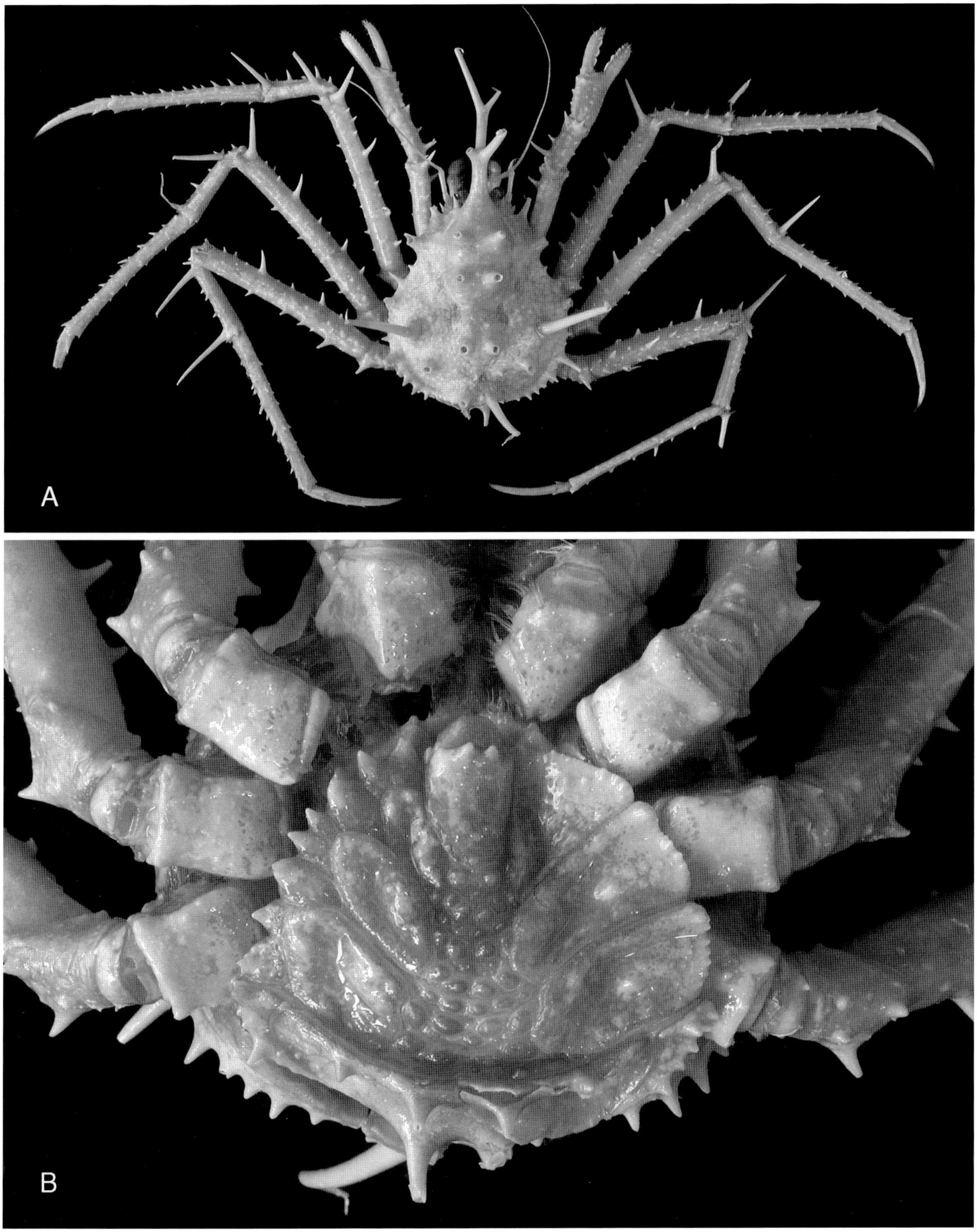

Figure 34. *Lithodes rachelae* sp. nov., female paratype, cl 80.0 mm, pcl 43.8 mm, cw 40.0 mm, SE of Albany, Western Australia (SAM C6853). A, dorsal habitus. C, ventral surface and abdomen.

spines. Dactylus curved, laterally compressed; proximally with 2 spines on either side adjacent to articulation; extensor margin with 2–4 slender corneous spines along length; flexor margin unarmed; apex corneous.

Pereopod 2 length 3.26–3.27 pcl (male), 2.85 pcl (female). Merus 1.23–1.24 pcl (male), 1.01 pcl (female); length:height ratio 8.1–9.82 (male), 7.60 (female). Carpus 0.46–0.51 merus length (male), 0.49 (female). Propodus 0.82–0.87 merus length (male), 0.85 (female); length:height ratio 13.13–14.16 (male), 11.23 (female). Dactylus 0.50–0.54 propodus length (male), 0.60 (female).

Pereopod 3 length 3.57–3.64 pcl (male), 3.06 pcl (female). Merus 1.33–1.36 pcl (male), 1.07 pcl (female); length:height ratio 8.92–9.09 (male), 7.73 (female). Carpus 0.47–0.51 merus length (male), 0.51 (female). Propodus 0.85–0.87 merus length (male), 0.87 (female); length:height ratio 14.72–16.61 (male), 11.84 (female). Dactylus 0.48–0.52 propodus length (male), 0.57 (female).

Pereopod 4 length 3.62–3.71 pcl (male), 3.10 pcl (female). Merus 1.29–1.30 pcl (male), 1.06 (female); length: height ratio 9.17–9.36 (male), 7.60 (female). Carpus 0.51–0.55 merus length (male), 0.51 (female). Propodus 0.88–0.91 merus length (male), 0.90 (female); length: height ratio 15.24–15.61 (male), 13.21 (female). Dactylus 0.51–0.53 propodus length (male), 0.59 (female).

Colour in life. Deep-red overall (Pl. 1D).

Etymology. Named for Rachel Ahyong, for her support during the course of the study.

Remarks. *Lithodes rachelae* sp. nov. most closely resembles *L. australiensis* sp. nov., *L. robertsoni* sp. nov., and *L. paulayi* Macpherson & Chan, 2008, sharing long carapace spines in adults; two long dorsal branchial spines; a three-segmented abdominal somite 2 in adults; and unarmed flexor margins of the dactyli of the walking legs. *Lithodes rachelae* is readily separated from *L. paulayi* in having proportionally longer primary branchial spines, with the 2nd primary branchial spine in the new species distinctly longer, rather than shorter than the outer oribital spine; and in longer walking legs with the third leg of males 3.6–3.7 versus less than 3.2 pcl. *Lithodes rachelae* differs from *L. australiensis* and *L. robertsoni* in having proportionally longer, more slender walking legs, being most pronounced in adult males in which the merus of the third walking leg is 1.3 pcl versus less than 1.2 pcl in the latter two species (further features distinguishing *L. rachelae* from *L. robertsoni* are given under the account of the latter). The third walking leg of male *L. rachelae* is 3.6–3.7 versus less than 3.5 pcl, with a merus length:height ratio 9.2–9.4 versus a maximum of 7.8 in *L. australiensis*. The third walking leg of female *L. rachelae* is 3.1 versus 2.6 pcl or less and the merus length:height ratio 7.6 versus a maximum of 7.3. It should also be noted that the differences between *L. rachelae* and *L. australiensis* in relative length and slenderness of the walking legs can be expected to be more pronounced when larger specimens of the former become available (largest known *L. rachelae* pcl 117.3 mm; largest *L. australiensis* pcl 133.8 mm), and size-matched specimens can be directly compared. The proportional length of the merus and overall length of the third walking leg of the smallest male *L. rachelae* (1.3 pcl and 3.6 pcl, respectively) exceeds that of the largest male *L. australiensis* (1.2 pcl; 3.5 pcl). In particular, more accurate comparisons between female *L. australiensis* and *L. rachelae* must await capture of adult females of both species. The female paratype of *L. rachelae* is a subadult, with the abdomen not having reached full size or maximum asymmetry.

Lithodes rachelae is presently known only between the Great Australian Bight and Albany, Western Australia, whereas *L. australiensis* ranges from central New South Wales to Tasmania.

One of two juvenile *Lithodes*, reported and figured as *L.* aff. *longispina* by Poore *et al.* (2008) was collected from southeast of Albany, where adult *L. rachelae* is known to occur. The juvenile is tentatively referred to *L. rachelae*. The second specimen, from Perth Canyon (31°58.28′S, 115°06′W), was not available for examination at the time of the study.

Distribution. South Australia, from the Great Australian Bight, to southwestern Western Australia; 930–1500 m.

Lithodes richeri Macpherson, 1990

(Figs 26, 35–37, Pl. 1E)

Lithodes richeri. — Dawson, 1989: 317 [nomen nudum].
Lithodes richeri Macpherson, 1990: 219–221, figs. 1a, 2b [type locality: New Caledonia]; 2001: 798–799. — Zaklan, 2002: 767, 782. — ?Takeda & Nagai, 2004: 15–18, figs 5, 6.

Type material. *Holotype*: MNHN Pg 4269, male (cl 166.5 mm, pcl 89.0 mm, cw 81.0 mm), New Caledonia, outside coral reef, trap.

Other material examined. *Loyalty Islands*: IRD, 2 males (cl 179.5 mm, pcl 95.3 mm, cw 87.4 mm; pcl 106.6 mm, cw 100.0 mm), off Lifou, 20°54′N, 166°59′E, 980–1000 m, 17–18 Jun 1977.

Australia: AM P40778, 1 male (cl 100.9 mm, pcl 53.6 mm, cw 49.6 mm), E of Tuncurry, 32°02′S, 153°09′E, 1025–1080 m, FRV *Kapala*, K88-08-04, coll. K. Graham, 4 May 1988; AM P53233, 1 male (pcl 42.2 mm, cw 36.8 mm), E of Cape Hawke, 32°06′S, 153°08′E, 952–970 m, FRV *Kapala*, K84-10-06, coll. K. Graham, 18 Jul 1984; AM P35600, 1 male (cl 205.0 mm, pcl 112.1 mm, cw 105.8

Figure 35. *Lithodes richeri* Macpherson, 1990, male, pcl 53.6 mm, E of Tuncurry (AM P40778). A, dorsal habitus. B, ventral surface and abdomen. C, posterior carapace and abdominal somite 2.

mm), E of Cape Hawke, 32°06'S, 153°08'E, 952–970 m, FRV *Kapala*, K84-10-06, coll. K. Graham, 18 Jul 1984; AM P35593, 1 male (cl 99.0 mm, pcl 50.5 mm, cw 45.4 mm), E of Cape Hawke, 32°08–04'S, 153°09–10'E, 1006–1052 m, FRV *Kapala*, K84-10-04, coll. K. Graham; AM P52704, 4 males (pcl 57.6–97.3 mm, cw 52.7–90.1 mm), 3 juvenile females (pcl 27.4–52.6 mm, cw 24.2–46.3 mm), E of Tuncurry, 32°08'S, 153°09'E, 1080 m, FRV *Kapala*, K89-12-04, coll. K. Graham, 15 Jun 1989; AM P75312, 1 ovigerous female (cl 255.0 mm, pcl 143.0 mm, cw 127.0 mm), E of Broken Bay, 33°35'S, 152°05'E, 860 m, from fisherman, 6 Jan 1992; AM P35594, 1 male (cl 56.2 mm, pcl 26.8 mm, cw 23.7 mm), E of Shoalhaven Heads, 34°51–55′S, 151°15–14′E, 1042–1061 m, FRV *Kapala*, K83-14-05, coll. K. Graham, 26 Oct 1983; AM P35597, 4 males (pcl 21.6–25.2 mm, cw 18.8–23.1 mm), E of Shoalhaven Heads, 34°53–56′S, 151°15–13′E, 1079–1116 m, FRV *Kapala*, K83-14-06, coll. K. Graham, 26 Oct 1983; AM P35602, 1 female (cl 210.2 mm, pcl 101.3 mm, cw 94.4 mm), E of Brush Island, 35°28–31′S, 150°54–50′E, 905–960 m, FRV *Kapala*, K84-11-01, coll. K. Graham, 26 Jul 1984; AM P35605, 1 ovigerous female (cl 131.4 mm, pcl 125.1 mm, cw 132.9 mm), E of Brush Island, 35°38′S, 150°44′E, 887 m, FRV *Kapala*, K83-11-03, coll. K. Graham, 8 Sep 1983.

Diagnosis. Carapace dorsal surface and margins with slender spines, becoming reduced to short conical protuberances in large adults; gastric region with 4 upright spines, anterior pair longer than half length of hepatic spine, posterior pair half length of hepatic spine; cardiac and intestinal regions each with 2 upright spines; branchial surface with 2 long upright spines. First primary marginal branchial spine half length of hepatic spine or less, 2nd primary marginal branchial spine minute, not larger than posterior branchial spines; posterior branchial margin with 5–7 short spines or protuberances. Rostrum exceeding 0.8 pcl; proximal half strongly upraised by about 30°. Abdominal somite 2 comprising single fused plate. Antennal peduncle article 2 with slender outer spine not reaching beyond midlength of article 4. Chelipeds unequal; palms with low tubercles or spines on dorsal surface. Walking legs spinose; surface between major spines smooth or with very few, scattered spines. Pereopod 4 merus longer than pcl in both sexes; dactylus flexor margin unarmed, half propodus length.

Description. *Carapace*: Pyriform, 1.07–1.15 times longer than wide; regions indicated; dorsal spines slender, becoming reduced to short conical protrusions in large adults; dorsal surface otherwise smooth apart from scattered, widely separated granules. Gastric region convex, with 2 pairs of spines. Cardiac region with pair of anterior tubercles and posterior pair of spines. Hepatic spine longest of carapace spines (0.1 pcl in 89.0 mm male holotype), directed anterolaterally. Intestinal region with pair of spines on posterior margin. Branchial surface with 2 spines and 2 low tubercles as follows: 1 spine at level of pereopod 3 coxa, 1 spine at level of pereopod 4 coxa, 2 conical tubercles in transverse row slightly behind posterior spine. Branchial margins spinose; anterior branchial margin with 2 short spines, posterior spine subeqal to or longer than anterior; 1st primary marginal branchial spine half as long as hepatic spine or less; lateral branchial margin with 2 or 3 short spines; 2nd primary marginal branchial spine short, little longer than anterior branchial marginal spines; posterior branchial margin with 5–7 short spines. All marginal branchial spines reduced to small protrusions in largest specimens. Pterygostomian region granulate or tuberculate, with small, anterior, submarginal spine.

Rostrum as slightly shorter to slightly longer than pcl (0.82–1.08 pcl), comprising proximal and distal portions; proximal portion angled dorsally by at least 30°; with pair of divergent, dorsal spines at rostral midlength; distal portion subhorizontal, distally bifurcate for half to two-thirds length, forming pair of strongly divergent spines; ventral rostral spine elongate, extending anteriorly beyond apices of outer orbital spine. Posterior orbital margin concave; outer orbital spine directed anteriorly, subequal to or shorter than anterolateral spine, not reaching beyond cornea.

Ocular peduncle: Longer than cornea; unarmed.

Antennule: Peduncle unarmed, reaching anteriorly beyond antennal peduncle by three-quarters to full length of distal antennular peduncle article.

Antenna: Basal antennal article unarmed; outer margin of article 2 with slender spine, not reaching beyond midlength of article 4; article 3 unarmed; scaphocerite minute, blunt, shorter than article 4; article 4 unarmed, about half as long as article 5.

Abdomen: Somite 1 with low triangular teeth on posterior margin. Somite 2 rugose; plates fused into single unit; median surface with pair of submedian spines or angular prominences; posterior margin with 4–6 stout spines or blunt angular prominences; lateral margins irregularly dentate. Somites 3–5 with nodular median plates; surface of submedian plates irregular, unarmed; marginal plates of males subdivided, with angular or sharp apices; marginal plates of females with left margin smooth to crenulate (fused with submedians), right marginals subdivided, with rounded or pointed apices. Somite 6 of both sexes longer than wide, distal margin with 2 or 4 distal spines or prominences. Telson rounded, unarmed.

Pereopod 1 (chelipeds): Dimorphic but spination similar. Surfaces of both chelipeds smooth or with scattered granules in addition to major spines. Coxa with tufts of setae, unarmed; ischiobasis with 3 stout ventral spines and about 6 blunt tubercles. Merus inner margin, with stout subdistal spine and 2 or 3 smaller spines or acute tubercles; ventral margin with two rows of 2 or 3 low conical spines; dorsal and lateral surface sparsely spinose, spines largest distally. Carpus with 2 or 3 short spines on dorsal margin; lateral surfaces with distal spine and scattered acute tubercles; ventral margin with small, scattered, acute tubercles. Palms of both chelipeds in both sexes with similar ornamentation; with small spines and acute tubercles on dorsal, lateral, and ventral surfaces, inner surface with few

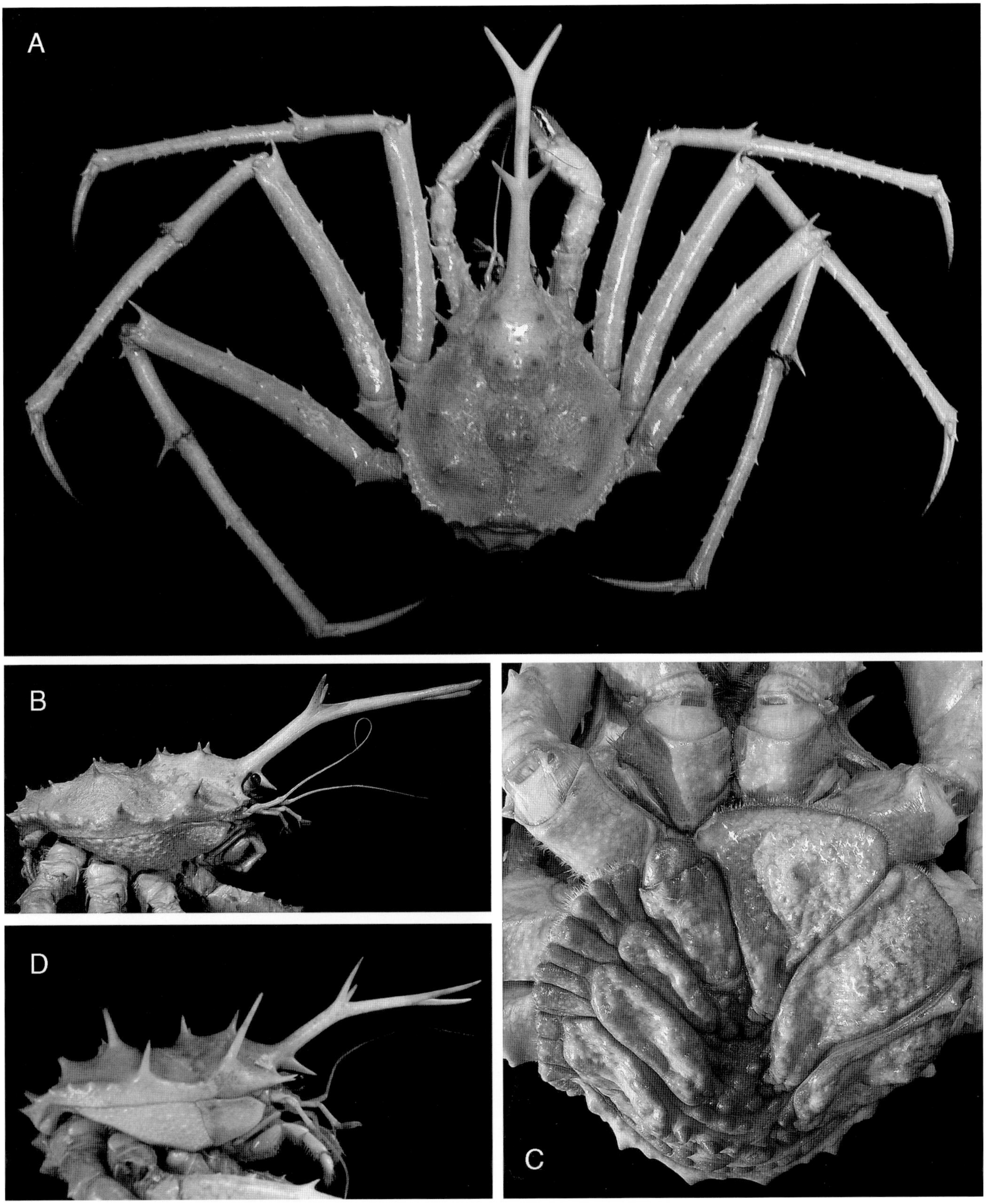

Figure 36. *Lithodes richeri* Macpherson, 1990. A–C, female, pcl 101.3 mm, E of Brush Island (AM P35602). D, male, pcl 53.6 mm, E of Tuncurry (AM P40778). A, dorsal habitus. B, D, carapace, right lateral view. C, ventral surface and abdomen.

low tubercles; dorsal margin with row of 3 or 4 small conical spines; midlateral surface and ventral surfaces with small scattered tubercles or small spines.

Major cheliped 1.25–1.38 pcl (male), 1.21–1.32 (female); upper palm length 1.25–1.48 times height (male), 1.24–1.27 (female); occlusal margins of fingers corneous for distal third, proximally with 3 calcareous nodules; dactylus dorsal margin broadly convex, with rows of tufts of golden setae and blunt proximal tooth, 1.31–1.79 times longer than dorsal margin of palm (male), 1.46–1.47 (female).

Minor cheliped 1.21–1.35 pcl (male), 1.22–1.28 (female); upper palm length 1.41–1.60 times height (male), 1.33–1.41 (female); occlusal margin corneous for slightly less than distal half, proximally crenulate; dactylus dorsal margin broadly convex, with rows of tufts of golden setae and blunt proximal tooth, 1.75–1.96 times longer than dorsal margin of palm (male), 1.88–1.90 (female).

Pereopods 2–4 (walking legs 1–3): Similar, slender, elongate; ornamentation similar in both sexes; segments sparsely spinose, surface between major spines smooth or with few, scattered spines or granules. Pereopod 4 longest. Distal margins of coxae unarmed; surface smooth. Ischiobasis with 3 or 4 non-setose distal spines. Merus subcircular to ovate in cross section; extensor margin with 4–10 spines of which 2 or 3 distinctly longer in addition to prominent distal spine; flexor margin with 2 rows of 4 or 5 small spines; upper surface with few, small, scattered spines. Carpus about half merus length; extensor margin with distal and second proximal spines longest, not exceeding one-third carpus length in adults; surface with few, small, scattered spines. Propodus ovate in cross section, with 6–9 small spines on extensor margin and few small spines on dorsal surface; flexor margin with 9–15 small spines. Dactylus curved, rounded in cross section distally; proximally with 2 spines on either side adjacent to articulation; with 0–4 small corneous spines along extensor margin; flexor margin smooth or with few minute, widely spaced setae; apex corneous.

Pereopod 2 length 2.96–3.17 pcl (male), 2.47–2.91 pcl (female). Merus 1.13–1.16 pcl (male), 0.91–1.07 pcl (female); length:height ratio 8.73–9.54 (male), 7.35–7.58 (female). Carpus 0.44–0.50 merus length (male), 0.50–0.52 (female). Propodus 0.80–0.87 merus length (male), 0.84–0.85 (female); length:height ratio 13.24–13.88 (male), 11.44–12.34 (female). Dactylus 0.49–0.55 propodus length (male), 0.50–0.51 (female).

Pereopod 3 length 3.36–3.57 pcl (male), 2.82–3.29 pcl (female). Merus 1.29–1.31 pcl (male), 1.03–1.22 pcl (female); length:height ratio 9.14–9.79 (male), 8.08–8.17 (female). Carpus 0.43–0.48 merus length (male), 0.48–0.50 (female). Propodus 0.78–0.86 merus length (male), 0.85–0.87 (female); length:height ratio 13.99–14.87 (male), 13.12–13.17 (female). Dactylus 0.45–0.52 propodus length (male), 0.46–0.49 (female).

Pereopod 4 length 3.54–3.81 pcl (male), 3.04–3.47 pcl (female). Merus 1.31–1.38 pcl (male), 1.10–1.25 pcl (female); length:height ratio 9.28–9.96 (male), 8.35–8.42 (female). Carpus 0.44–0.51 merus length (male), 0.50–0.51 (female). Propodus 0.81–0.91 merus length (male), 0.87–0.88 (female); length:height ratio 14.60–15.46 (male), 12.60–12.72 (female). Dactylus 0.50 propodus length (male), 0.46–0.49 (female).

COLOUR IN LIFE. Overall deep-red (Pl. 1E).

REMARKS. Poore's (2004) suggestion that previous reports of *L. longispina* from eastern Australia (Dawson & Yaldwyn 1985) could be based on *L. richeri* is confirmed by the present series.

The Australian specimens of *L. richeri* agree well with New Caledonian material and exhibit morphological variation as reported by Macpherson (1990). As in the holotype of *L. richeri*, the marginal and dorsal carapace spines of the largest Australian specimens are reduced or obsolete, but longer in smaller individuals. A distinctive feature of *L. richeri* is the considerably elongated rostrum, measuring almost as long as or longer than the pcl; the rostrum in adults of other Australian and New Zealand lithodids is always distinctly shorter than pcl. The dorsal spines in *L. richeri* are long and slender in juveniles, becoming reduced to short, conical protrusions in large adults. The hepatic spine is as long as the pcl in the 26.8 mm male (AM P35594), about half pcl in the 53.6 mm male (AM P40778), and about 0.1 pcl in the 89.0 mm male holotype. In the 143 mm female (AM P75312), the hepatic spine is less than 0.05 pcl.

Lithodes richeri is sometimes sympatric with *Lithodes australiensis* and distinguishing between juveniles of the two species can be difficult. Adults are readily separated by the marked difference in rostral length (almost as long as pcl in *L. richeri*; less than half pcl in *L. australiensis*), marked difference in dorsal spine length (reduced to short conical projections in *L. richeri*; prominent in *L. australiensis*), slenderness of the walking legs (pereopod 4 merus length:height ratio of males exceeds 9 in *L. richeri*; less than 8 in *L. australiensis*), spination of the branchial margins of the carapace (at most one prominent primarly marginal branchial spine in *L. richeri*; two in *L. australiensis*), and fusion of the plates of abdominal somite 2 (fused in *L. richeri*; separate in *L. australiensis*). Juvenile *L. richeri* and *L. australiensis* are more difficult to separate because of their similarly long dorsal spines and rostra. As in adults, however, the fused plates of abdominal somite 2 and the presence of only one prominent primary marginal branchial carapace spine will separate juvenile *L. richeri* from *L. australiensis*. In addition, the

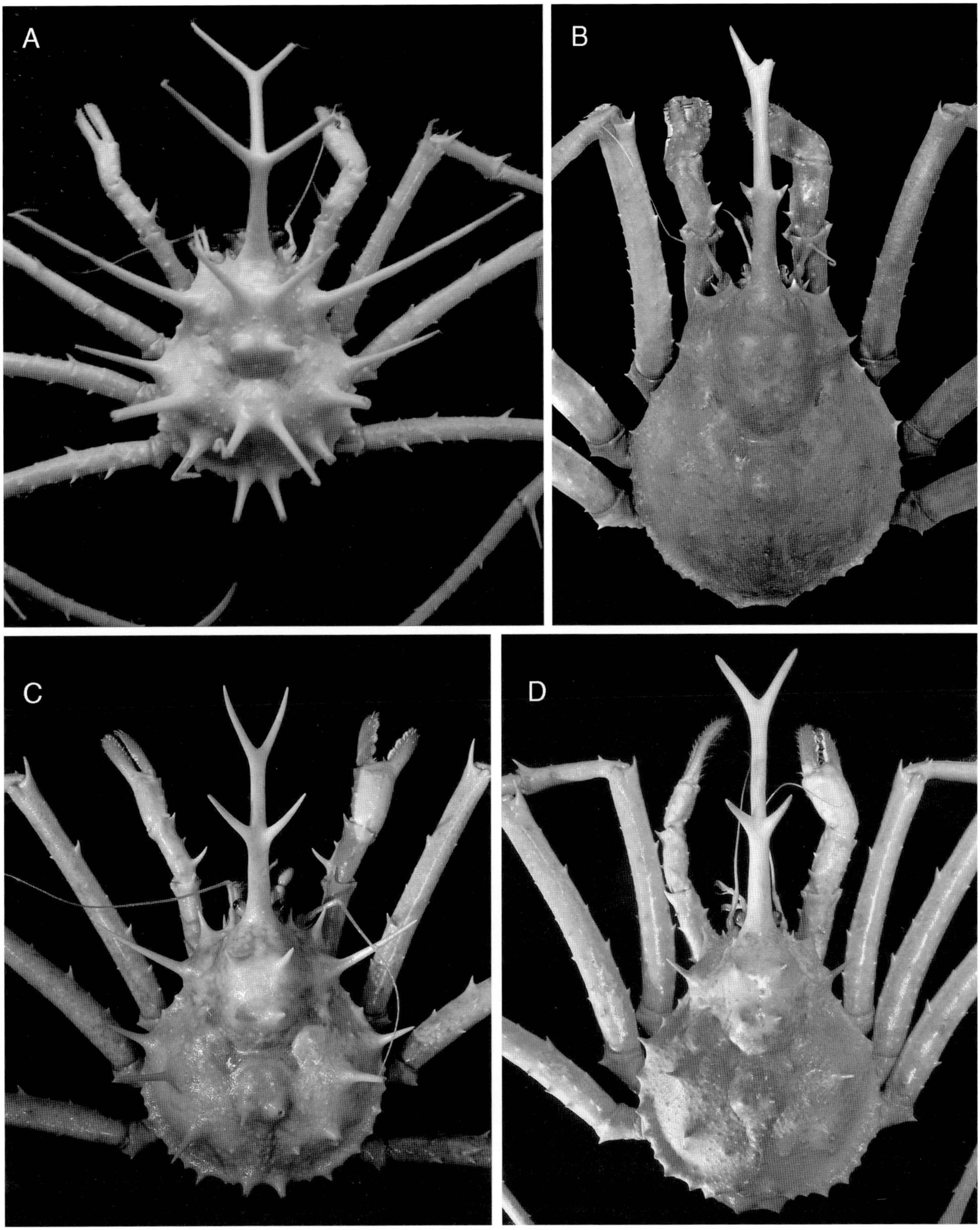

Figure 37. *Lithodes richeri* Macpherson, 1990. Carapace. A, juvenile male, pcl 26.8 mm, E of Shoalhaven Heads (AM P35594). B, female, pcl 143.0 mm, E of Broken Bay (AM P75312). C, male, pcl 53.6 mm, E of Tuncurry (AM P40778). D, female, pcl 101.3 mm, E of Brush Island (AM P35602).

flexor margins of the walking leg dactyli of juvenile *L. australiensis* up to about 60 mm pcl are armed with a row of corneous spinules, but are unarmed in *L. richeri* of all sizes examined (smallest with pcl 21.6 mm).

Takeda & Nagai (2004) tentatively identified a large female specimen (pcl 175 mm) from the Ceram Sea, Indonesia, as *L. richeri*, which differed from the type description in carapace shape — the carapace of the Ceram specimen is widest across the anterior, rather than mid-portion of the branchial region, with angular rather than evenly rounded anterior branchial margins, and the junction of the hepatic and branchial regions forming an angular rather than evenly curved concavity. Takeda & Nagai (2004) speculated that the differences in carapace shape might be attributable to allometry, with the Ceram specimen considerably larger than the holotype (pcl 81 mm). The specimens in series of *L. richeri* reported here, however, span a wide size range (pcl 26.8–143.0 mm), and the carapace outline is consistently pyriform, exhibiting no trend towards the condition of the Ceram specimen. Additionally, the dorsal carapace spines are distinctly more prominent than in the largest specimen examined here (pcl 143.0 mm). The Ceram specimen possibly represents an undescribed species.

Although Australian records of *L. richeri* are limited to New South Wales, it can be expected to occur off Queensland given its occurrence at Vanuatu, Lifou, and New Caledonia.

Distribution. Vanuatu, Lifou, New Caledonia, and eastern Australia south to at least 35°S; possibly from the Ceram Sea, Indonesia; 860–1220 m (Macpherson, 2001; present study).

Lithodes robertsoni sp. nov.

(Figs 31, 38–42, Pl. 1F)

Type material. *Holotype*: NIWA 42889, male (cl 186.6 mm, pcl 128.1 mm, cw 117.0 mm), near Pitt Island, Chatham Rise, 44°41.6′S, 177°01.4′W, 1002 m, trip 2521/69, FV *Thomas Harrison*, coll. T. Turton, 15 Nov 2007.

Paratypes: NIWA 34914, 1 male (cl 122.7 mm, pcl 81.7 mm, cw 76.4 mm), The Snares, 48°02.01′S, 166°06.00′E, 935 m, trip 1171/12, Z9583, 25 Nov 1998; NIWA 34908, 1 female (cl 119.0 mm, pcl 71.2 mm, cw 65.2 mm), Bounty Trough, 46°27.0′S, 172°01.8′E, 1030–1236 m, trip 1650, tow 58, FV *Amital Explorer*, 18 May 2002; NIWA 61239, 1 female (pcl 70.2 mm, cw 64.6 mm), Bounty Trough, 47°53.1′S, 175°02.5–02.7′E, 1252–1259 m, trip 2494, tow 27, 4 Sep 2007.

Other material examined. *Chatham Rise:* NIWA 60564, 1 badly damaged male (pcl 103.7 mm, cw 100.2 mm), Chatham Rise, FV *San Waitaki*, SWA0901/-. *Challenger Plateau*: NIWA 62808, 1 badly damaged male (pcl 114.2 mm, cw 113.3 mm), 39°45.97–46.29′S, 167°15.49–15.14′E, 880–1044 m, FV *Thomas Harrison*, THH0601/15, 26 Jun 2006.

Diagnosis. Carapace dorsal surface with slender spines in adults; gastric region with 4 upright spines, anterior pair longer than half length of hepatic spine, posterior pair half length of hepatic spine; cardiac and intestinal regions each with 2 long upright spines; branchial surface with 2 long upright spines. Branchial margins spinose; 1st primary marginal branchial spine about half length of hepatic spine, 2nd primary marginal branchial spine shorter than 1st primary marginal branchial spine; posterior branchial margin with 7 or 8 short spines and 1 or 2 low tubercles. Rostrum not exceeding 0.8 pcl, upraised by about 30°; almost straight or only slightly angled near midlength. Abdominal somite 2 comprising 3 plates in adults. Antennal peduncle article 2 with slender outer spine not reaching beyond article 4. Chelipeds unequal; palms with low tubercles or spines on dorsal surface. Walking legs spinose; surface between major spines with very few, scattered spines. Pereopod 4 merus 1.1 pcl or less, length less than 8.2 times height; propodus length 11–14 times height (male), 11 (female); dactylus flexor margin unarmed.

Description. *Carapace*: Pyriform, 1.03–1.09 times longer than wide; regions indicated; dorsal surface armed with long, slender spines and numerous, well-spaced granules. Gastric region convex, with 2 pairs of long, upright spines. Cardiac region with pair of anterior tubercles and posterior pair of long upright spines. Hepatic spine longest of carapace spines (0.3 pcl in holotype; 0.5 pcl in 70.2 mm pcl female paratype), directed anterolaterally. Intestinal region with pair of long upright spines on posterior margin. Branchial surface with 2 long upright spines and 2 small conical spines as follows: 1 spine at level of pereopod 3 coxa, 1 spine at level of pereopod 4 coxa, 2 small conical spines in transverse row slightly behind posterior spine. Branchial margins spinose; anterior branchial margin with 2 small spines, subequal or posterior spine longer than anterior; 1st primary marginal branchial spine about half length of hepatic spine; lateral branchial margin with 1 short spine and 1 short granule or 3 spines (NIWA 60564); 2nd primary marginal branchial spine shorter than two-thirds length of 1st primary marginal branchial spine; posterior branchial margin with 7 or 8 spines and 1 or 2 low tubercles. Pterygostomian region granulate or tuberculate, with small, anterior, submarginal spine.

Rostrum half to two-thirds pcl, upraised by about 30°; almost straight or only slightly angled near midlength; with pair of long, divergent, dorsal spines that overreach midlength of distal portion of rostrum; distally bifurcate for about half length, forming pair of strongly divergent spines; ventral rostral spine

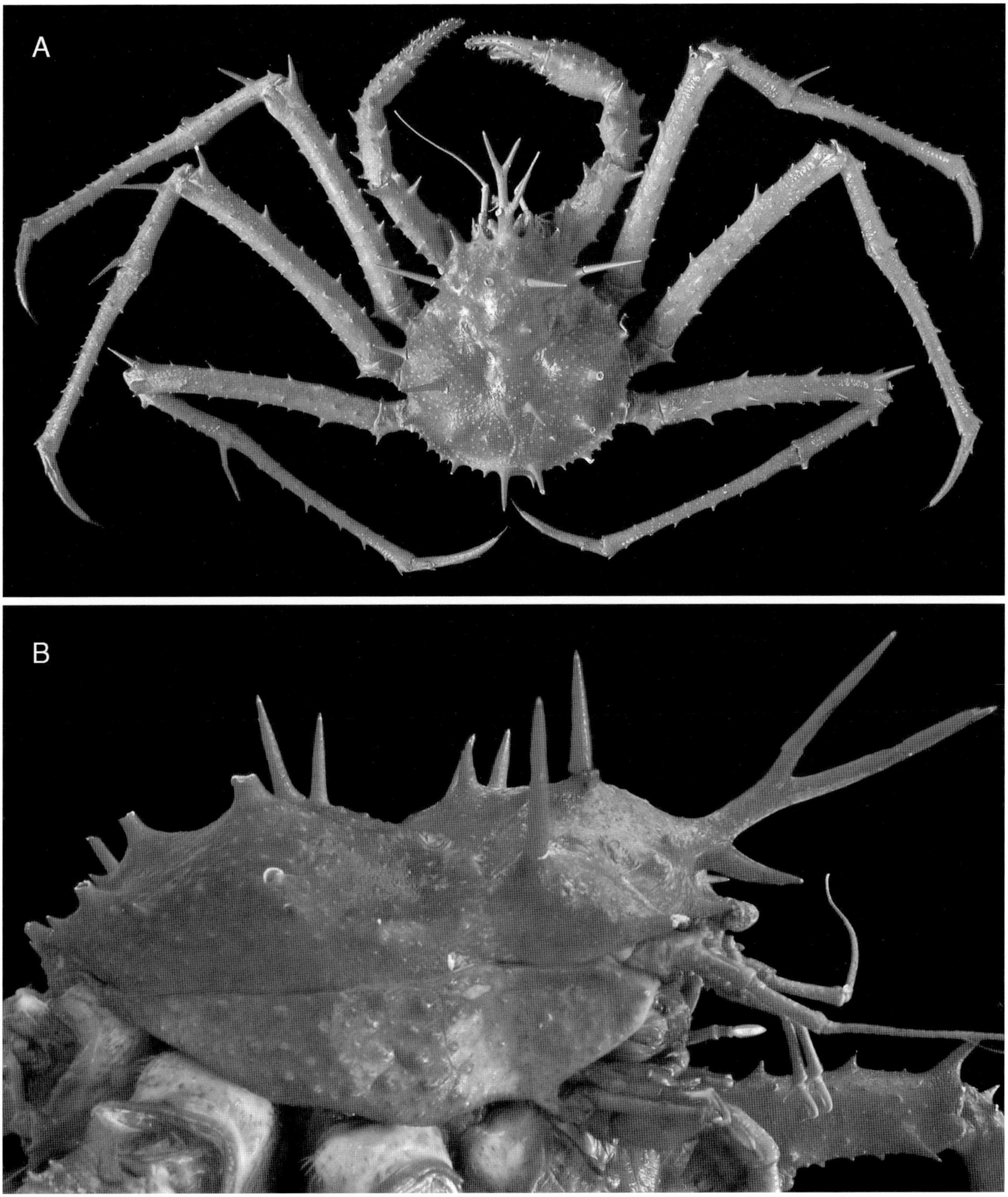

Figure 38. *Lithodes robertsoni* sp. nov., male holotype, cl 186.6 mm, pcl 128.1 mm, cw 117.0 mm, Chatham Rise (NIWA 42889). A, dorsal habitus. B, carapace, right lateral view.

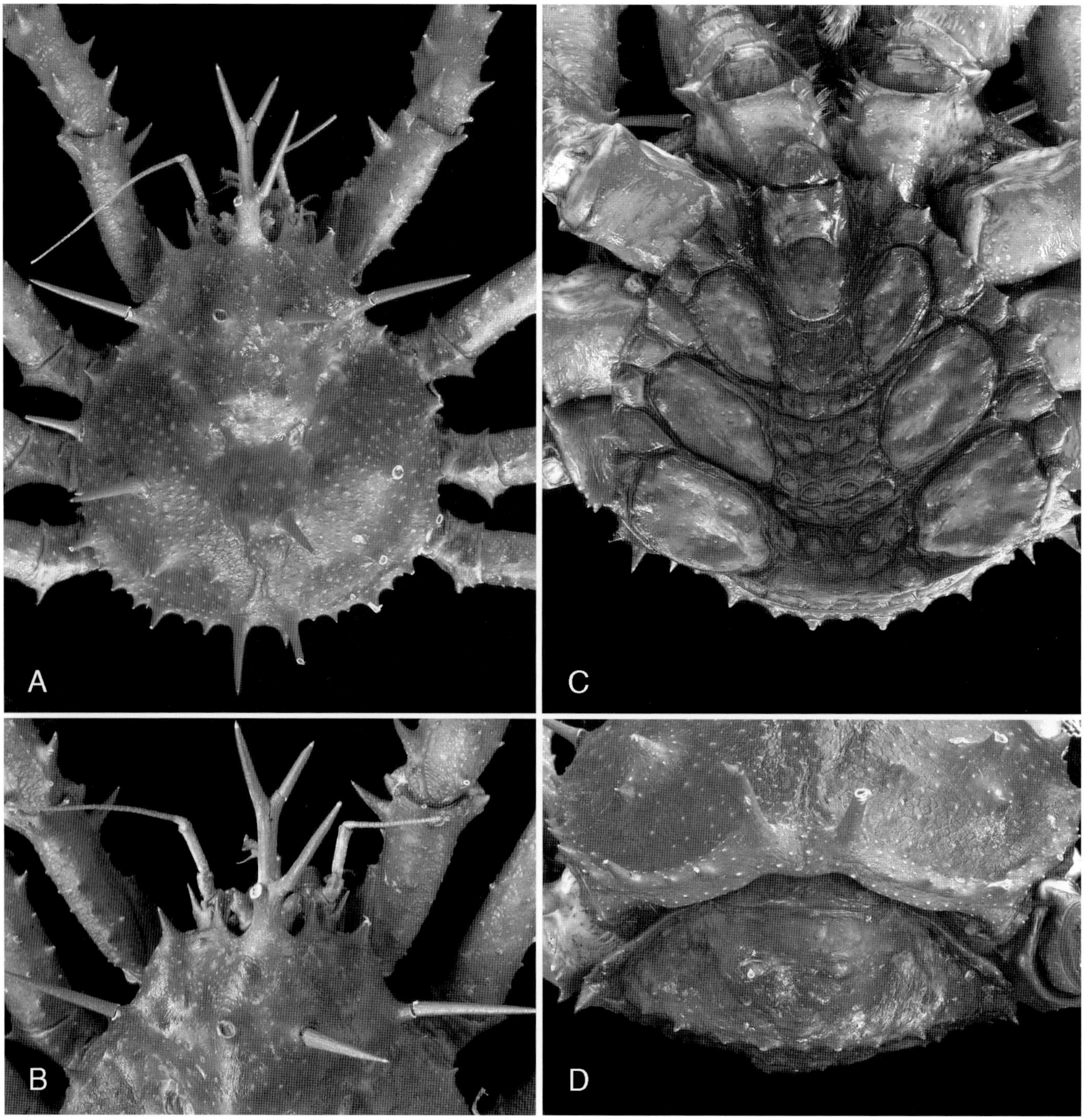

Figure 39. *Lithodes robertsoni* sp. nov., male holotype, cl 186.6 mm, pcl 128.1 mm, cw 117.0 mm, Chatham Rise (NIWA 42889). A, carapace, dorsal view. B, anterior carapace, dorsal view. C, abdomen. D, posterior carapace and abdominal somite 2.

elongate, extending anteriorly beyond apices of outer orbital spine. Posterior orbital margin concave; outer orbital spine directed anteriorly, slightly longer than anterolateral spine, reaching to or beyond cornea.

Ocular peduncle: Longer than cornea; unarmed.

Antennule: Peduncle unarmed, reaching anteriorly beyond antennal peduncle by about three-quarters length of distal antennular peduncle article.

Antenna: Basal antennal article unarmed; outer margin of article 2 with slender spine, not reaching apex of article 4; article 3 unarmed; scaphocerite minute, blunt, shorter than article 4; article 4 unarmed, about half as long as article 5.

Abdomen: Somite 1 with low triangular teeth on posterior margin. Somite 2 rugose; composed of 3 distinct plates separated by narrow groove; median

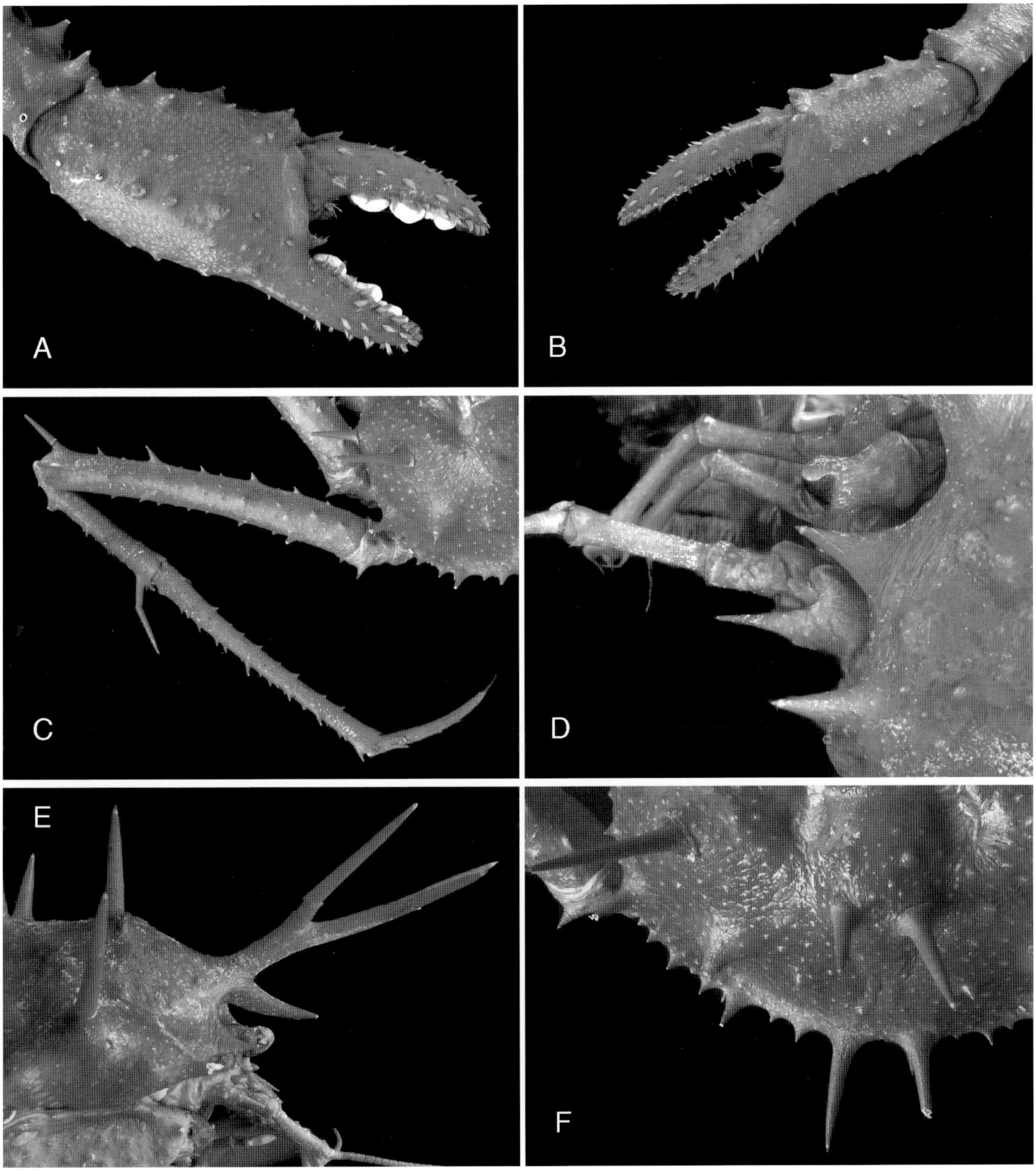

Figure 40. *Lithodes robertsoni* sp. nov., male holotype, cl 186.6 mm, pcl 128.1 mm, cw 117.0 mm, Chatham Rise (NIWA 42889). A, right chela. B, left chela. C, left pereopod 4. D, left antenna and orbit. E, anterior carapace, right lateral view. F, carapace, left posterior.

surface with pair of stout submedian spines; posterior margin with 4–8 stout teeth or angular prominences; lateral margins irregularly dentate. Somites 3–5 with nodular median plates; surface of submedian plates irregular, unarmed except for 1 or 2 small distolateral teeth; marginal plates of males subdivided, with angular or sharp apices; marginal plates of females with left margin spinose (fused with submedians), right marginals subdivided, with angular to spiniform apices. Somite 6 of both sexes longer than wide, distal margin with 4 distal spines or prominences. Telson rounded, unarmed.

Pereopod 1 (chelipeds): Dimorphic but spination similar. Surfaces of both chelipeds smooth or with scattered granules in addition to major spines. Coxa with tufts of setae, unarmed; ischiobasis with 3 stout ventral spines and about 6 blunt tubercles. Merus inner margin with stout subdistal spine and 2 or 3 smaller spines or acute tubercles; ventral margin with two rows of 2 or 3 low conical spines; dorsal and lateral surface spinose, spines largest distally. Carpus with prominent spines on dorsal and lateral surfaces; dorsal margin with row of 3 stout spines; ventral margin with small, scattered, acute tubercles. Palms of both chelipeds in both sexes with similar ornamentation; with small spines and acute tubercles on dorsal, lateral and ventral surfaces, inner surface with few low tubercles; dorsal margin with row of 3 or 4 small conical spines; midlateral surface with 2 rows of 3 or 4 small spines or tubercles, of similar size to dorsal row; ventral surface with scattered tubercles, smaller than lateral and dorsal spines.

Major cheliped 1.27–1.52 pcl (male), 1.33–1.34 (female); upper palm length 1.22–1.35 times height (male), 1.29–1.40 (female); occlusal margins of fingers corneous for distal quarter, proximally with 3 calcareous nodules; dactylus dorsal margin broadly convex, with rows of tufts of golden setae and blunt proximal tooth, 1.18–1.39 times longer than dorsal margin of palm (male), 1.32–1.36 (female).

Minor cheliped 1.36–1.50 pcl (male), 1.32–1.33 (female); upper palm length 1.36–1.53 times height (male), 1.43–1.52 (female); occlusal margin corneous for slightly less than distal half, proximally crenulate; dactylus dorsal margin broadly convex, with rows of tufts of golden setae and blunt proximal tooth, 1.61–1.70 times longer than dorsal margin of palm (male), 1.68–1.73 (female).

Pereopods 2–4 (walking legs 1–3): Similar, slender, elongate; ornamentation similar in both sexes; segments spinose, surface between major spines with minute granules and few, scattered spines. Pereopod 3 and 4 subequal, longer than pereopod 2. Distal margins of coxae unarmed; surface smooth. Ischiobasis with 3 or 4 non-setose distal spines. Merus subcircular to ovate in cross section; extensor margin with 7–10 spines of which 2 or 3 are distinctly longer in addition to long, prominent, distal spine; flexor margin with 2 rows of 5–7 small spines; upper surface with irregular row of 7–9 small spines. Carpus half merus length of merus; extensor margin with distal and second proximal spines longest, exceeding half carpus length; surface with scattered small spines. Propodus compressed; with 7–11 small spines on extensor margin and 4–7 small spines on dorsal surface; flexor margin with 4–7 small spines. Dactylus curved, rounded in cross section distally; proximally with 2 spines on either side adjacent to articulation; with 3 or 4 small corneous spines along extensor margin; flexor margin smooth or with about 4 soft, minute, widely spaced setae; apex corneous.

Pereopod 2 length 2.76–2.95 pcl (male), 2.63–2.66 pcl (female). Merus 0.99–1.07 pcl (male), 0.90–0.92 pcl (female); length:height ratio 6.26–7.50 (male), 5.70–6.15 (female). Carpus 0.51–0.53 merus length (male), 0.52–0.54 (female). Propodus 0.84–0.85 merus length (male), 0.86–0.90 (female); length:height ratio 10.26–13.22 (male), 9.89–10.71 (female). Dactylus 0.52–0.54 propodus length (male), 0.59–0.62 (female).

Pereopod 3 length 3.04–3.18 pcl (male), 2.84–2.87 pcl (female). Merus 1.08–1.13 pcl (male), 0.97–0.98 pcl (female); length:height ratio 6.92–7.87 (male), 6.05–6.58 (female). Carpus 0.52 merus length (male), 0.53–0.55 (female). Propodus 0.85–0.86 merus length (male), 0.89–0.90 (female); length:height ratio 10.78–13.76 (male), 10.71–11.37 (female). Dactylus 0.48–0.52 propodus length (male), 0.57–0.61 (female).

Pereopod 4 length 3.00–3.21 pcl (male), 2.91–2.92 pcl (female). Merus 1.05–1.10 pcl (male), 0.96–0.98 pcl (female); length:height ratio 6.70–8.16 (male), 6.11–6.76 (female). Carpus 0.55 merus length (male), 0.55–0.58 (female). Propodus 0.90–0.92 merus length (male), 0.92–0.95 (female); length:height ratio 11.43–14.08 (male), 11.55–11.87 (female). Dactylus 0.51–0.53 propodus length (male), 0.54–0.62 (female).

Colour in life. Overall deep-red (Pl. 1F).

Etymology. Named for Don Robertson, NIWA, on occasion of his retirement, and for his strong support and advocacy for biodiversity research.

Remarks. *Lithodes robertsoni* sp. nov. closely resembles *L. australiensis* sp. nov., *L. rachelae* sp. nov., and *L. paulayi* Macpherson & Chan, 2008, species sharing long dorsal spines in adults, two long dorsal branchial spines, unarmed flexor margins of the walking leg dactyli, and a three-segmented abdominal somite 2 in adults. Unfortunately, the aforementioned species exhibit few meristic differences, and are best separated via morphometrics of the walking legs. The walking legs of *L. robertsoni* are proportionally shorter and less slender than in *L. rachelae*. The overall pereopod 4 length in *L. robertsoni* does not exceed 3.2 pcl, compared to 3.6 in male *L. rachelae*; and in females, less than 3.0 versus 3.1 or greater, respectively. The pereopod 4 merus length of male *L. robertsoni* is shorter than 1.1 pcl with a length:height ratio <8.2 versus about 1.3 pcl and >9.1, respectively, in male *L. rachelae*. *Lithodes robertsoni* can be readily distinguished from *L. paulayi*

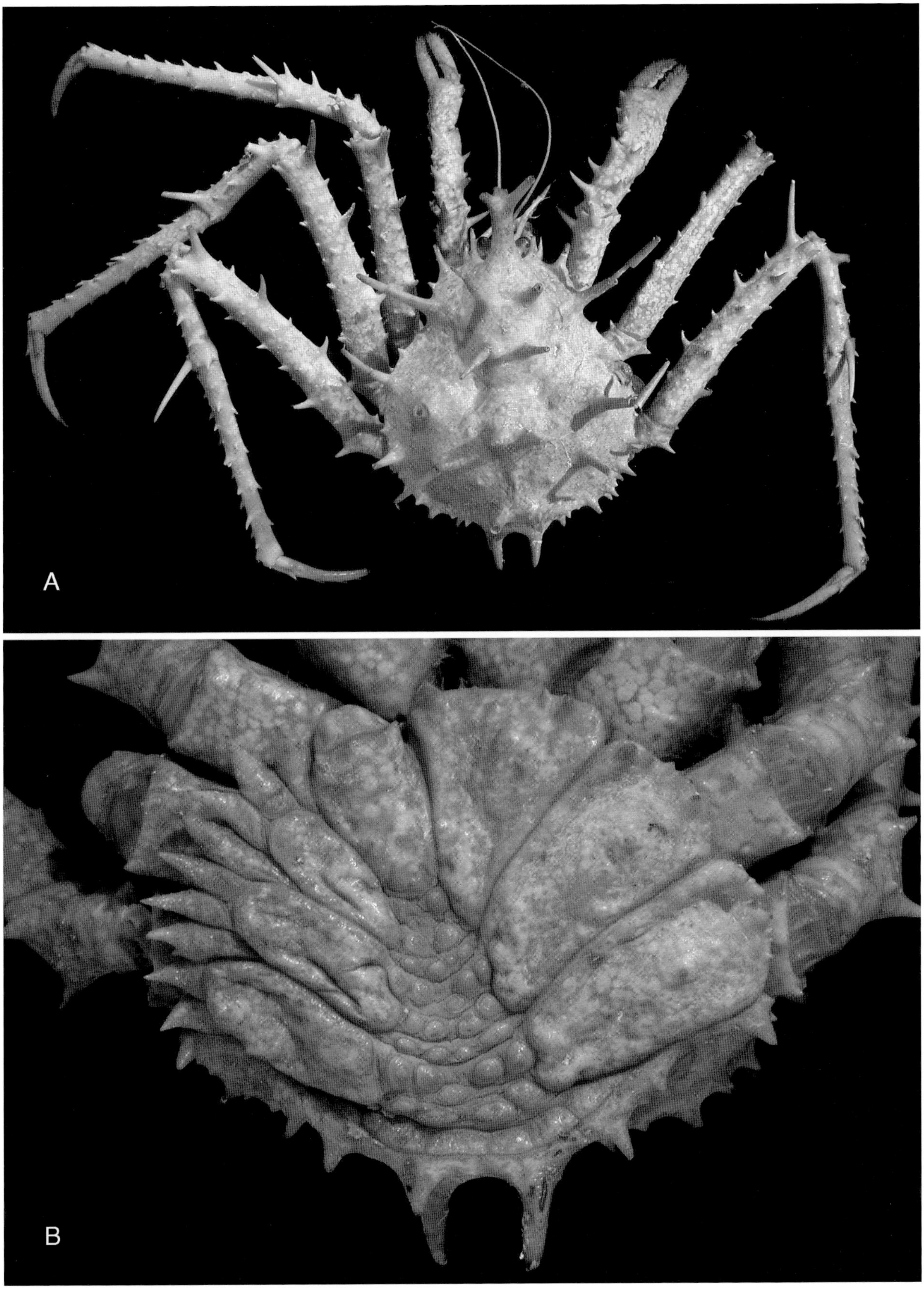

Figure 41. *Lithodes robertsoni* sp. nov., female paratype, pcl 70.2 mm, cw 64.6 mm, Bounty Trough (NIWA 61239). A, dorsal habitus. B, abdomen.

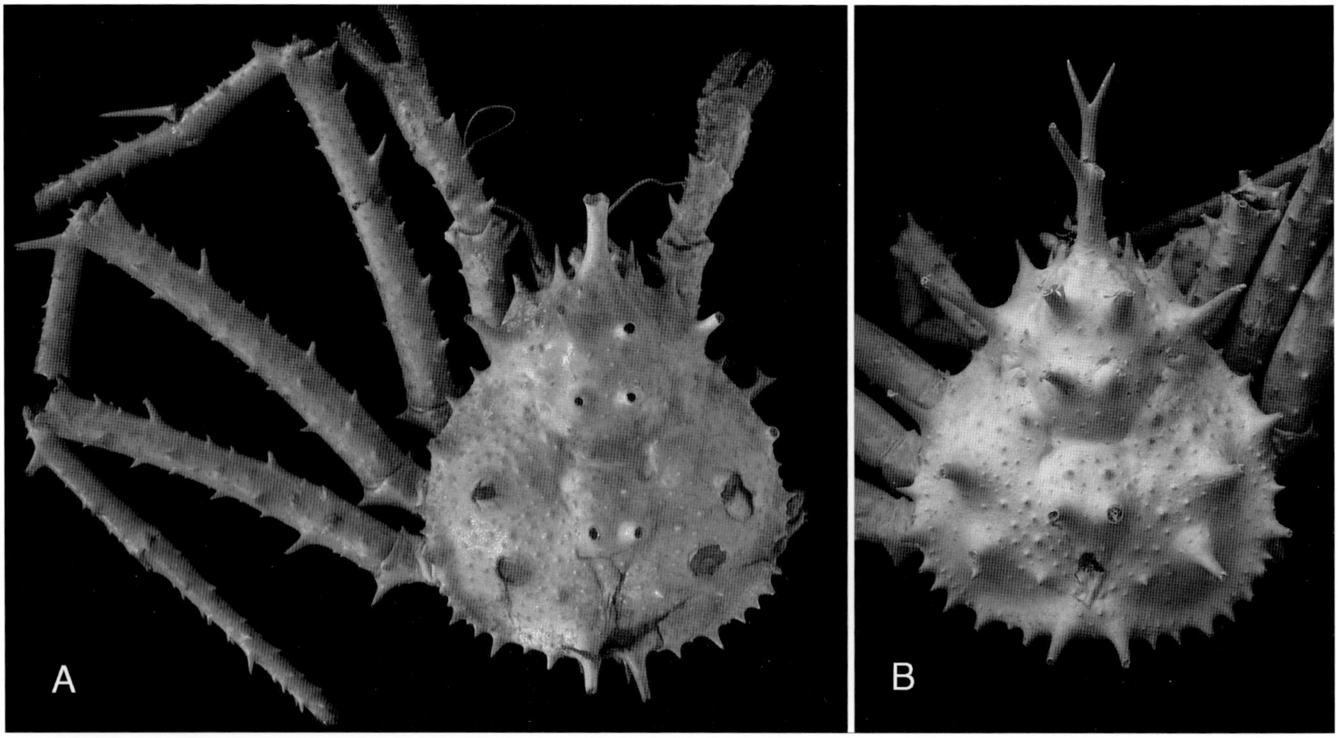

Figure 42. *Lithodes robertsoni* sp. nov. A, male, pcl 103.7 mm, Chatham Rise (NIWA 60564). B, male paratype, cl 122.7 mm, pcl 81.7 mm, cw 76.4 mm, The Snares (NIWA 34914).

by the longer outer spine on antennal article 2, which reaches to, or almost to, the end of article 4, rather than to the midlength in *L. paulayi*; and the less slender pereopod 4 propodi (length 11–14 times height versus about 17).

Lithodes robertsoni is morphologically closest to *L. australiensis*. The two species are difficult to distinguish and it is with some hesitation that both are recognised as separate new species. Comparisons are confounded by the limited available material of *L. robertsoni*, in which the largest and most complete specimen is the male holotype. The two forms, however, exhibit subtle but consistent morphological differences in addition to disjunct distributions (southeastern New Zealand versus southeastern Australia). The number of posterior marginal branchial spines occupies a higher range in *L. robertsoni* (7 or 8) than in *L. australiensis* (5 or 6), but occasional overlap seems likely given the closeness of the ranges. Adult, size-matched male *L. robertsoni* differ from adult male *L. australiensis* in having proportionally shorter walking legs. Pereopod 4 is 3.0 pcl at 81.7 mm pcl in *L. robertsoni* compared to 3.1 pcl in 78.0 mm pcl *L. australiensis*. At 128.1 mm pcl, pereopod 4 is 3.2 times pcl in *L. robertsoni* compared to 3.5 for *L. australiensis* at 133.8 mm pcl. The pereopod 4 merus is proportionally shorter in *L. robertsoni* than in *L. australiensis* (1.05–1.10 pcl vs 1.10–1.17 pcl across the adult size range), with the largest specimens exbiting the most marked differences (1.10 pcl versus 1.17 pcl for 128.1 mm and 133.8 mm specimens of *L. robertsoni* and *L. australiensis*, respectively).

Mature female *L. robertsoni* and *L. australiensis* are currently known only from very few specimens, and no large females of either species have yet been collected. Based on current data, the propodus of pereopod 4 of female *L. robertsoni* appears to be slightly more slender than in *L. australiensis* (length exceeding 11.5 times height versus about 11). Unfortunately, *L. robertsoni* is apparently an uncommon species, so its full range of morphological variation remains to be documented. Moreover, most specimens are damaged and have broken rostra. More 'satisfactory' characterisation of differences between *L. robertsoni* and *L. australiensis* must await capture of more specimens of both species, especially females. A badly damaged male (NIWA 60564, pcl 103.7 mm), with a broken rostrum and broken carapace spines, and with only the proximal segments of the walking legs intact, is tentatively assigned to *L. robertsoni*. It agrees well with other males in most aspects that can be studied, but differs in having three stout spines, instead of one, on the lateral branchial margin.

In New Zealand waters, the only species that *L. robertsoni* is likely to be mistaken for is *L. aotearoa*, especially because the geographic and bathymetric ranges of the species overlap. Adult *L. robertsoni* differs from *L. aotearoa* in abdominal somite 2 comprising three separate plates rather than a single fused plate; in usually having only one spine and one or two granules on the lateral branchial margin rather than two or three spines; the surface of the submedian plates on the male abdominal somites 2–5 do not bear acute tubercles except distolaterally; the pereopod 4 merus is always longer than the pcl in males rather than usually shorter than pcl in *L. aotearoa* (except in some of the largest specimens); and the dorsal spines are proportionally longer. By 120 mm pcl, the dorsal spines are reduced to short conical protrusions in *L. aotearoa* of about 0.1 pcl, rather than slender spines of at least 0.2 pcl in *L. robertsoni*. *Lithodes robertsoni* might also mature at a smaller size than *L. aotearoa*. The 71.2 mm pcl female of *L. robertsoni* (NIWA 34908) appears to be mature, based on abdomen development. *Lithodes aotearoa* matures by about 90 mm pcl.

Distribution. New Zealand, from the Challenger Plateau, Chatham Rise and Bounty Trough to The Snares; 935–1259 m.

Neolithodes A. Milne-Edwards & Bouvier, 1894

Neolithodes A. Milne-Edwards & Bouvier, 1894b: 120 [type species: *Lithodes grimaldii* A. Milne-Edwards & Bouvier, 1894a, by monotypy. Gender: masculine].

Diagnosis. Carapace pyriform, not covering bases of walking legs; regions indicated; gastric region elevated above other regions; cardiac region triangular, separated from gastric region by deep groove; cervical groove shallow, indistinct. Rostrum comprised of median spine and one pair of dorsal spines at base. Abdominal somite 2 comprising 5 plates (median plate, paired submedian and paired marginal plates). Abdominal somites 3–5 of males composed of small spiniform nodules embedded in arthrodial membrane; female somites composed either of well-developed plates or spiniform nodules on right side, and well-developed plates on left side. Sternite 5 (between pereopods 2) with deep median fissure. Scaphocerite absent or a simple spine. Walking legs (pereopods 2–4) similar in form, third walking leg longest, always longer than pcl; dactyli of adults without row of corneous flexor spines.

Composition

- *N. agassizi* (Smith, 1882) [western Atlantic Ocean]
- *N. asperrimus* Barnard, 1947 [southeastern Atlantic Ocean to southwestern Indian Ocean]
- *N. brodiei* Dawson & Yaldwyn, 1970 [New Zealand, southwestern Pacific Ocean]
- *N. bronwynae* sp. nov. [New Zealand and Tasman Sea, southwestern Pacific Ocean]
- *N. capensis* Stebbing, 1905 [South Africa, western Indian Ocean]
- *N. diomedeae* (Benedict, 1895) [southeastern Pacific to southwestern Atlantic oceans]
- *N. duhameli* Macpherson, 2004 [Crozet Islands, southwestern Indian Ocean]
- *N. flindersi* Ahyong, 2010 [southeastern Australia, southwestern Pacific Ocean and Southern Ocean]
- *N. grimaldii* (A. Milne-Edwards & Bouvier, 1894a) [northern Atlantic Ocean]
- *N. nipponensis* Sakai, 1971 [Japan to Taiwan, northwestern Pacific Ocean]
- *N. vinogradovi* Macpherson, 1988b [southeastern Indian Ocean]
- *N. yaldwyni* Ahyong & Dawson, 2006 [Ross Sea, Antarctica]

Remarks. *Neolithodes* currently includes 12 species including *N. bronwynae* sp. nov., described herein from New Zealand. Five species are known from the Atlantic Ocean, of which two also range into the southeastern Pacific and southwestern Indian Ocean, respectively. Two species (*N. vinogradovi* Macpherson, 1988b, *N. duhameli* Macpherson, 2004) are known from the southern Indian Ocean, two from New Zealand (*N. brodiei* Dawson & Yaldwyn, 1970 and *N. bronwynae* sp. nov.), one from the Ross Sea Antarctica (*N. yaldwyni* Ahyong & Dawson, 2006), one from southern Australia (*N. flindersi* Ahyong, 2010), and one from the northwestern Pacific (*N. nipponensis* Sakai, 1971). Records assigned to known *Neolithodes* species from other localities, especially in the Indo-Pacific, require verification. In particular, records of *N. agassizi* (Smith, 1882) from the northwestern Indian Ocean (Anderson 1896; Alcock & Anderson 1899; Alcock 1901) are based on an undescribed species (Dawson 1989). Similarly, records of *N. nipponensis* from the southwestern Pacific and records of *N. brodiei* from Vanuatu appear to be based on undescribed species. Records of *N. vinogradovi* from New Caledonia appear referable to *N. bronwynae*. Macpherson (2001) recorded an unidentifiable juvenile *Neolithodes* from the Marquesas Islands (French Polynesia); as yet, adult *Neolithodes* have not been identified from the central Pacific.

Neolithodes includes some of the deepest occurring lithodids, down to 3200 m for *N. capensis* (see Macpherson 1988b).

Key to species of *Neolithodes* from New Zealand, Australia, and the Ross Sea

1. Carapace surface between major spines with sparsely distributed granules or secondary spinules. Pereopod 2–4 merus subcylindrical......................2
- Carapace surface between major spines densely covered with granules or secondary spinules. Pereopod 2–4 merus compressed, flattened.................3

2. Carapace covered with long, slender dorsal spines in adults; longest spines exceeding 0.2 pcl [New Zealand, northern Tasman Sea] .... *N. bronwynae* sp. nov.

- Carapace spines short, conical in adults; longest spines shorter than 0.2 pcl [Ross Sea, Antarctica].. .... *N. yaldwyni*

3. Pereopod 2–4 coxae of males and juvenile females covered with short conical spines. Outline of extensor margin of walking legs generally even in adults, without several prominent spines protruding above level of general spines [Australia]......... *N. flindersi*

- Pereopod 2–4 coxae in both sexes covered with low, blunt nodules or tubercles (occasionally subacute in juveniles). Outline of extensor margin of walking legs uneven in adults, with several prominent spines protruding above level of general spines [New Zealand] .... *N. brodiei*

Neolithodes brodiei Dawson & Yaldwyn, 1970 (Figs 43–48, Pl. 2A)

Neolithodes brodiei Dawson & Yaldwyn, 1970: 227–228 [type locality: E of Campbell Plateau, 50°58'S, 173°57'E, 832 m]. — Dawson & Yaldwyn, 1985: 70 [New Zealand occurrences]. — McLay, 1988: 36–38, fig. 3. — Dawson, 1989: 3, frontispiece, 318. — Takeda *in* Amaoka *et al.*, 1990: 361. — Webber, 1997: 81, figs 1–3, 5. — O'Driscoll *et al.*, 2003: 65. — Davie, 2002: 73 [New Zealand occurrences]. — Zaklan, 2002: 768, 785, 804 [New Zealand occurrences]. — Batson, 2003: 136–137. — Webber & Naylor, 2004a: 79. — Thatje & Lörz, 2005: fig. 2B [note: figure caption transposed]. — Ahyong *et al.*, 2007: 155.

NOT *Neolithodes brodiei*. — Thatje & Lörz, 2005: 335–336, fig. 2A. [= *N. yaldwyni* Ahyong & Dawson, 2006] [note: figure caption transposed].

NOT *Neolithodes brodiei*. — Macpherson, 2001: 799 [= undescribed species].

NOT *Neolithodes brodiei*. — Dawson & Yaldwyn, 1985: 70. — Davie, 2002: 73 [Tasmania and Victoria]. — Zaklan, 2002: 768 [Australian occurrences]. — Poore, 2004: 268, fig. 75. [= *N. flindersi* Ahyong, 2010].

TYPE MATERIAL. *Holotype*: NIWA 735, ovigerous female (pcl 111.5 mm, cw 103.9 mm), Campbell Plateau, 50°58.00'S, 173°56.99'E, 832 m, *Globigerina* ooze, small Agassiz trawl, HMNZ *Endeavour*, F135, 30 Jan 1965.

OTHER MATERIAL EXAMINED. *West Norfolk Ridge*: NMNZ Cr11150, 1 male (pcl 111.4, cw 101.0 mm), 34°34.26–33.37'S, 168°56.53–54.76'E, 1013–1350 m, TAN0308/156, 4 Jun 2003.

Challenger Plateau: NIWA 3919, 1 damaged juvenile female, 37°27.30–28.80'S, 168°14.29–29.89'E, 953–992 m, MFish SOP 1630/23, FV *Baldur*, coll. S. Smith, 24 Mar 2002; NMNZ Cr3747, 1 male (pcl 86.2 mm, cw 78.4 mm), 38°33.6'S, 167°00.2'E, 1168–1179 m, FV *Arrow* A02/143/83, 16 Sep 1983; NMNZ Cr7351, 1 male (pcl 124.7 mm, cw 121.5 mm), W Challenger Plateau, 39°51.90'S, 167°09.00'E, 996 m, FV *Oyang* 7, trip 284, coll. Sharples, 9 Sep 1988; NMNZ Cr3746, 1 male (pcl 137.5 mm, cw 136.6 mm), 39°27.9'S, 167°13.6'E, 1153–1155 m, FV *Arrow*, A02/108/83, 16 Sep 1983; NIWA 31729, 1 juvenile, 39°55.57'S, 167°41.60'E, 1139–1144 m, TAN0707/67, 1 Jun 2007; NMNZ Cr3748, 1 male (pcl 123.8 mm, cw 122.3 mm), 41°39.6'S, 169°20.9'E, 1025 m, FV *Arrow*, A03/49/83, 7 Oct 1983.

Bay of Plenty: NIWA 6713, 1 juvenile male (cl 24.8 mm, pcl 13.7 mm, cw 11.6 mm), Otara Seamount, 36°56.81–57.09'S, 177°20.09–19.90'E, 1323–1346 m, TAN0413/41, 10 Nov 2004; NIWA 6712, 1 indet. juvenile (cl 13.7 mm, pcl 8.6 mm, cw 7.3 mm), Otara Seamount, 36°57.32–57.26'S, 177°20.93–20.67'E, 1080–1081 m, TAN0413/29, 9 Nov 2004; NIWA 34911, 1 male (pcl 97.4 mm, cw 96.1 mm), 37°05.80–03.40'S, 176°43.80–38.59'E, 1071–1049 m, Z10064, 27 Mar 2000; NIWA 34912, 1 male (pcl 112.7 mm, cw 106.2 mm), Nukuhou Knoll, 37°16.0'S, 177°17.3'E, 1130 m, FV *Ocean Fresh*, Z10063, 29 Mar 2000; NMNZ Cr11786, 1 ovigerous female (crushed), Bay of Plenty, 37°17.5'S, 176°58.67'E, 1000 m, crab pot, FV *Savannah*, coll. G. Gibbs, 8 Oct 2001.

Hawkes Bay: NMNZ Cr8589, 1 male (cl 177.6 mm, pcl 155.5 mm, cw 149.9 mm), Ritchie Bank, 39°37'S, 178°25'E; NMNZ Cr4812, 1 ovigerous female (cl 129.9 mm, pcl 105.8 mm, cw 100.1 mm), off Portland Island, 39°43.0'S, 177°53.3'E, 1120 m, Orange Roughy trawl, FV *Arrow*, A01/17/87, 22 Jun 1987; NIWA 34910, 1 male (cl 137.0 mm, pcl 112.9 mm, cw 107.5 mm), 39°46.59–45.76'S, 178°09.28–10.05'E, 1250–1200 m, TAN0109/22, Z10873, 4 Jul 2001; NMNZ Cr3753, 1 ovigerous female (cl 142.6 mm, pcl 127.4 mm, cw 111.8 mm), 39°47.7–40.8'S, 177°58.4–00.6'E, 1160–1240 m, FV *Kalinovo*, K01/22/81, 24 Nov 1981; NMNZ Cr3754, 1 male (pcl 146.0 mm, cw 144.3 mm), 39°47.7–40.8'S, 177°58.4–00.6'E, 1160–1240 m, FV *Kalinovo*, K01/22/81, 24 Nov 1981; NMNZ Cr3755, 1 ovigerous female (cl 151.5 mm, pcl 123.9 mm, cw 110.0 mm), 39°47.7–40.8'S, 177°58.4'E–178°00.6'E, 1160–1240 m, FV *Kalinovo*, 24 Nov 1981; NIWA 25531, 1 male (pcl 48.8 mm, cw 44.0 mm), 39°49.99'S, 177°39.00'E, 1000–1100 m, Z8371, 10 Aug 1995; NIWA 34909, 1 male (pcl 98.8 mm, cw 96.0 mm), 39°49.99'S, 177°39.00'E, 1000–1100 m, Z8371, 10 Aug 1995; NMNZ Cr3751 P404, 1 male (pcl 134.0 mm, cw 126.9 mm), 40°47'S, 177°05'E, 1055–1180 m, FV *Kalinovo*, K01/147/81, 18 Dec 1981.

Wairarapa coast: NMNZ Cr3793, 1 male (cl 130.0 mm, pcl 101.5 mm, cw 95.0 mm), 30.5 nautical miles [56 km] off Cape Turnagain, 1006–1189 m, FV *Twofold Bay*, coll. A.D. Carruthers, 28 Dec 1984; NMNZ Cr3840–3843, 3 males (pcl 48.1–115.6 mm, cw 44.5–111.3 mm), 1 female (cl 123.4 mm, pcl 103.8 mm, cw 102.5 mm), off Cape

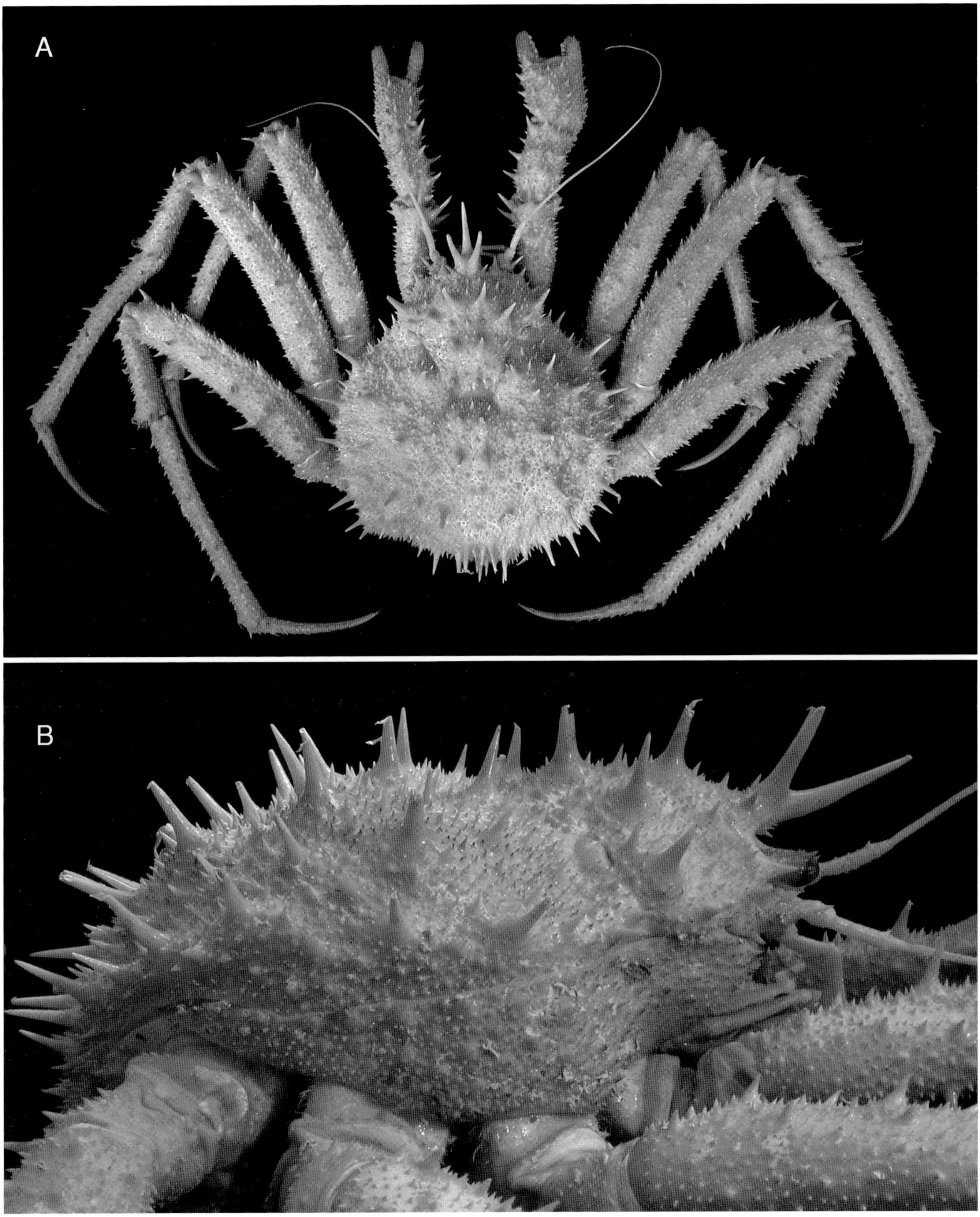

Figure 43. *Neolithodes brodiei* Dawson & Yaldwyn, 1970, male, pcl 115.6 mm, off Cape Palliser (NMNZ Cr3840). A, dorsal habitus. B, carapace, right lateral view.

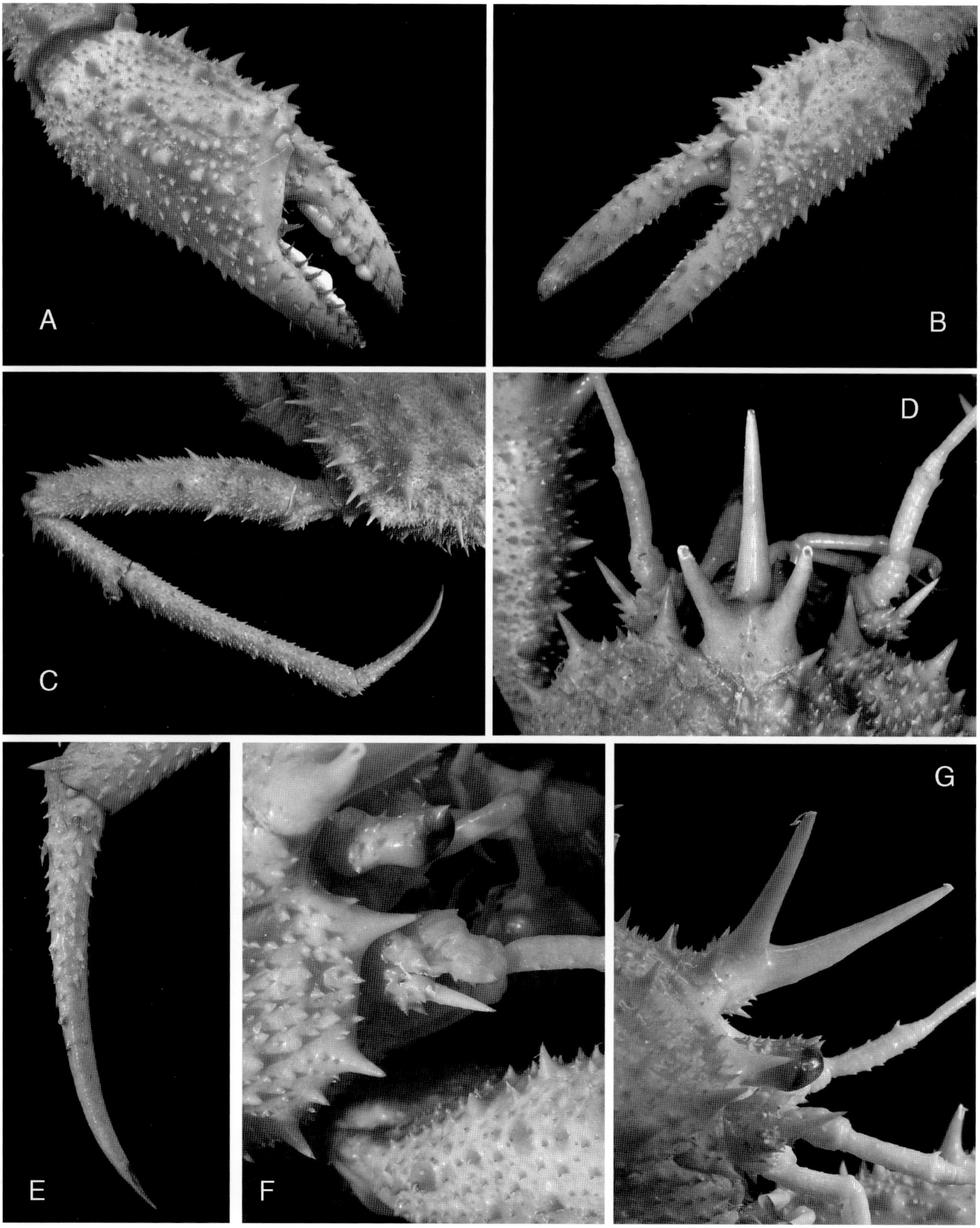

Figure 44. *Neolithodes brodiei* Dawson & Yaldwyn, 1970, male, pcl 115.6 mm, off Cape Palliser (NMNZ Cr3840). A, right chela. B, left chela. C, left pereopod 4. D, anterior carapace, dorsal view. E, left pereopod 4 dactylus. F, right orbit and antenna. G, anterior carapace, right lateral view.

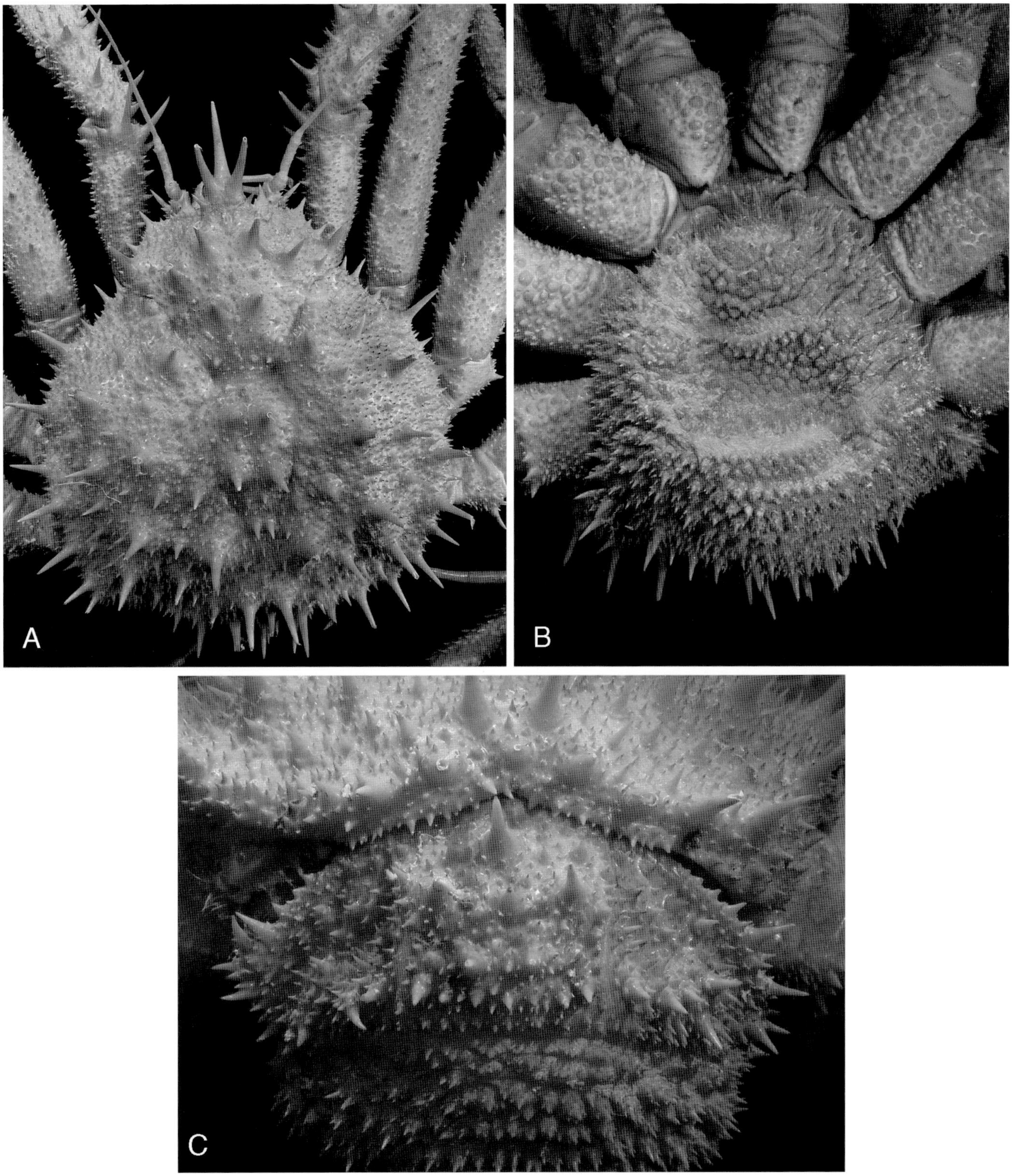

Figure 45. *Neolithodes brodiei* Dawson & Yaldwyn, 1970, male, pcl 115.6 mm, off Cape Palliser (NMNZ Cr3840). A, carapace. B, ventral surface and abdomen. C, posterior carapace and abdominal somite 2.

Palliser, Orange Roughy depth, Apr/May 1985; NMNZ Cr3791, 1 male (cl 172.0 mm, pcl 135.1 mm, cw 135.2 mm), off Cape Palliser, 40°00′S–41°30′S, 176°00–31′E, 803–1120 m, RV *James Cook*, J18/7/84, 16–18 Oct 1984; NMNZ Cr3837, 1 male (pcl 40.2 mm, cw 38.0 mm), off Castle Point, 40°51.2–49.1′S, 176°55.9–55.7′E, 1030–1060 m, FV *Kalinovo*, K1/40/81, 25 Nov 1981; NMNZ Cr3838, 1 male (cl 83.7 mm, pcl 53.0 mm, cw 48.7 mm), off Castle Point, 40°51.5–53.6′S, 176°57.9–56.3′E, 1125–1150 m, FV *Kalinovo*, K1/40/81, 28 Nov 1981; NMNZ Cr3789, 1 female (cl 142.1 mm, pcl 122.3 mm, cw 108.1 mm), off Cape Palliser, 41°00–30′S, 176°00–31′E, 803–1107 m, RV *James Cook*, J18/-/84, 16–18 Oct 1984; NMNZ Cr3745, 1 male (pcl 129.3 mm, cw 126.1 mm), off Wairarapa, 41°20.4′S, 176°16.3′E, 1070–1080 m, RV *James Cook*, J6/9/84, 2 Apr 1984; NMNZ Cr3839, 1 juvenile female (pcl 32.3 mm), E of Cape Palliser, 41°29.7–23.8′S, 176°08.4–12.1′E, 1103–1071 m, RV *James Cook*, J18/4/84, 16 Oct 1984; NMNZ Cr3752, 1 damaged male (pcl 116.2 mm), SE of Castle Point, 41°13.7–15.4′S, 176°36.1–30.9′E, 1040–1100 m, FV *Kalinovo*, K01/35/81, 27 Nov 1981; NMNZ Cr3775, 1 female (pcl 105.0 mm, cw 95.4 mm), off Wairarapa, 41°21.5′S, 176°20.9′E, 1073–1112 m, RV *James Cook*, J6/14/84, 3 Apr 1984; NMNZ Cr3776, 1 male (pcl 66.7 mm, cw 61.4 mm), off Wairarapa coast, 41°23.6′S, 176°21.7′E, 995–1082 m, RV *James Cook*, J6/13/84, 3 Apr 1984; NMNZ Cr3777, 1 male (cl 146.2 mm, pcl 119.6 mm, cw 118.8 mm), off Wairarapa coast, 41°23.6′S, 176°21.7′E, 995–1082 m, RV *James Cook*, J6/13/84, 3 Apr 1984; NMNZ Cr3774, 1 male (cl 170.6 mm, pcl 141.5 mm, cw 141.3 mm), off Wairarapa, 41°27.5′S, 176°02.4′E, 1007–1075 m, RV *James Cook*, J6/7/84, 2 Apr 1984; NMNZ Cr3792, 1 ovigerous female (pcl 145.1 mm, cw 138.1 mm), off Wairarapa coast, 41°28.4–26.1′S, 176°05.2–08.2′E, 1121–1120 m, RV *James Cook*, J18/3/84, 3 16 Oct 1984; NMNZ Cr11570, 1 male (pcl 143.0 mm, cw 143.5 mm), off Wairarapa coast, 700 m, coll. D. Wylie, 17 Oct 1986; NMNZ Cr11621, 1 male (pcl 133.2 mm, cw 133.1 mm), 2 females (cl 149.8–159.5 mm, pcl 127.6–132.1 mm, cw 117.4–120.0 mm), off Wairarapa coast, 700 m, coll. D. Wylie, 13 Oct 1983.

Chatham Rise: NMNZ Cr7353, 1 male (pcl 145.9 mm, cw 142.4 mm), NW of Chatham Islands, 42°42.0–41.7′S, 178°01.0–03.7′W, 1025–1055 m, COR/223/89, 13 Aug 1989; NIWA 33773, 1 ovigerous female (pcl 88.5 mm, cw 85.2 mm), 42°49.44–49.38′S, 175°29.87–31.93′W, 1105–1095 m, TAN0709/24, 9 Jul 2007; NMNZ Cr3758–3759, 2 males (pcl 107.9–117.1 mm, cw 104.8–109.6 mm), NE of Chathams, 42°55.4′S, 174°31.9′W, 900 m, FV *Kalton*, KTN61/82; NIWA 34588, 1 female (pcl 99.0 mm, cw 92.5 mm), 42°55.62′S, 174°42.01–39.93′W, 1076–1074 m, TAN0709/86, 17 Jul 2007; NIWA 34923, 1 ovigerous female (cl 145.0 mm, pcl 120.9 mm, cw 113.8 mm), NW Chatham Rise off Mernoo Bank, TAN0709, no specific station; NMNZ Cr14893, 1 juvenile male (pcl 34.5 mm, cw 30.1 mm), E of Chatham Islands, 43°01.06′S–42°59.62′S, 174°33.46–37.20′W, 979–988 m, TAN9206/153, 6 Jul 1992; NIWA 41608, 1 female (pcl 81.5 mm, cw 74.1 mm), 1 ovigerous female (pcl 97.3 mm, cw 94.6 mm), 44°05.49–06.33′S, 178°32.04–32.10′E, 952–982 m, TAN9812, Z9386, 30 Sep 1998; NIWA 60581, 1 male (cl 97.0 mm, pcl 76.9 mm, cw 70.9 mm), 44°08.20–08.19′S, 178°47.90–46.61′E, 921–922 m, FV *San Waitaki*, SWA0901/88, 17 Nov 2009; NIWA 34895, 1 male (cl 79.0 mm, pcl 57.7 mm, 51.8 mm), 44°12.39–12.99′S, 179°03.54–0.46′E, 959–961 m, TAN9812/15, Z9468, 2 Oct 1998; NIWA 60580, 1 female (pcl 64.6 mm, cw 59.3 mm), 44°22.83–22.66′S, 179°54.68–56.27′E, 952–957 m, FV *San Waitaki*, SWA0901/82, 16 Nov 2009; NIWA 34613, 1 female (pcl 70.2 mm, cw 64.2 mm), 44°25.26–26.28′S, 174°59.30′W–175°00.75′W, 913–936 m, TAN0709/115, 22 Jul 2007; NIWA 60578, 1 male (pcl 149.4 mm, cw 149.2 mm), 44°32.78–33.62′S, 175°23.73–23.77′W, 939–985 m, FV *San Waitaki*, SWA0901/61, 13 Nov 2009; NIWA 60556, 1 ovigerous female (cl 151.6 mm, pcl 122.4 mm, cw 113.2 mm), 44°34.38–35.02′S, 175°20.27–20.28′W, 1070–1106 m, FV *San Waitaki*, SWA0901/060, 13 Nov 2009; NIWA 34896, 1 female (pcl 31.5 mm, cw 28.5 mm), 44°35.28–34.90′S, 177°27.43–29.03′W, 852–858 m, TAN9812/75, Z9461, 21 Oct 1998; NMNZ Cr3772, 1 female (pcl 95.3 mm, cw 96.1 mm), off Canterbury, 44°45.1′S, 173°05.2′E, 922–929 m, RV *James Cook*, J10/16/84, 10 Jun 1984; NMNZ Cr9211, 1 juvenile female (cl 52.6 mm, pcl 35.4 mm, cw 32.9 mm), 44°50.6′S, 175°29.4′E, 1044–1050 m, RV *James Cook*, JCO90/015/5, 20 Sep 1990; NIWA, 1 male (pcl 148.6 mm, cw 151.2 mm), Chatham Rise, Tow 123, NEB.

Off West Coast: NMNZ Cr3786, 1 ovigerous female (cl 178.1 mm, pcl 153.0 mm, cw 145.6 mm), 43°15.3′S, 168°27.2′E, 1027–1041 m, RV *James Cook*, J16/10/83, 13 Dec 1983;

Louisville Ridge: NIWA 34893, 1 male (pcl 62.9 mm, cw 58.3 mm), southern Louisville Ridge, 45°33.99′S, 156°40.00′W, 870–900 m, Z8370, FV *Peterson*, Sep 1995.

Campbell Plateau: NMNZ Cr3779, 1 male (pcl 73.9 mm, cw 66.8 mm), Campbell Plateau (Pukaki Rise or near Campbell Island), RV *Shinkai Maru*, Stn 241; NMNZ Cr3778, 1 parasitised female (cl 128.8 mm, pcl 113.7 mm, cw 106.1 mm), Campbell Plateau (Pukaki Rise or near Campbell Island), RV *Shinkai Maru*, Stn 241; NMNZ Cr3750, 1 male (pcl 139.4 mm, cw 130.3 mm), 47°27′S, 169°26′E, 500 m, FV *Wessermunde*, WO 2B/165/79, 2 Mar 1979; NIWA 34894, 1 male (pcl 76.8 mm, cw 70.7 mm), 47°57.30′S–48°00.19′S, 169°55.83–57.01′E, 910 m, TAN0012/79, Z10849, 15 Dec 2000; NIWA 60552, 1 male (pcl 80.5 mm, cw 73.6 mm), 1 ovigerous female (pcl 104.7 mm, cw 98.3 mm), E of Pukaki Rise, 48°36.79′S, 175°15.03′E, 820–815 m, TAN0911/15, 1 Dec 2009; NIWA 44075, 1 male (pcl 109.8 mm, cw 102.5 mm), Pukaki Rise, 50°02–03′S, 174°41–45′E, 855–975 m, with *Briarosaccus callosus* (Rhizocephala), SOP TRIP2468/70, 28 Aug 2007; NIWA 60554, 2 ovigerous females (pcl 104.1–111.7 mm, cw 99.1–102.5 mm), SE of Pukaki Rise, 50°15.14′S, 174°05.49′E, 772–776 m, TAN0911/19, 2 Dec 2009;

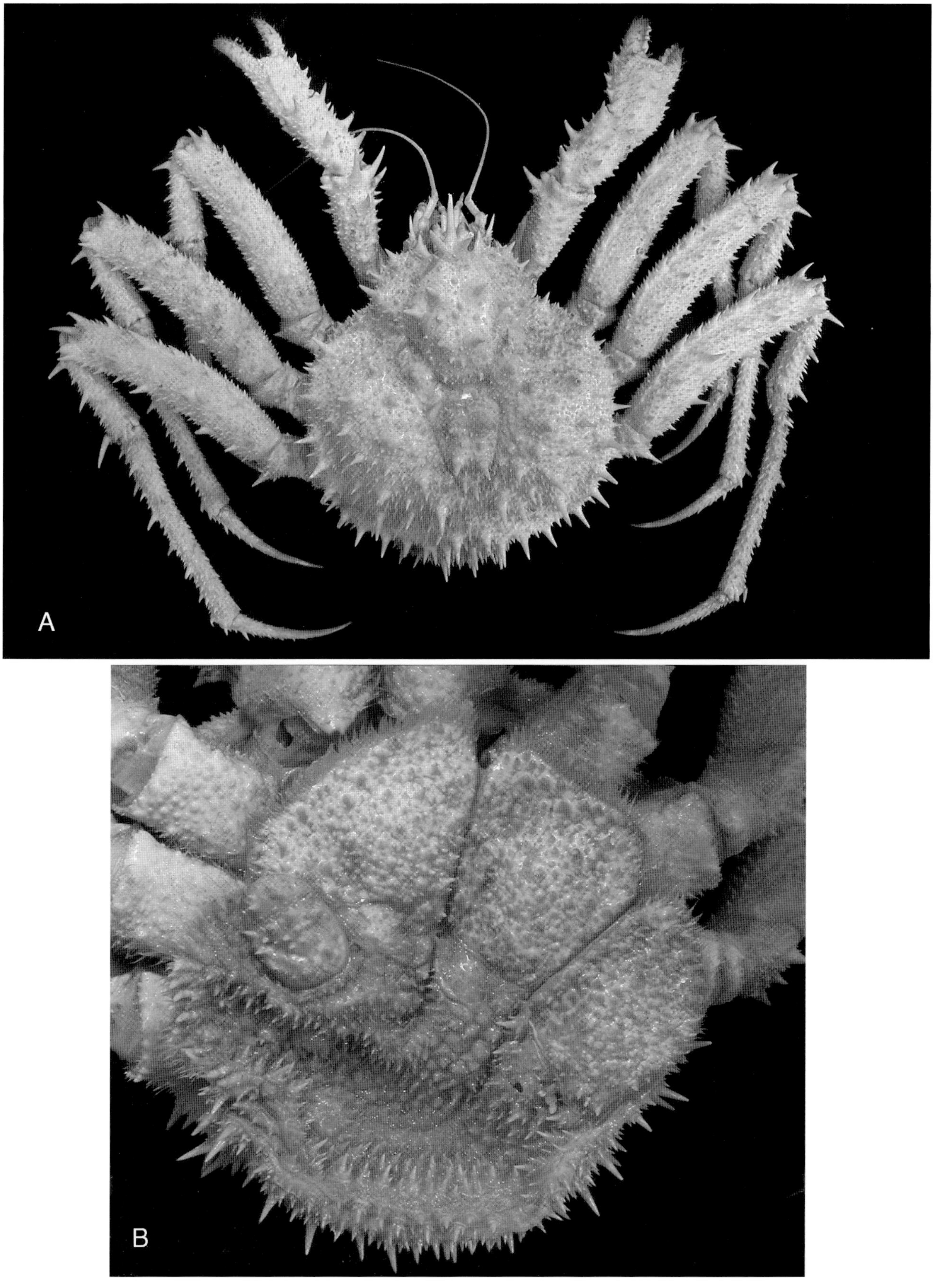

Figure 46. *Neolithodes brodiei* Dawson & Yaldwyn, 1970, female, pcl 103.8 mm, off Cape Palliser (NMNZ Cr3841). A, dorsal habitus. B, ventral surface and abdomen.

NIWA 34892, 1 male (pcl 66.2 mm, cw 61.0 mm), Auckland Islands, 51°38.80'S, 164°35.28'E, 968–973 m, TAN0306/8, 14 Apr 2003.

New Zealand, no specific locality: NMNZ Cr3831, 1 male (cl 188.0 mm, pcl 165.5 mm, cw 161.6 mm), northern New Zealand, FV *Wanaka*, 1984; NMNZ Cr3833, 1 male (cl 83.8 mm, pcl 55.5 mm, cw 50.6 mm), northern New Zealand waters, FV *Wanaka*, 1984; NMNZ Cr3832, 1 female (pcl 121.4 mm, cw 112.1 mm), northern New Zealand waters, FV *Wanaka*, 1984; NMNZ Cr3835, 1 male (cl 89.4 mm, pcl 75.1 mm, cw 68.8 mm), New Zealand.

Diagnosis. Carapace with numerous small secondary spinules on the carapace and pereopods in addition to the major spines. Cheliped dactyli with convex dorsal margins, rounded in cross section. Antennal peduncle with few, small, scattered granules or minute spinules, not distinctly spinose. Walking leg meri compressed, flattened; extensor margin with several upright major spines protruding above level of secondary spines; pereopod 4 dactylus covered with spinules over surfaces of proximal half; ventral surfaces of coxae with blunt tubercles or granules.

Description. *Carapace*: Pyriform, 0.99–1.14 times longer than wide; dorsal surface armed with scattered, longer spines amongst dense, uniform covering of granules or minute secondary spinules. Gastric region bearing 6 large spines forming transverse hexagon; with 2 near transverse rows of 3 smaller spines, first across centre of hexagon and second slightly anterior to hexagon. Hepatic spine strong, directed anterolaterally; margin between outer orbital spine and hepatic spine with 1 or 2 spines. Branchial regions with 8–12 large dorsal spines in addition to scattered smaller secondary spines; margins with 12–16 major spines in addition to numerous minute spines. Cardiac region with 4 equal spines of similar size to gastric spines and 1 or 2 smaller spines near intestinal region. Intestinal region with 2 upright spines in transverse row and 2 or 3 smaller posterior directed spines. Pterygostomian region spinose or tuberculate, with small, anterior, submarginal spine.

Rostrum 0.12–0.28 pcl; median spine inclined dorsally, with pair of divergent dorsal spines and smaller pair of basal spines; ventral proximal surface with cluster of spinules. Posterior orbital margin concave in lateral view, margin usually unarmed, occasionally with 1–4 minute spinules; outer orbital spine slender, reaching to but not beyond cornea (when eyes directed forwards). Anterolateral spine shorter than outer orbital spine.

Ocular peduncle: Longer than cornea; with scattered dorsal granules or spinules.

Antennule: Peduncle unarmed, reaching anteriorly beyond antennal peduncle by less than one-quarter distal antennular peduncle article.

Antenna: Basal antennal article with small anterolateral spine; outer margin of article 2 with cluster of 4–7 small spines basally and long slender spine that reaches to end of article 4; article 3 with sharp to angular inner distal tooth; scaphocerite minute, often with bifid or trifid apex, shorter than article 4; article 4 with mesial and lateral distal spinule and occasionally 1–3 other minute spinules, about half as long as article 5; article 5 unarmed or with 1 or 2 minute acute granules.

Abdomen: Ornamentation similar in both sexes. Somite 2 densely spinose, with about 10 long spines and numerous smaller spines on median plate; submedian and marginal plates densely spinose, spines largest laterally; surface and margins of remaining somites and telson multispinose.

Pereopod 1 (chelipeds): Dimorphic but spination similar. Entire surface of both chelipeds, except for distal half or two-thirds of fingers, covered with numerous minute secondary spines in addition to major spines. Coxa ventral surface with blunt tubercles and tufts of setae; ischiobasis with 4 or 5 stout ventral spines. Merus inner margin with stout subdistal spine; ventral margin with two rows of stout spines, mesial row with 2 spines, lateral row with 3 or 4 spines; dorsal and lateral surface spinose, spines largest distally. Carpus with prominent spines on dorsal and lateral surfaces, with 3 or 4 irregular rows of 3–5 spines. Palm of both chelipeds in both sexes with similar ornamentation; all surfaces spinose, though less pronounced mesially and ventrally; dorsal margin with 2 irregular rows of about 3 or 4 conical spines; midlateral surface with 2 rows of 4 or 5 spines of similar size to dorsal row; ventral surface irregularly spinose.

Major cheliped 1.19–1.31 pcl (male), 1.16–1.23 (female); upper palm length 1.07–1.11 times height (male), 1.05–1.15 (female); occlusal margins of fingers corneous for distal third to half, proximally with 3 calcareous nodules; dactylus dorsal margin broadly convex, with rows of tufts of golden setae and proximal cluster of 5–7 small spines, 1.29–1.35 times longer than dorsal margin of palm (male), 1.33–1.35 (female).

Minor cheliped 1.19–1.29 pcl (male), 1.16–1.20 pcl (female); upper palm length 1.13–1.28 times height (male), 1.13–1.18 (female); occlusal margin corneous in distal half, proximally crenulate; dactylus dorsal margin broadly convex, with rows of tufts of golden setae and proximal cluster of 5–8 small spines proximally, 1.70–1.83 times longer than dorsal margin of palm (male), 1.65–1.78 (female).

Pereopods 2–4 (walking legs 1–3): Similar, slender, elongate; ornamentation similar in both sexes; segments spinose, surface between major spines densely covered with numerous minute secondary spinules or granules. Pereopod 4 longest. Coxae with well-spaced,

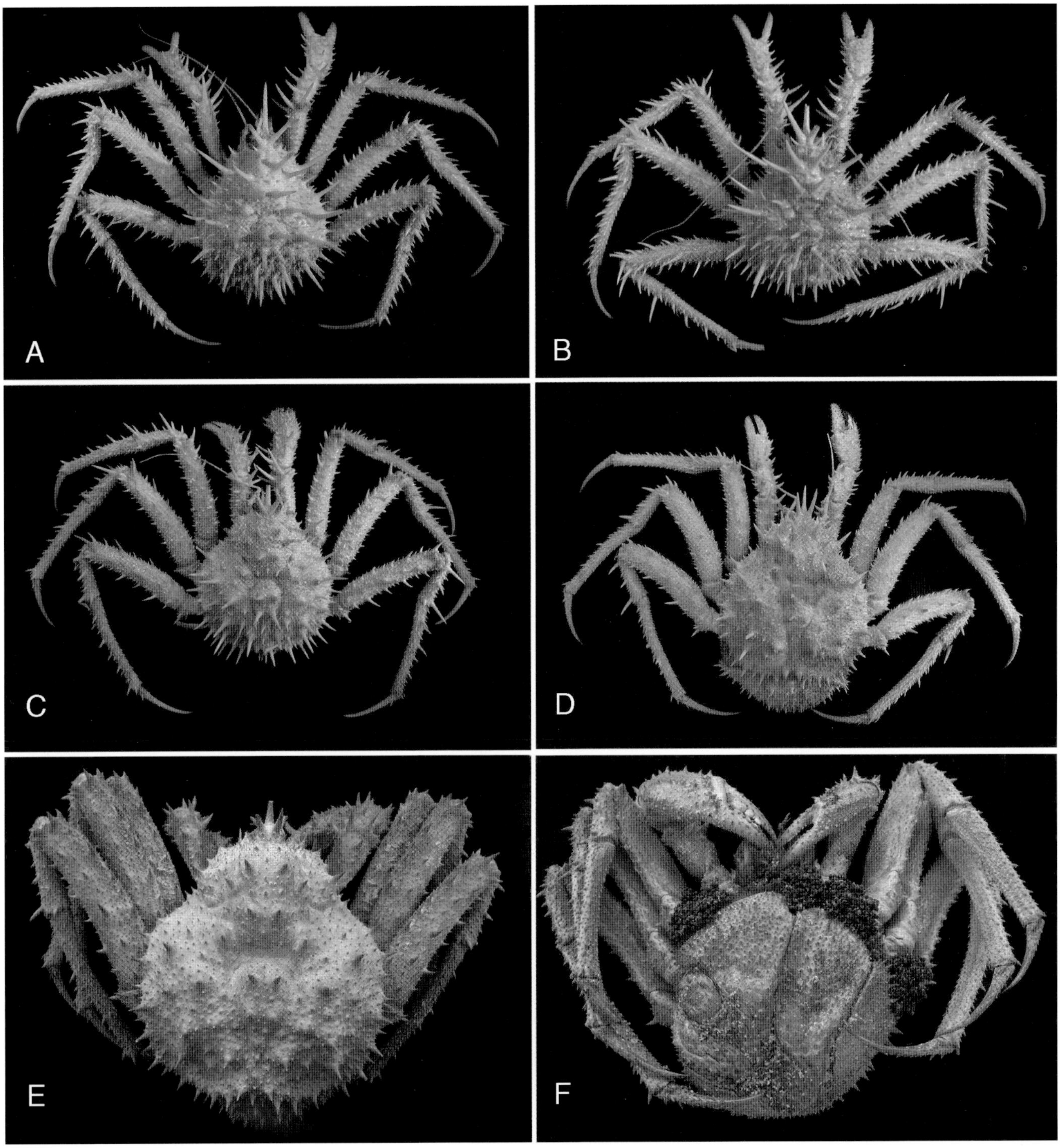

Figure 47. *Neolithodes brodiei* Dawson & Yaldwyn, 1970. A, juvenile female, pcl 35.4, Chatham Rise (NMNZ Cr9211). B, juvenile male, pcl 48.1 (NMNZ Cr3843). C, male, pcl 86.2, Challenger Plateau (NMNZ Cr3747). D, ovigerous female, pcl 105.8 mm, Hawkes Bay (NMNZ Cr4812). E–F, ovigerous female holotype, pcl 111.5 mm, dorsal and ventral view, Campbell Plateau (NIWA 735).

blunt, rounded to flat tubercles, distal margins crenulate to bluntly dentate. Ischiobasis with 3 or 4 stout spines around distal margins and numerous smaller ventral spinules. Merus dorsoventrally compressed, flattened ovate in cross section, longer than propodus; extensor margin with 6–9 major spines in addition to paired distal spines; dorsal surface with row of 5–7 major spines of similar size to extensor spines; flexor margin with 2 rows of 4–7 major spines; major spines on surface of merus distinctly standing out from

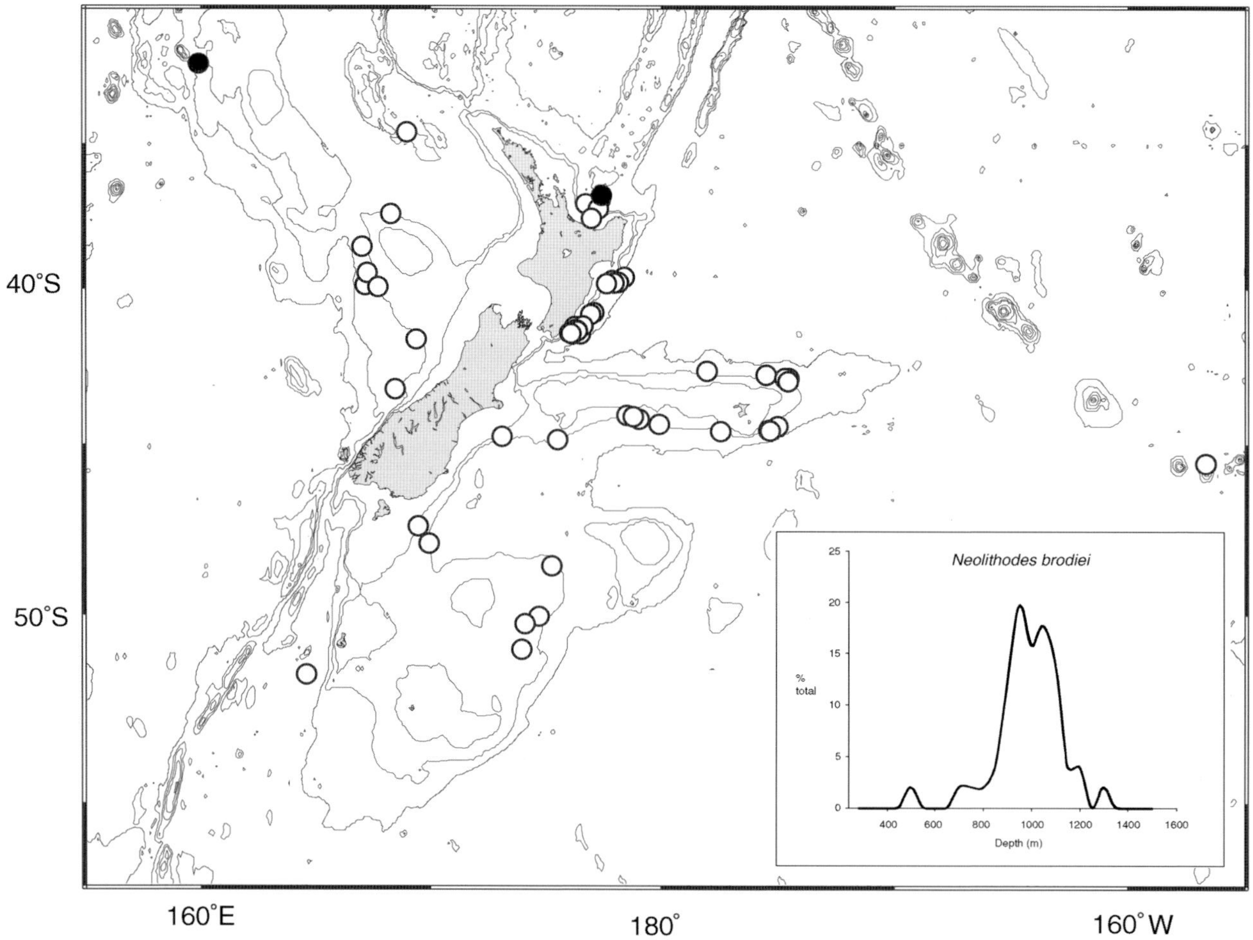

Figure 48. Geographic distributions of *Neolithodes brodiei* Dawson & Yaldwyn, 1970 (○) and *N. bronwynae* sp. nov. (●). Inset: bathymetric distribution of *N. brodiei.*

secondary spinules. Carpus slightly longer than half merus length; with 5–7 spines on extensor margin and 3 or 4 other major spines distributed on carpal surface, distal and second proximal spines longest, about twice as long as other spines. Propodus dorsoventrally compressed; with 5–9 spines on extensor margin and 3–5 similar spines on dorsal surface; flexor margin with numerous smaller spines. Dactylus curved, rounded in cross section, with 4 or 5 small proximal spines and numerous secondary spines covering proximal half; apex corneous.

Pereopod 2 length 2.21–2.41 pcl (male), 2.01–2.09 pcl (female). Merus 0.77–0.86 pcl (male), 0.66–0.68 pcl (female); length:height ratio 4.41–5.04 (male), 3.80–4.05 (female). Carpus 0.51–0.60 merus length (male), 0.56–0.63 (female). Propodus 0.83–088 merus length (male), 0.91–0.92 (female); length:height ratio 8.00–8.55 (male), 6.43–7.71 (female). Dactylus 0.60–0.64 propodus length (male), 0.62–0.63 (female).

Pereopod 3 length 2.48–2.77 pcl (male), 2.30–2.43 pcl (female). Merus 0.85–0.99 pcl (male), 0.76–0.77 pcl (female); length:height ratio 4.60–4.92 (male), 4.11–4.38 (female). Carpus 0.50–0.53 merus length (male), 0.55–0.59 (female). Propodus 0.85–0.87 merus length (male), 0.92–0.93 (female); length:height ratio 8.52–9.14 (male), 7.72–8.50 (female). Dactylus 0.61–0.68 propodus length (male), 0.59–0.60 (female).

Pereopod 4 length 2.56–2.94 pcl (male), 2.36–2.62 pcl (female). Merus 0.84–0.99 pcl (male), 0.75–0.80 pcl (female); length:height ratio 4.54–5.06 (male), 4.26–4.31 (female). Carpus 0.52–0.55 merus length (male), 0.56–0.58 (female). Propodus 0.89–0.93 merus length (male), 0.98–1.03 (female); length:height ratio 8.76–9.14 (male), 8.03–9.33 (female). Dactylus 0.61–0.66 propodus length (male), 0.59–0.63 (female).

Colour in life. Deep-red overall.

Remarks. *Neolithodes brodiei* Dawson & Yaldwyn, 1970 belongs to the group within the genus characterised by bearing numerous secondary spinules on the carapace and pereopods in addition to the major spines; convex dorsal margins of the cheliped dactylus; and compressed, flattened meri of the walking legs:

N. brodiei Dawson & Yaldwyn, 1970, *N. nipponensis* Sakai, 1971, and *N. flindersi* Ahyong, 2010. Of these, *N. brodiei* most closely resembles *N. flindersi* Ahyong, 2010 from southeastern Australia, previously misidentified as *N. brodiei* from that region.

Neolithodes brodiei is best distinguished from *N. flindersi* by patterns of spination: the ventral surfaces of the coxae of the walking legs in *N. brodiei* are covered in low, blunt tubercles rather than short, conical spines (in males and juvenile female *N. flindersi*); the spinulation covering the surfaces of the walking leg dactyli of *N. brodiei* usually extends only slightly beyond the proximal half, rather than beyond the distal quarter; and in *N. brodiei*, the extensor margins of the pereopod 2–4 meri are lined with short, closely-spaced spines that are interspersed by 6–9 distinctly longer spines, whereas in *N. flindersi* and *N. nipponensis*, the meral extensor spines are of similar size. *Neolithodes brodiei* also differs from *N. flindersi* in more subtle features: the rostrum in specimens exceeding 80 mm pcl is proportionally longer (0.12–0.28 versus 0.10–0.19 pcl); the pereopod 4 propodus of *N. brodiei* is generally less slender than in *N. flindersi* (usually less than 10 times as long as high, with the merus longer than the propodus; versus usually 10 times as long as high or greater, and longer than the merus). Note that specimens with regenerating limbs may be difficult to distinguish using pereopod 4 characters. Overall dorsal spination of juvenile *N. brodiei* is similar to juvenile *N. flindersi* and *N. nipponensis* but, as in adults, can be distinguished by the ornamentation of the pereopod 2–4 coxae. The strong morphological similarities and adjacent, but discrete, distributions suggest that *N. brodiei* and *N. flindersi* might be sibling species.

As with other lithodids, rostral and general spine length in *N. brodiei* is inversely proportional to body size, and the walking legs of females are proportionally shorter than those of males. The median rostral spine reaches up to 0.6 pcl in juveniles smaller than about 50 mm pcl, falling to about 0.1–0.3 pcl in specimens exceeding about 80 mm pcl. Bluntness of the tubercles on the ventral surface of the pereopod 2–4 coxae also varies allometrically, being flattest and bluntest in adults. Occasionally, the tubercles of the coxae of pereopods 3–4 may be slightly pointed in juveniles, approaching the condition of *N. flindersi*. The smallest ovigerous female measures 88.5 mm pcl (NIWA 33773).

Neolithodes brodiei is the most common and widespread lithodid in New Zealand waters, ranging from the southern Norfolk Ridge south to the Campbell Plateau, and east to the Louisville Ridge (156°W; outside of the New Zealand Exclusive Economic Zone). Reports of *N. brodiei* from outside of the New Zealand region are based on other species: Antarctic records of *N. brodiei* (see Thatje & Lörz 2005) are referable to *N. yaldwyni* Ahyong & Dawson, 2006; southern Australian and Vanuatu records (Davie 2002; Poore 2004; Macpherson 2001) represent *N. flindersi* Ahyong, 2010, and an undescribed species, respectively. *Neolithodes brodiei* is not known from depths much beyond 1200 m, so its distribution appears to be largely dictated by the extent of contiguous continental shelf and slope habitat in the New Zealand region.

The only other species of *Neolithodes* known from New Zealand, *N. bronwynae* sp. nov., markedly differs from *N. brodiei* by the very long dorsal spines in adults, sparse rather than dense covering of secondary spinules on the carapace and pereopods, and subcylindical rather than flattened meri of the walking legs.

Neolithodes brodiei and *L. aotearoa* sp. nov. are the two largest species of lithodid in New Zealand waters, both being subject to fishery management under the Quota Management System implemented by the New Zealand Ministry of Fisheries.

Batson (2003) reported the association between *N. brodiei* and the liparid fish, *Careproctus* sp., which incubates its eggs in the branchial chambers of the crab.

Distribution. New Zealand and adjacent waters, from the southern Norfolk Ridge to the Campbell Plateau, and the southern Louisville Ridge; 500–1240 m, usually from 950–1150 m.

Neolithodes bronwynae sp. nov.

(Figs 48–52, Pl. 2B)

Neolithodes vinogradovi. — ?Macpherson, 1990: 218, fig. 2a [not *Neolithodes vinogradovi* Macpherson, 1988b].

Type material. *Holotype*: NIWA 49026, male (cl 171.4 mm, pcl 123.3 mm, cw 114.5 mm), Whakatane Seamount, 36°47.71–47.28′S, 177°25.53–25.65′E, 1515–1530 m, TAN0413/19, 8 Nov 2004.

Paratype: NMNZ Cr11147, 1 male (cl 220.0 mm, pcl 170.0 mm, cw 155.9 mm), Lord Howe Rise, 32°03.97–02.26′S, 159°52.80–51.10′E, 1934–1920 m, NORFANZ, TAN0308/071, 24 May 2003.

Diagnosis. Carapace dorsal surface with long, slender spines and scattered, widely separated granules or minute secondary spines; longest spine exceeding 0.2 pcl. Rostrum 0.3–0.4 pcl; ventral surface smooth. Posterior orbital margin near vertical; outer orbital spine not reaching beyond cornea. Outer spine of antennal article 2 reaching beyond article 4; scaphocerite reduced to a short, acute point, shorter than antennal article 4. Major cheliped palm with prominent dorsal and lateral spines; dactylus with convex dorsal margin, longer than dorsal margin of palm. Walking legs spinose, surface between major spines finely granulate, secondary

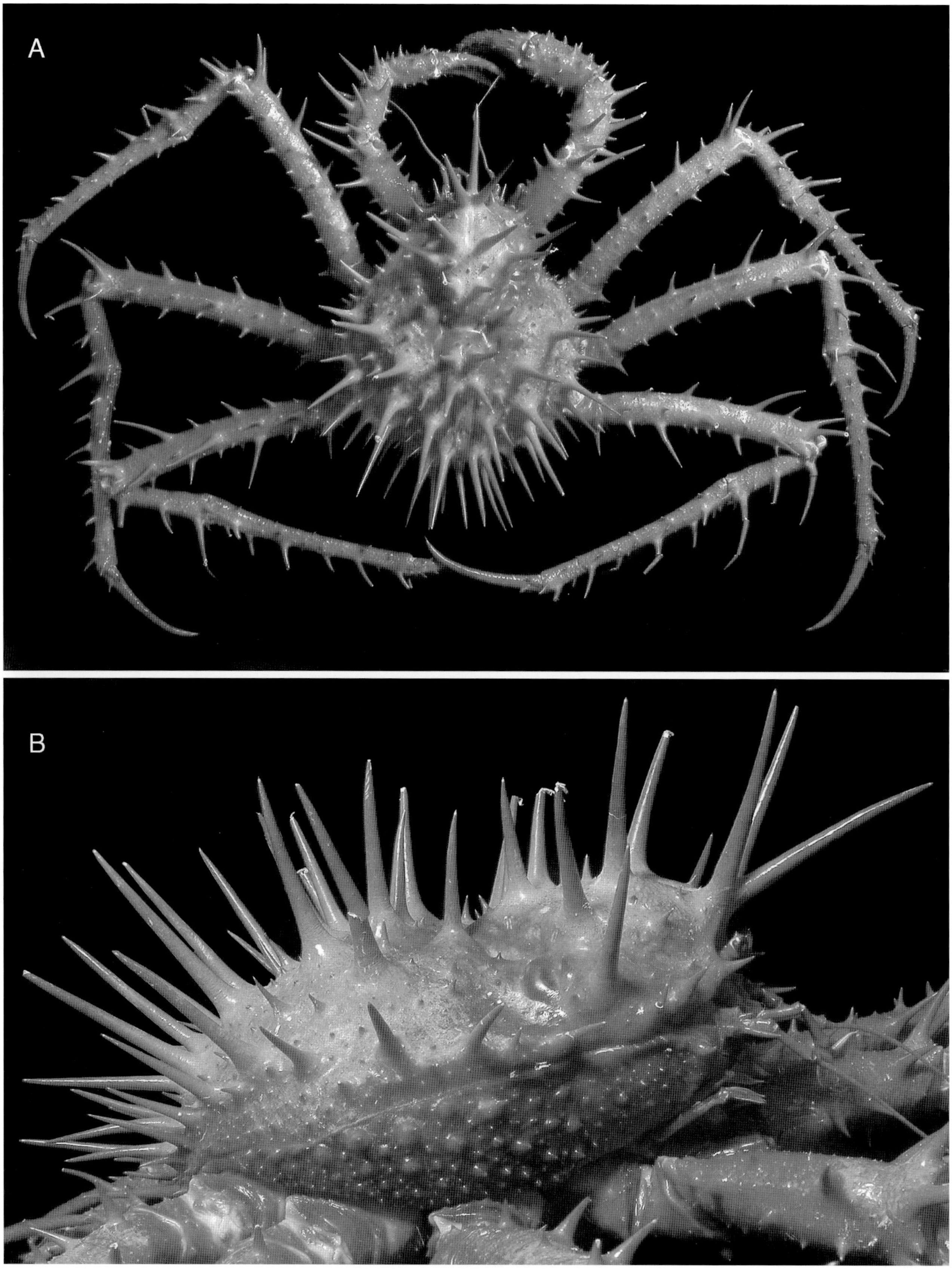

Figure 49. *Neolithodes bronwynae* sp. nov., male holotype, cl 171.4 mm, pcl 123.3 mm, cw 114.5 mm, Whakatane Seamount (NIWA 49026). A, dorsal habitus. B, carapace, right lateral view.

Figure 50. *Neolithodes bronwynae* sp. nov., male holotype, cl 171.4 mm, pcl 123.3 mm, cw 114.5 mm, Whakatane Seamount (NIWA 49026). A, carapace. B, ventral surface and abdomen. C, right antenna. D, posterior carapace and abdominal somite 2.

spines absent or sparsely distributed; merus ovate in cross section, shorter than or subequal to pcl.

Description. *Carapace*: Pyriform, 1.07–1.09 times longer than wide; dorsal surface armed with long, slender spines and scattered, widely separated granules or minute secondary spines; longest spine 0.2–0.4 pcl (0.37 pcl in holotype). Gastric region bearing 6 long spines forming transverse hexagon; with 2 near transverse rows of 3 and 4 smaller spines, first across centre of hexagon and second slightly anterior to hexagon. Hepatic spine strong, anterodorsally directed; margin between outer orbital spine and hepatic spine with 2 spines. Branchial regions with 6 long dorsal spines in addition to scattered, minute, secondary spines; margins with 10 or 11 long spines in addition to several markedly shorter spines. Cardiac region with 4 long, equal spines of similar size to posterior gastric spines;

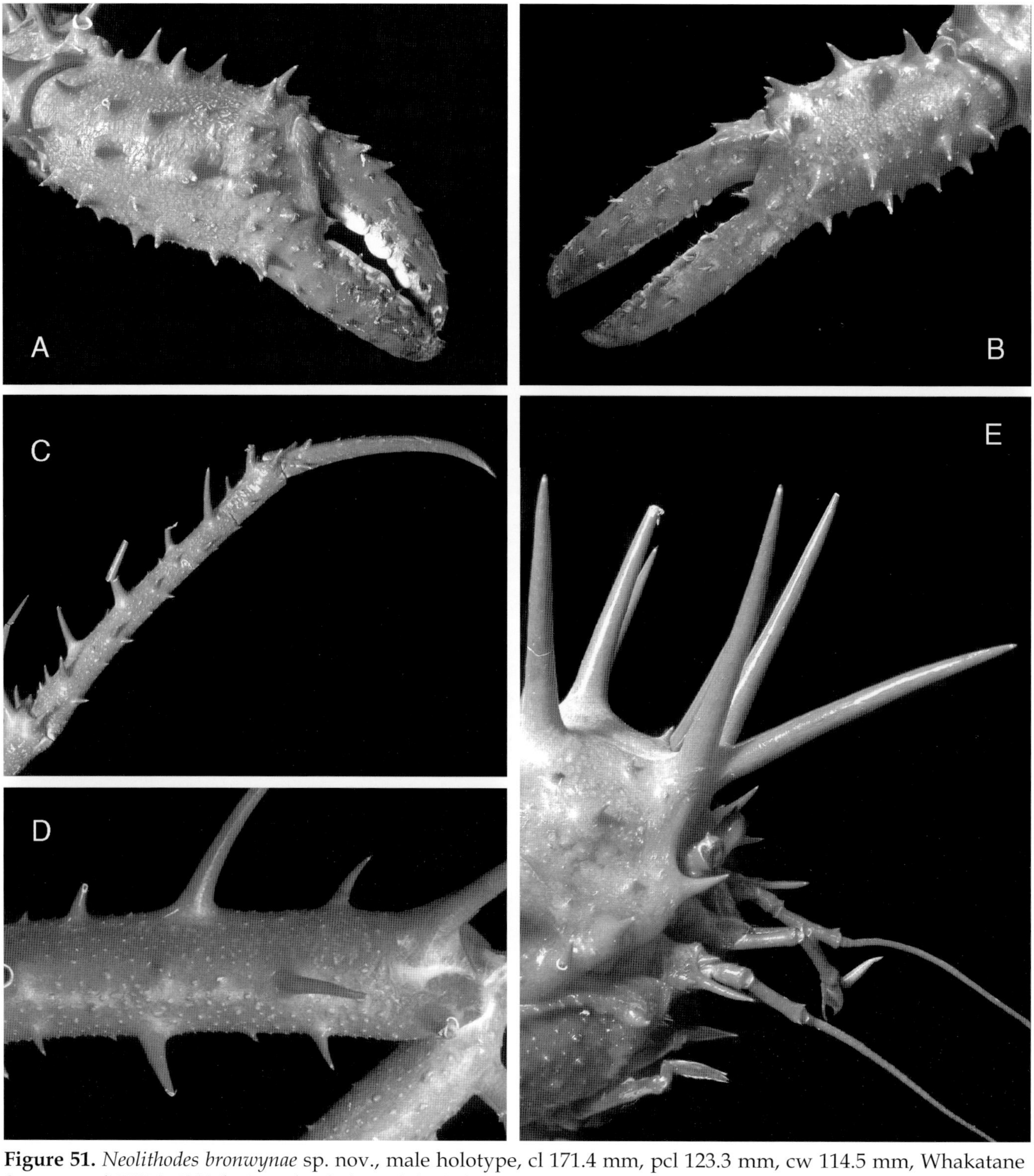

Figure 51. *Neolithodes bronwynae* sp. nov., male holotype, cl 171.4 mm, pcl 123.3 mm, cw 114.5 mm, Whakatane Seamount (NIWA 49026). A, right chela. B, left chela. C, right pereopod 4 dactylus and propodus. D, right pereopod 4 merus, distal end. E, anterior carapace, right lateral view.

with single shorter spine near intestinal region. Intestinal region with 2 long upright spines in transverse row and 2, much shorter, posterior directed spines. Pterygostomian region spinose or tuberculate, with stout, conical subdistal spine.

Rostrum 0.29–0.39 pcl; median spine straight, angled dorsally, with pair of long, divergent dorsal spines; ventral surface smooth. Posterior orbital margin shallowly concave to almost straight in lateral view; outer orbital spine slender, reaching to but not beyond cornea (when eyes directed forwards). Anterolateral spine subequal to or shorter than outer orbital spine.

Ocular peduncle: Longer than cornea; with 5 dorsal granules or spinules.

Antennule: Peduncle unarmed, reaching anteriorly beyond antennal peduncle by half length of distal antennular peduncle article.

Antenna: Basal antennal article with small anterolateral spine; outer margin of article 2 with small basal spine and long slender spine that reaches proximal quarter to third of article 5; article 3 with angular inner distal tooth; scaphocerite minute, shorter than article 4, apex sharp; article 4 unarmed, about half length of article 5.

Abdomen: Somite 2 with 8–10 long spines of similar length to cardiac spines, and several much shorter spines on median plate; submedian plate with 4 or 5 long spines near posterior margin and 5–7 shorter spines on surface; lateral border of marginal plates with 7 or 8 low, graded teeth proximally and 3 or 4 slender spines distally; surface and margins of remaining somites multispinose or nodulose; telson spinose.

Pereopod 1 (chelipeds): Dimorphic but spination similar; surface of segments minutely granular. Coxa with blunt tubercles and tufts of setae, unarmed; ischiobasis with 5 or 6 stout ventral spines. Merus outer surface spinose, longest spine distally; ventral margin with 2 rows stout spines, mesial row with 2 spines, lateral row with 3 spines; dorsal and lateral surface spinose, spines largest distally. Carpus dorsal margin with irregular row of 4 or 5 slender spines; lateral margin with 2 irregular rows of 3 or 4 slender spines of similar size to dorsal row; ventrally with small, scattered, acute tubercles. Palm prominently spinose on dorsal, lateral and ventral surfaces, inner surface with acute tubercles and small spines; dorsal margin with 2 irregular rows of 3 or 4 conical spines; midlateral surface with 2 irregular rows of 4 or 5 spines of similar size to dorsal rows; ventral surface with 2 rows of 4 or 5 spines, smaller than lateral and dorsal spines.

Major cheliped 1.44–1.40 pcl (male); upper palm length 1.27–1.21 times height (male); occlusal margins corneous for slightly less than distal half, proximally with 3 calcareous nodules; dactylus dorsal margin broadly convex, with rows of tufts of golden setae and proximal cluster of 4–7 small spines, 1.29–1.07 times longer than dorsal margin of palm (male).

Minor cheliped 1.45–1.34 pcl (male); upper palm length 1.39–1.35 times height (male); occlusal margins corneous for slightly more than distal half, proximally crenulate; dactylus dorsal margin broadly convex, with rows of tufts of golden setae and proximal cluster of 4–6 small spines, 1.70–1.51 times longer than dorsal margin of palm (male).

Pereopods 2–4 (walking legs 1–3): Similar; segments prominently spinose, surface between major spines granulate; length increasing posteriorly, pereopod 4 longest. Distal margins of coxae with short triangular spines. Ischiobasis with 4 distal spines (dorsal 2 shortest) and 2 or 3 smaller ventral spines. Merus ovate in cross section, shorter than carapace; extensor margin with 4 or 5 prominent spines increasing in length distally and intervening shorter spines, in addition to long, paired distal spines; dorsal surface with row of 5 spines of similar length to proximal extensor spines (except on pereopod 4 merus with spine near midlength as long as cardiac carapace spines); flexor margin with 2 rows of 5–7 spines (longest on pereopod 4). Carpus slightly longer than half merus length; extensor margin with 5 or 6 spines, distal and second proximal spines longest, about twice length of other spines; dorsal surface with 4 or 5 spines, distal and proximal spine longest. Propodus ovate to subcircular in cross section; with 8 or 9 spines on extensor margin and 9–11 similar spines on dorsal surface; flexor margin with 6–10 smaller spines. Dactylus curved, rounded in cross section, with 4 small proximal spines and several scattered spinules on proximal surfaces; apex corneous.

Pereopod 2 length 2.51–2.52 pcl (male). Merus 0.86–0.91 pcl (male); length:height ratio 6.41–6.59 (male). Carpus 0.51–0.52 merus length (male). Propodus 0.86–0.89 merus length (male); length:height ratio 10.10–10.18 (male). Dactylus 0.66–0.50 propodus length (male).

Pereopod 3 length 2.70–2.73 pcl (male). Merus 0.92–0.98 pcl (male); length:height ratio 6.52–6.31 (male). Carpus 0.52–0.50 merus length (male). Propodus 0.91–0.85 merus length (male); length:height ratio 10.29 (male). Dactylus 0.64–0.52 propodus length (male).

Pereopod 4 length 2.81–2.79 pcl (male). Merus 0.91–0.99 pcl (male); length:height ratio 6.84–6.62 (male). Carpus 0.54–0.52 merus length (male). Propodus 0.96–0.91 merus length (male); length:height ratio 10.14–11.10 (male). Dactylus 0.66–0.51 propodus length (male).

Colour in life. Deep-red overall (Pl. 2B).

Etymology. Named for Bronwyn Ahyong.

Remarks. On the basis of the long dorsal spines retained in adults, *Neolithodes bronwynae* sp. nov. resembles *N. vinogradovi* Macpherson, 1988b [type locality: southeast Indian Ocean] and *N. duhameli* Macpherson, 2004 [type locality: Crozet Islands]. *Neolithodes bronwynae* differs from *N. duhameli* in its proportionally longer, more slender walking legs that lack a dense covering of secondary spinules and acute tubercles. The third walking leg in *N. bronwynae* is 2.8 pcl versus 1.9–2.3 pcl in *N. duhameli*, with the merus about 0.9 rather than 0.7 pcl., and the propodi about 10 rather than 5–6 times longer than high. The dense secondary spinulation

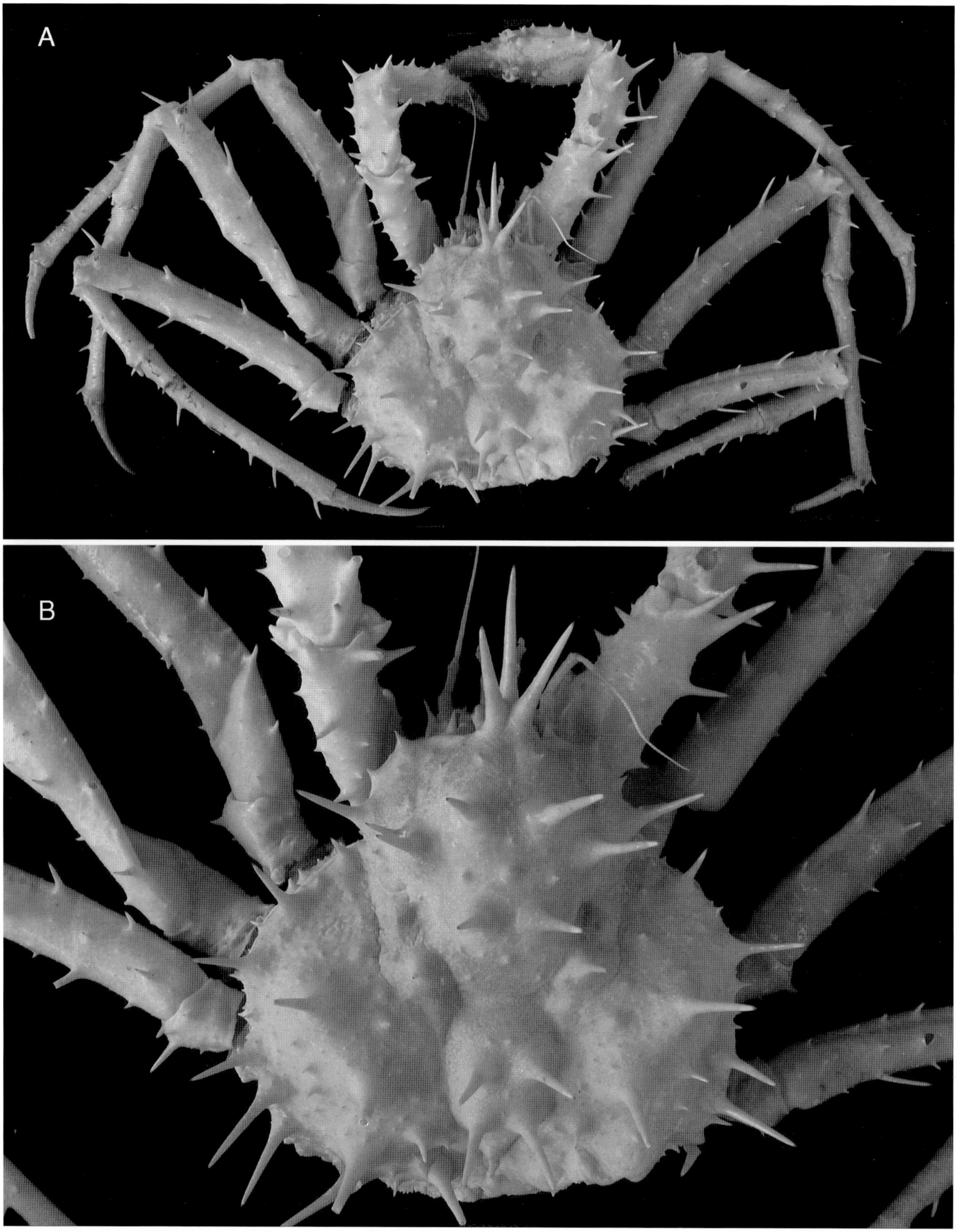

Figure 52. *Neolithodes bronwynae* sp. nov., male paratype, cl 220.0 mm, pcl 170.0 mm, cw 155.9 mm, Lord Howe Rise (NMNZ Cr11147). A, dorsal habitus. B, carapace.

of the walking legs in *N. duhameli* is replaced in *N. bronwynae* by a rather even granulation (Fig. 51D). Although the secondary ornamentation of the walking legs differs between *N. bronwynae* and *N. duhameli*, both have similar secondary spinulation on the carapace, distinguishing them both from *N. vingradovi*, which bears very few, scattered secondary spinules. *Neolithodes bronwynae* further differs from *N. vinogradovi* in its shorter walking legs with a granular rather than smooth surface between the major spines. The third walking leg of *Neolithodes bronwynae*, in comparison to *N. vinogradovi*, measures 2.8 versus 3.5 pcl; the merus is shorter than or almost as long as pcl versus 1.2 pcl.

The holotype and paratype agree well, differing chiefly in size-related features: the dorsal spines of the latter are proportionally shorter, the granules on the walking legs are slightly less distinct, and the dactyli of the walking legs are proportionally shorter (0.5 versus 0.6–0.7 propodus length). The posterior margin and right branchial surface of the carapace of the paratype exhibits evidence of prior damage, perhaps from moulting.

Macpherson (1990) reported a large female (pcl 105 mm, cw 91 mm) from New Caledonia as *N. vinogradovi* that, apart from sexual dimorphism, differed from the male holotype in more numerous secondary spines. In view of the close geographic proximity and similar secondary spination, the New Caledonian material might be referable to *N. bronwynae*. *Neolithodes bronwynae* is readily distinguished from the only other New Zealand congener, *N. brodiei*, by its long dorsal spines and the granular, rather than densely spinose, surfaces between the major spines of the walking legs.

Distribution. Presently known from the Whakatane Seamount (Bay of Plenty) and the Lord Howe Rise near Lord Howe Island (Tasman Sea); possibly also from New Caledonia; 1515–1920 m.

Neolithodes flindersi Ahyong, 2010

(Figs 53–58, Pl. 2C)

Neolithodes brodiei. — Dawson & Yaldwyn, 1985: 70. — Davie, 2002: 73 [Tasmania and Victoria]. — Zaklan, 2002: 768 [Australian occurrences]. — Poore, 2004: 268, fig. 75. [Not *N. brodiei* Dawson & Yaldwyn, 1970].
Neolithodes flindersi Ahyong, 2010: 56–62, figs 1–3 [type locality: NW Tasmania, about 900 m].

Type material. *Holotype*: NMV (ex NMNZ Cr011763), male (cl 183.3 mm, pcl 126.1 mm, cw 115.8 mm), NW Tasmania, about 900 m, otter trawl, FV *Margaret Philippa*, coll. W. Nicholls, 1983.

Paratypes: NMNZ Cr011764, female (pcl 117.0 mm, cw 107.3 mm), NW Tasmania, about 900 m, otter trawl, FV *Margaret Philippa*, coll. W. Nicholls, 1983; TM G2973, 1 male (cl 52.4 mm, pcl 40.3 mm, cw 35.6 mm), 1 female (pcl 56.1 mm, cw 48.4 mm), off W coast of Tasmania, 40°45′S, 143°31′E, 950 m, FRV *Challenger*, coll. M. Wilson, 16 Dec 1981; TM G2975, 1 female (pcl 55.8 mm, cw 49.9 mm), off W coast of Tasmania, 40°45′S, 143°30′E, 860–1060 m, FRV *Challenger*, coll. M. Wilson, 9 Jan 1982; TM G3480, 1 juvenile male (cl 41.1 mm, pcl 27.6 mm, cw 22.7 mm), E coast of Tasmania, 41°27′S, 148°40′E, FRV *Soela*, SO4/86/27, coll. P. Last, 13 Jul 1986; NMV J12942, 1 female (pcl 101.0 mm, cw 91.5 mm), off NW Tasmania, 43°45′S, 143°40′E to 41°12′S, 143°55′E, 500–650 fm [915–1189.5 m], FV *Derwent Venture*, coll. G. Switzer, 4–8 May 1986.

Other material examined. *New South Wales*: AM P35596, 1 female (cl ~76 mm, badly damaged), E of Broken Bay, 33°31–28′S, 152°12–14′E, 1079–951 m, FRV *Kapala*, K83-15-03, 3 Nov 1983; AM P35604, 1 male (cl 161.6 mm, pcl 135.5 mm, cw 132.9 mm), E of Brush Island, 35°38′S, 150°44′E, 887 m, FRV *Kapala*, K83-11-03, coll. K. Graham, 8 Sep 1983; AM P81161, 1 male (cl 124.1 mm, pcl 105.8 mm, cw 93.0 mm), off Bermagui, 36°10.8–15.0′S, 150°24.4–22.4′E, 457.5–503.3 m, FV *Shelley H*, coll. K. Graham, 1 Sep 1999; AM P35598, 1 male (pcl 149.2 mm, cw 149.2 mm), E of Gabo Island, 37°37′S, 150°21′E, 960 m, FRV *Kapala*, K84-04-04, coll. K. Graham, 5 Apr 1984; AM P35603, 1 male (pcl 112.7 mm, cw 109.5 mm), E of Brush Island, 35°31–28′S, 150°50–53′E, 942–978 m, FRV *Kapala*, K83-11-04, coll. K. Graham, 8 Sep 1983; AM P35606, 1 male (cl 124.0 mm, pcl 110.3 mm, cw 101.1 mm), E of Brush Island, 35°31′S, 150°50′E, 978 m, on *Chrysogorgia orientalis*, FRV *Kapala*, K83-11-04, coll. K. Graham, 8 Sep 1983.

Victoria: AM P52067, 1 female (cl 146.9 mm, pcl 126.2 mm, cw 112.5 mm), E of Gabo Island, 37°42.1–39.4′S, 150°16.6–17.6′E, 604–617 m, FRV *Kapala*, K97-01-02, coll. K. Graham, 16 Apr 1997; AM P35592, 1 female (pcl 38.6 mm, cw 32.4 mm), E of Gabo Island, 37°43–41′S, 150°20′E, 960–978 m, FRV *Kapala*, K83-12-01, 26 Sep 1983; SAM C6854, 1 male (cl 86.3 mm, pcl 75.3 mm, cw 68.8 mm), 65 nautical miles [120 km] SE of Portland, 39°21.55–20.66′S, 142°40.54–39.75′E, demersal trawl, RV *Soela*, coll. K. Gowlett-Holmes, 4 Mar 1989.

South Australia: SAM C6855, 1 female (cl 110.0 mm, pcl 96.2 mm, cw 89.0 mm), Great Australian Bight, approx. 120 nautical miles [222 km] SW of Cape Adieu, 33°58′S, 131°22′E, 1000 m, trawled, FV *Saxon Progress*, coll. D. Wheenan, Nov 1989; SAM C6856, 1 male (pcl 83.9 mm, cw 78.4 mm), Great Australian Bight, approx. 120 nautical miles [222 km] SSE of Eucla, 33°39′S, 129°50′E, 984–1015 m, FV *Longva III*, coll. K. Gowlett-Holmes, 12 Dec 1989; SAM C6857, 1 female (cl 110.6 mm, pcl 99.23 mm, cw 91.3 mm), Great Australian Bight, approx. 135 nautical miles [250 km] SSW of Cape Adieu, 34°13′S, 131°30′E, 1122–1330 m, from black coral 'A', FV *Longva III*, coll. K. Gowlett-Holmes, 19

Figure 53. *Neolithodes flindersi* Ahyong, 2010, male holotype, cl 183.3 mm, pcl 126.1 mm, cw 115.8 mm, NW Tasmania (NMV, ex NMNZ Cr011763). A, dorsal habitus. B, carapace.

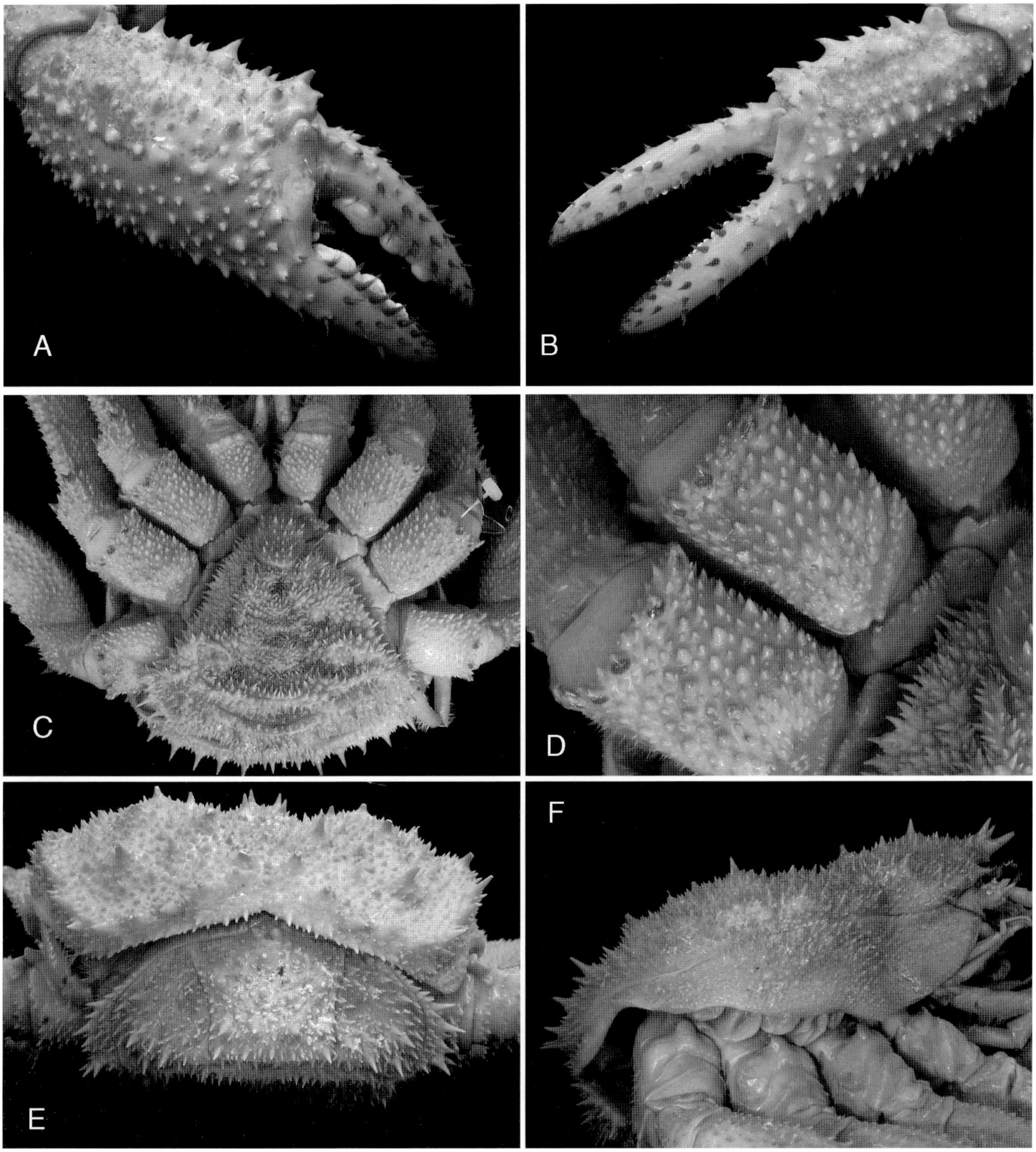

Figure 54. *Neolithodes flindersi* Ahyong, 2010, male holotype, cl 183.3 mm, pcl 126.1 mm, cw 115.8 mm, NW Tasmania (NMV, ex NMNZ Cr011763). A, right chela. B, left chela. C, ventral surface and abdomen. D, right pereopods 2–3 coxae. E, posterior carapace and abdominal somite 2. F, carapace, right lateral view.

Dec 1989; TM G2977, 1 male (cl 75.4 mm, pcl 55.0 mm, cw 48.8 mm), 20 miles [32 km] SW of Port MacDonnell, 38°19–23'S, 140°20–32'E, 800–1000 m, FV *Tuna Endeavour*, coll. R.M. Green, 5 Aug 1983; SAM C6355, 1 female (pcl 97.3 mm, cw 96.1 mm), Great Australian Bight, approx. 120 nautical miles [222 km] SSW of Cape Adieu, 33°57'S, 131°20'E, 1004–1030 m, FV *Longva III*, coll. K. Gowlett-Holmes, 30 Nov 1989; SAM C6356, 1 male (cl 105.0 mm, pcl 92.8 mm, cw 82.0 mm), 1 female (cl 81.0 mm, pcl 67.1 mm, cw 57.5 mm), Great Austral-

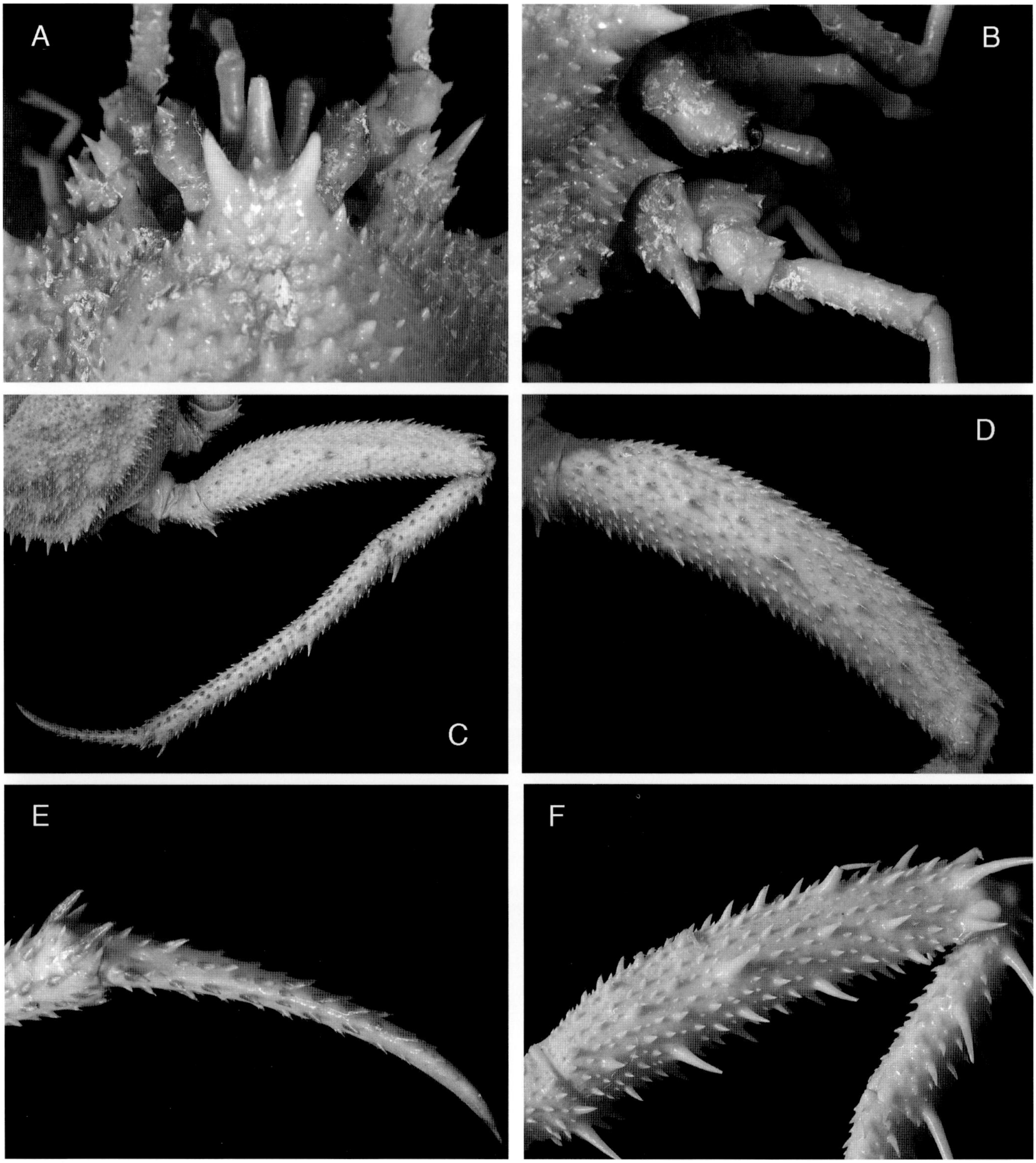

Figure 55. *Neolithodes flindersi* Ahyong, 2010. A–E, male holotype, cl 183.3 mm, pcl 126.1 mm, cw 115.8 mm, NW Tasmania (NMV, ex NMNZ Cr011763). F, male, pcl 55.0 mm, SW of Port MacDonnell (TM G2977). A, anterior carapace, dorsal view. B, right orbit and antenna. C, right pereopod 4. D, right pereopod 4 merus. E, right pereopod 4 dactylus. F, right pereopod 4 merus and carpus.

ian Bight, approx. 120 nautical miles [222 km] SSW of Cape Adieu, 34°01′S, 131°18′E, 994–1065 m, FV *Longva III*, coll. K. Gowlett-Holmes, 14 Dec 1989; SAM C6360, 1 male (pcl 108.6 mm, cw 96.2 mm), 1 female (cl 93.0 mm, pcl 82.3 mm, cw 74.1 mm), Great Australian Bight, approx. 120 nautical miles [222 km] SSW of Cape Adieu,

Figure 56. *Neolithodes flindersi* Ahyong, 2010, female paratype, pcl 117.0 mm, cw 107.3 mm, NW Tasmania (NMNZ Cr11764). A, dorsal habitus. B, abdomen.

33°59′S, 131°27′E, 1020 m, FV *Longva III*, stn 136, coll. K. Gowlett-Holmes, 3 Dec 1989; SAM C6359, 1 female (cl 105.2 mm, pcl 94.1 mm, cw 87.4 mm), Great Australian Bight, approx. 120 nautical miles [222 km] SSW of Cape Adieu, 33°57′S, 131°19′E, 1020–1055 m, FV *Longva III*, coll. K. Gowlett-Holmes, 22 Nov 1989; SAM C6365, 1 male (cl 114.7 mm, pcl 99.4 mm, cw 92.3 mm), 32–35 miles [51–56 km] off Beachport, 1333 m, trawl, FV *Silent Venture*, coll. W. Rumball, 27–29 Nov 1987; SAM C6362, 1 male (cl 144.9 mm, pcl 128.8 mm, cw 125.9 mm), Great Australian Bight, approx. 120 nautical miles [222 km] SSW of Cape Adieu, 33°57′S, 131°20′E, 954–1016 m, FV *Longva III*, coll. K. Gowlett-Holmes, 30 Nov 1989; SAM C6363, 1 male (pcl 154.1 mm, cw 151.5 mm), Great Australian Bight, approx. 120 nautical miles [222 km] SSW of Cape Adieu, 33°57′S, 131°22′E, 1012–1033 m, FV *Longva III*, coll. K. Gowlett-Holmes, 30 Nov 1989; SAM C6364, 1 ovigerous female (cl 165.3 mm, pcl 151.1 mm, cw 143.8 mm), Great Australian Bight, approx. 95 nautical miles [176 km] W of Cape Wiles, 34°57′S, 133°20′E, 850–970 m, FV *Longva III*, coll. K. Gowlett-Holmes, 11 Nov 1989; NMNZ Cr20314–20315 (ex SAM), 2 males (cl 89.0–94.3 mm, pcl 75.3–77.0 mm, cw 70.7–74.8 mm), 30 nautical miles [56 km] SW of Cape Martin Lighthouse, 37°48.6′S, 139°30.99′E, 1000–1060 m, trawl, FV *Silent Victory*, K. Gowlett-Holmes *et al.*, 15 Dec 1987; SAM TC15379, 1 juvenile male (cl 31.9 mm, pcl 21.6 mm, cw 18.4 mm), 31 nautical miles [57 km] SW of Cape Martin lighthouse, Beachport, 37°48.61′S, 139°29.74′E, 933–1098 m, FV *Silent Victory*, coll. K. Gowlett-Holmes *et al.*, 16 Dec 1987.

Diagnosis. Carapace with numerous small secondary spinules on the carapace and pereopods in addition to major spines. Antennal peduncle with few, small, scattered granules or minute spinules, not distinctly spinose. Cheliped dactyli with convex dorsal margins, rounded in cross section. Walking leg meri compressed, flattened ovate in cross section; larger spines on extensor margin not protruding above level of secondary spines; dactyli of pereopod 4 covered with spinules onto proximal three-quarters; ventral surfaces of coxae with short conical spines in males and juvenile females (obsolete in adult females).

Description. *Carapace*: Pyriform, 1.01–1.15 times longer than wide; dorsal surface armed with scattered major spines amongst dense, uniform covering of small secondary spinules. Gastric region bearing 6 large spines forming transverse hexagon; with 2 near transverse rows of 3 smaller spines, first across centre of hexagon and second slightly anterior to hexagon. Hepatic spine anterolaterally directed; margin between outer orbital spine and hepatic spine with 2 or 3 larger spines in additional to small secondary spines. Branchial regions with 8–10 larger dorsal spines in addition to scattered smaller secondary spines; margins with 13–15 major spines in addition to numerous minute spines. Cardiac region with 4 equal spines of similar size to gastric spines and 1 or 2 smaller spines near intestinal region. Intestinal region with 2 upright spines in transverse row and 2 or 3 smaller, posteriorly directed spines. Pterygostomian region spinose or tuberculate, with small, anterior, submarginal spine.

Rostrum 0.09–0.19 pcl; median spine inclined dorsally, with pair of divergent dorsal spines and smaller pair of basal spines; ventral proximal surface with cluster of spinules. Posterior orbital margin concave in lateral view, occasionally with 2–4 spinules on lower margin; outer orbital spine slender, not reaching cornea (when eyes directed forwards). Anterolateral spine as long as or shorter than outer orbital spine.

Ocular peduncle: Longer than cornea; with scattered dorsal granules or spinules.

Antennule: Peduncle unarmed, reaching anteriorly beyond antennal peduncle by less than half distal antennular peduncle article.

Antenna: Basal antennal article with small anterolateral spine; outer margin of article 2 with or without cluster of small spinules basally and long slender spine that reaches to end of article 4; article 3 with 1 or 2 sharp to angular inner distal teeth; scaphocerite minute, often with bifid or trifid apex, shorter than article 4; article 4 with mesial and lateral distal spinule and occasionally 1 or 2 other minute spinules, about half as long as article 5; article 5 with scattered spinules or granules.

Abdomen: Ornamentation similar in both sexes. Somite 2 densely spinose, with about 10 long spines and numerous smaller spines on median plate; submedian and marginal plates densely spinose, spines largest laterally; surface and margins of remaining somites and telson multispinose.

Pereopod 1 (chelipeds): Dimorphic but spination similar. Entire surface of both chelipeds, except for distal half or two-thirds of fingers, covered with numerous minute secondary spines in addition to major spines. Coxa ventral surface with blunt tubercles and tufts of setae; ischiobasis with 4 or 5 stout ventral spines. Merus inner margin with stout subdistal spine; ventral margin with two rows of stout spines, mesial row with 2 spines, lateral row with 3 or 4 spines; dorsal and lateral surface spinose, spines largest distally. Carpus with prominent spines on dorsal and lateral surfaces, with 3 or 4 irregular rows of 3–5 spines. Palm of both chelipeds in both sexes with similar ornamentation; all surfaces spinose, although less pronounced mesially and ventrally; dorsal margin with 2 irregular rows of about 3 or 4 conical spines; midlateral surface with 2 rows of 4 or 5 spines of similar size to dorsal row; ventral surface irregularly spinose.

Major cheliped 1.27–1.35 pcl (male), 1.13–1.14 (female); upper palm length 1.06–1.14 times height (male),

Figure 57. *Neolithodes flindersi* Ahyong, 2010. A, juvenile male paratype, E Tasmania, pcl 27.6 mm (TM G3480). B, male, pcl 55.0 mm, SW of Port MacDonnell (TM G2977).

1.14–1.16 (female); occlusal margins of fingers corneous for distal third to half, proximally with 3 calcareous nodules; dactylus dorsal margin broadly convex, with rows of tufts of golden setae and proximal cluster of 4–7 small spines, 1.23–1.33 times longer than dorsal margin of palm (male), 1.26–1.43 (female).

Minor cheliped 1.24–1.32 pcl (male), 1.09–1.21 pcl (female); upper palm length 1.17–1.33 times height (male), 1.17–1.31 (female); occlusal margin corneous in distal half, proximally crenulate; dactylus dorsal margin broadly convex, with rows of tufts of golden setae and proximal cluster of 4–6 small spines proximally, 1.69–1.98 times longer than dorsal margin of palm (male), 1.65–1.86 (female).

Pereopods 2–4 (walking legs 1–3): Similar, slender, elongate; ornamentation similar in both sexes; segments spinose, surface between major spines densely covered with small spines. Pereopod 4 longest. Coxae covered with small conical spinules in males, in females becoming obsolete or reduced to blunt granules beneath overlap of abdomen (covered in conical spinules in juvenile females); distal margins crenulate to bluntly dentate. Ischiobasis with 3–5 stout spines around distal margins and numerous, smaller, ventral spinules. Merus dorsoventrally compressed; extensor margin lined with numerous, short, subequal spines in addition to paired distal spines; dorsal surface with irregular row of 4–7 major spines of similar size to distal extensor spines; flexor margin lined with short, subequal spines in addition to 2–5 larger spines; major spines on surfaces of merus distinctly standing out from secondary spinules. Carpus with 4 or 5 strong spines in addition to numerous small spines. Propodus dorsoventrally compressed; as long as or longer than merus; with 4–7 spines on extensor margin and 3–5 similar spines on dorsal surface; flexor margin with numerous smaller spines. Dactylus curved, rounded in cross section, with numerous small spines covering proximal half (pereopod 2) to three-quarters (pereopods 3–4); apex corneous.

Pereopod 2 length 2.20–2.47 pcl (male), 1.84–2.24 pcl (female). Merus 0.74–0.81 pcl (male), 0.60–0.73 pcl (female); length:height ratio 4.42–4.59 (male), 3.21–3.98 (female). Carpus 0.54–0.55 merus length (male), 0.56–0.60 (female). Propodus 0.93–0.99 merus length (male), 0.95–0.99 (female); length:height ratio 9.20–10.63 (male), 7.08–8.93 (female). Dactylus 0.62 propodus length (male), 0.61–0.73 (female).

Pereopod 3 length 2.85–2.89 pcl (male), 2.26–2.74 pcl (female). Merus 0.84–0.94 pcl (male), 0.71–0.85 pcl (female); length:height ratio 4.56–4.91 (male), 3.50–4.09 (female). Carpus 0.53–0.54 merus length (male), 0.56–0.59 (female). Propodus 0.96–1.01 merus length (male), 1.01–1.02 (female); length:height ratio 9.94–10.65 (male), 7.94–10.02 (female). Dactylus 0.58–0.59 propodus length (male), 0.62–0.68 (female).

Pereopod 4 length 3.05–3.14 pcl (male), 2.41–2.87 pcl (female). Merus 0.86–0.94 pcl (male), 0.70–0.86 pcl (female); length:height ratio 3.98–5.01 (male), 3.49–4.22 (female). Carpus 0.56–0.57 merus length (male), 0.59–0.62 (female). Propodus 1.03–1.10 merus length (male), 1.08–1.14 (female); length:height ratio 9.67–12.27 (male), 8.41–11.42 (female). Dactylus 0.58–0.60 propodus length (male), 0.62–0.70 (female).

COLOUR IN LIFE. Deep-red overall (Pl. 2C).

REMARKS. Recognition of *N. brodiei* from Australia stems from Dawson & Yaldwyn's (1985: 70) remark that the species "occurs in appreciable numbers off south-eastern Australia" based on collections in the Australian Museum. All of the Australian Museum specimens, however, as well as other Australian specimens identified as *N. brodiei* in other collections, are referable to *N. flindersi*. Records of *N. brodiei* from Vanuatu and *N. nipponensis* from the Solomon Islands (Macpherson 2001, 2003) may represent undescribed species under study by P. Davie.

Neolithodes flindersi most closely resembles *N. brodiei* Dawson & Yaldwyn, 1970, from New Zealand, and *N. nipponensis* Sakai, 1971, from Taiwan and Japan. It differs from *N. nipponensis* chiefly in being less prominently spinose: the secondary spines covering the surfaces of the walking legs are distinctly shorter, and the antennal peduncle is only sparsely granulate or spinulate, rather than prominently spinulose.

Neolithodes flindersi differs from *N. brodiei* in patterns of spination, particularly of the ventral surfaces of the coxae of the walking legs, spination of the walking leg dactyli, spination of the pereopod 2–4 meri and in morphometric features. These differences between *N. flindersi* and *N. brodiei* are discussed under the account of the latter.

As in other lithodids, dorsal spination of *N. flindersi* is most pronounced in smallest specimens, becoming progressively shorter with increasing size. Similarly, the proportional length of the walking leg dactyli decreases with increasing size, ranging from about 0.8 propodus length in the smallest specimen (juvenile male, pcl 21.6 mm) to 0.6 in the largest specimens (male, pcl 154.1 mm). The only known ovigerous female is 151.1 mm pcl, although females appear to be mature by about 95 mm pcl based on development of the abdomen and presence of setae on the coxal surfaces.

One specimen of *N. flindersi* (AM P35606) was collected from *Chrysogorgia orientalis*, and another (SAM C6857) from an unidentified black coral. The male holotype (NMNZ Cr11763) has an egg mass on the inside of the carapace in the left branchial chamber (egg diameter 3.98–4.08 mm). These eggs probably belong to a species of liparid fish (*Careproctus* sp.), which uses the crab as a mobile 'home' and uses the branchial chambers of various lithodids for egg incubation.

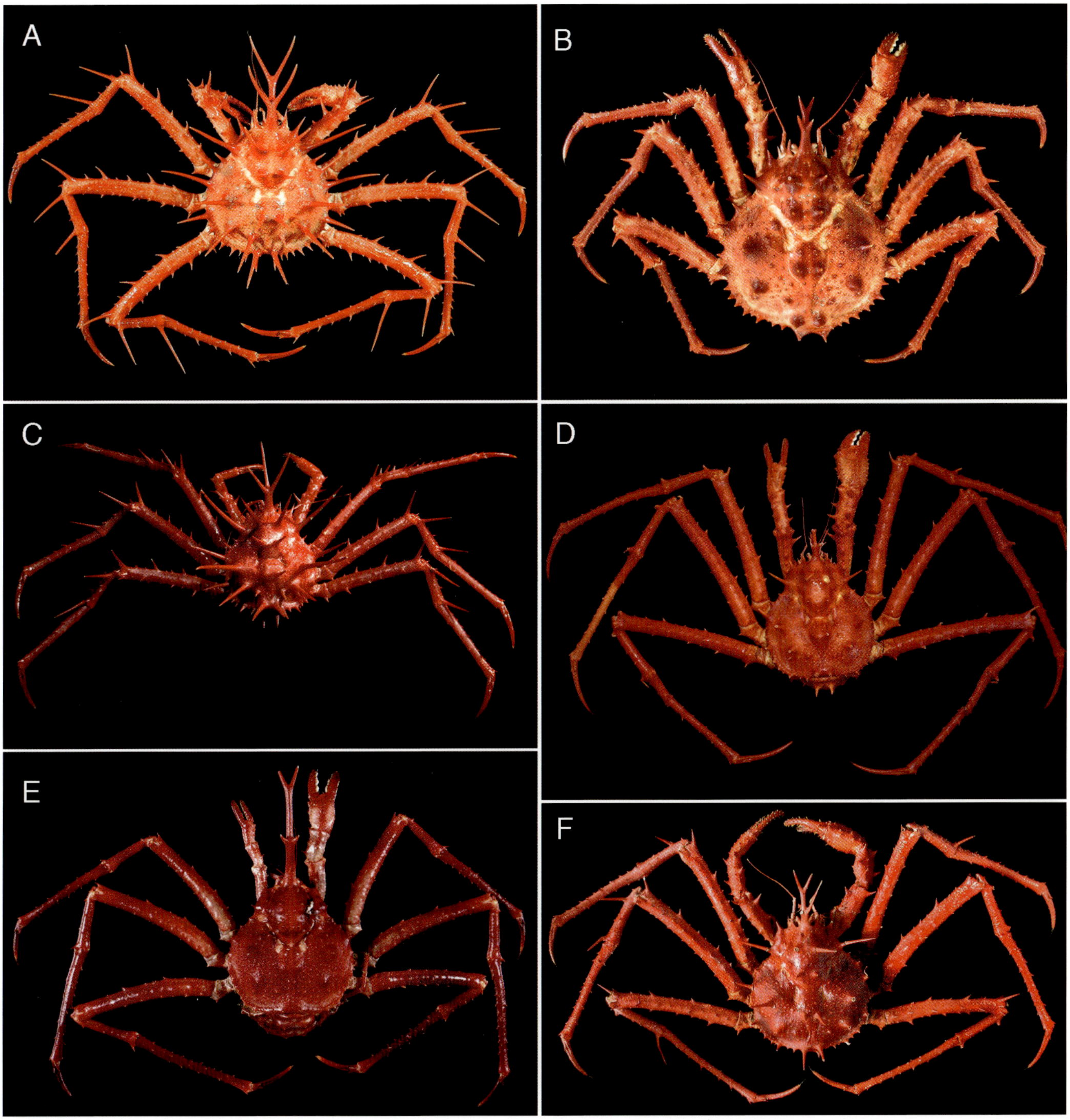

Plate 1. A, *Lithodes aotearoa* sp. nov., male holotype, pcl 88.6 mm, off Hawkes Bay (NIWA 34924). B, *Lithodes aotearoa* sp. nov., ovigerous female, pcl 118.9 mm, Chatham Rise (NIWA 42893). C, *Lithodes australiensis* sp. nov., male, pcl 60.3 mm, SSE of Southeast Cape (TM G3659). D, *Lithodes rachelae* sp. nov., male paratype, pcl 117.3 mm, Great Australian Bight (SAM C6367). E, *Lithodes richeri* Macpherson, 1991, ovigerous female, pcl 125.1 mm, SE of Brush Island (AM P35605). F, *Lithodes robertsoni* sp. nov., male holotype, pcl 128.1 mm, Chatham Rise (NIWA 42889). (Photo credits: A, P. Shearer, NIWA; C, K. Gowlett-Holmes, CSIRO; E, K. Lowe, Australian Museum; F, P. Marriott, NIWA).

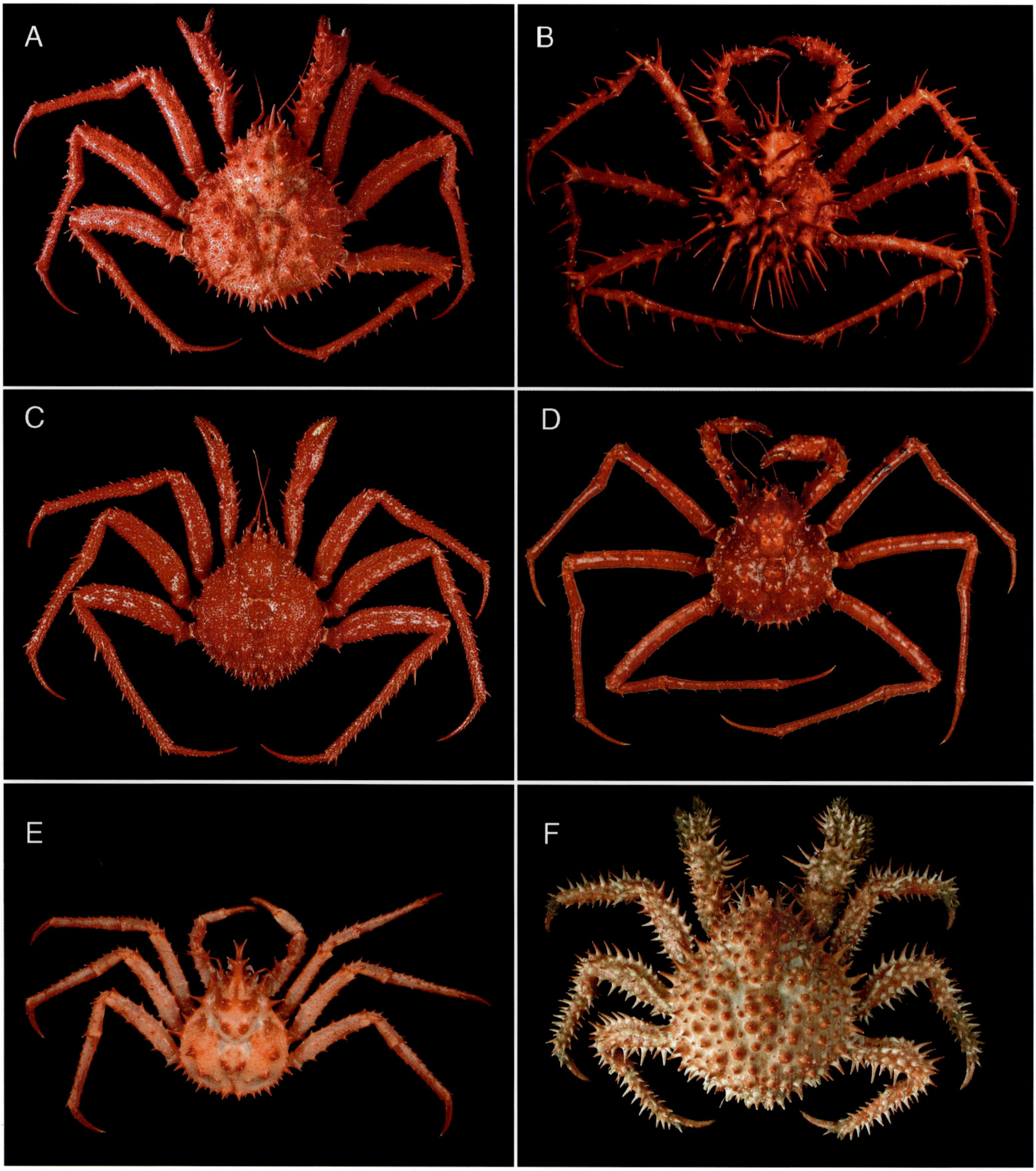

Plate 2. A, *Neolithodes brodiei* Dawson & Yaldwyn, 1970, ovigerous female, pcl 120.9 mm, Chatham Rise (NIWA 34923). B, *Neolithodes bronwynae* sp. nov., male holotype, pcl 123.3 mm, Whakatane Seamount (NIWA 49026). C, *Neolithodes flindersi* Ahyong, 2010, female, pcl 126.2 mm, E of Gabo Island (AM P52067). D, *Neolithodes yaldwyni* Ahyong & Dawson, 2006, male, pcl 116.9 mm, Ross Sea (NIWA 38209). E, *Lithodes macquariae* sp. nov. juvenile female, pcl 43.8 mm, Macquarie Ridge (NIWA 40902). F, *Paralomis zealandica* Dawson & Yaldwyn, 1971, female, pcl 79.4 mm, Chatham Rise (NIWA 42891). (Photo credits: C, K. Lowe, Australian Museum; D,E, P. Marriott, NIWA).

Plate 3 *(opposite)*. A, *Paralomis birsteini* Macpherson, 1988a, male, pcl 69.5 mm, Ross Sea (NIWA 38505). B, *Paralomis dawsoni* Macpherson, 2001, male, pcl 120.8 mm, Challenger Plateau (NIWA 48671). C, *Paralomis gowlettholmes* sp.

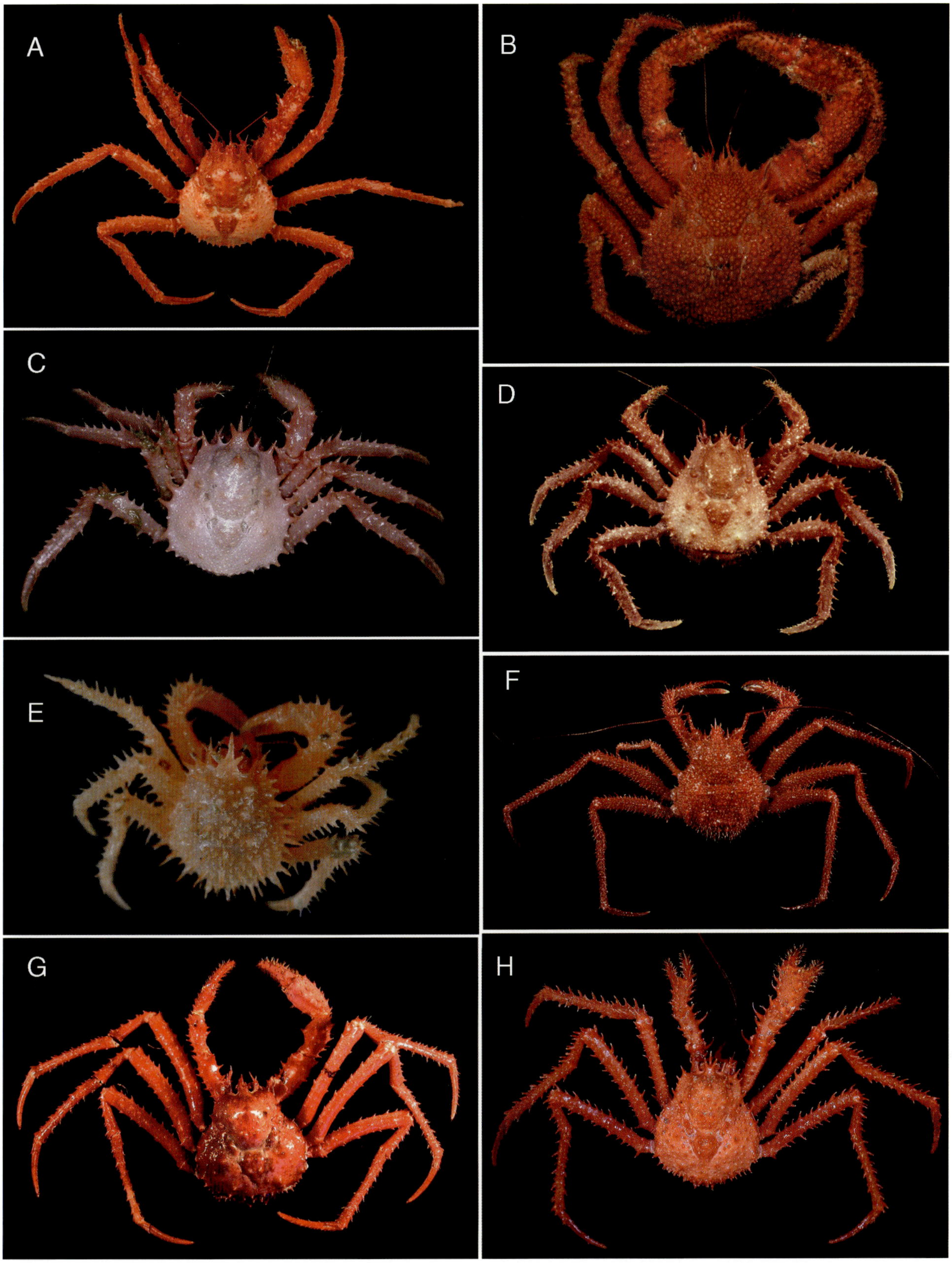

nov., juvenile female paratype, pcl 15.9 mm, Tasmania (NMV J61054). D, *Paralomis gowlettholmes* sp. nov., ovigerous female, pcl 30.5 mm, Tasmania (TM G3607). E, *Paralomis poorei* sp. nov., juvenile male paratype, pcl 11.3 mm, Chatham Rise (NIWA 29526). F, *Paralomis taylorae* sp. nov., female holotype, pcl 39.8 mm, Tasmania (TM G3609). G, *Paralomis stevensi* Ahyong & Dawson, 2006, male, pcl 81.7 mm, Ross Sea (NIWA 27835). H, *Paralomis staplesi* sp. nov., male holotype, pcl 50.0 mm, Tasmania (NMV J61052). (Photo credits: A, E, P. Marriott, NIWA; B, S. O'Shea; C–D, F, H, K. Gowlett-Holmes, CSIRO).

Plate 4. A, *Lithodes chaddertoni* sp. nov., holotype. B, *Paralomis hirtella* de Saint Laurent & Macpherson, 1998, Monowai Caldera, 25°48.237′S, 177°10.071′W, 1165 m. C, *Lithodes aotearoa* sp. nov., Kaikoura canyon, 42°35.01–33.37′S, 173°43.81–43.77′E, 1379 m, TAN0616/95, 15 Nov 2006. D, *Neolithodes yaldwyni* Ahyong & Dawson, 2006, Balleny Islands, Ross Sea, 67°47.28–47.43′S, 179°45.71–47.33′W, 1072–1282 m, TAN0802/219, 4 Mar 2008. E, *Paralomis birsteini* Macpherson, 1988a, Ross Sea, 68° 05.47–05.61′S, 179°16.63–14.96′W, 630–784 m, TAN0802/200, 2 Mar 2008. F, *Paralomis zealandica* Dawson & Yaldwyn, 1971, Chatham Rise, 44°00.37–00.31′S, 177°09.48–08.49′E, 637–630 m, TAN0705/54, 7 Apr 2007. (Photo credits: A, NMNZ; B, A. Rowden, NIWA; C–F, NIWA).

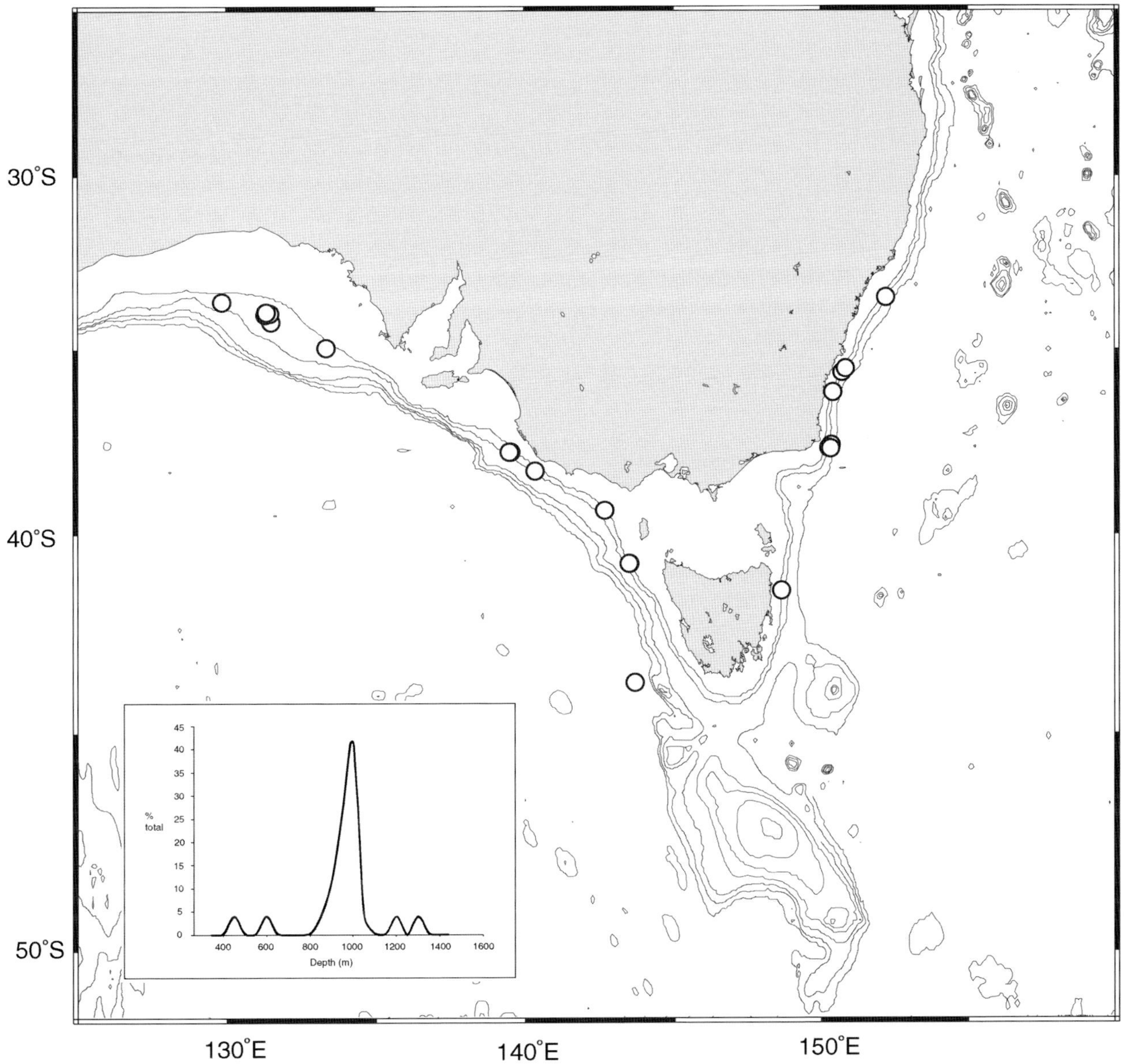

Figure 58. Geographic and bathymetric distribution of *Neolithodes flindersi* Ahyong, 2010.

Neolithodes flindersi occurs on soft substrates in outer shelf and slope depths, and does not appear to range beyond continental margins. It is frequently taken by deep-water trawlers in the central Great Australian Bight and is common between the Tasmanian seamounts (Poore 2004; as *N. brodiei*).

DISTRIBUTION. Off Broken Bay, New South Wales, to Victoria, Tasmania and the Great Australian Bight, South Australia; 887–1333 m, usually 950–1050 m.

Neolithodes yaldwyni Ahyong & Dawson, 2006 (Figs 59–62, Pl. 2D, 4D)

Neolithodes brodiei. — Thatje & Lörz, 2005: 335–336, fig. 2A [NIWA 3433, 3434 only; not *N. brodiei* Dawson & Yaldwyn, 1970].

Neolithodes yaldwyni Ahyong & Dawson, 2006: 46–54, figs 1–4 [type locality: N of Sturge Island, Balleny Islands, 67°13.22–11.97'S, 164°17.78–14.87'E, 522–538 m]. — Frankham, 2009: 66.

Type material. *Holotype*: NIWA 3434, male (cl.159 mm, pcl 141.5 mm, cw 124.1 mm), N of Sturge Island, Balleny Islands, 67°13.22–11.97'S, 164°17.78–14.87'E, 522–538 m, TAN0402/226, RV *Tangaroa*, 3 Mar 2004.

Paratypes: NMNZ Cr11002, 2 males (cl 124.1–131.8 mm, pcl 101.7–103.6 mm, cw 91.6–95.2 mm), between Scott and Balleny Islands, Ross Sea, 67°08–06'S, 170°54–47'E, 900–1160 m, RV *San Aotea II*, trip 2010, haul 180, coll. E. Winslade, 14 Feb 2005; NIWA 3433, 1 male (cl 144.0 mm, pcl 120.0 mm, cw 108.7 mm), seamount W of Sturge Island, Balleny Islands, 67°26.09–26.37'S, 163°52.98–51.79'E, 124–170 m, TAN0402/235, 4 Mar 2004; NMNZ Cr11003, 1 male (cl 183.3 mm, pcl 136.6 mm, cw 131.1 mm), near Scott Island, Ross Sea, 68°23–21'S, 179°53–57'E, 1337–1050 m, RV *San Aotea II*, trip 2010, haul 137B, on hook, B, coll. E. Winslade, 31 Jan 2005.

Other material examined. *Ross Sea, Antarctica*: NMNZ, 2 males (cl +140–140.5 mm, pcl 101.7–112.7 mm, cw 99.1–97.6 mm), 65°S, 178°W, area 88.1C, FV *San Aotea*; NIWA 42890, 1 male (pcl 115.5 mm, cw 99.4 mm), Ross Sea, area 88.1C, 65°07.20–08.40'S, 178°29.40–25.80'W, 1669–1950 m, trip 2530, set 11, coll. G. Higgins, 7 Dec 2007; NMNZ Cr11780, 1 male (cl 136.4 mm, pcl 109.00 mm, cw 95.3 mm), S of Scott Island, 68°02.00'S, 179°44.00'W, 1312–1564 m, trip 1728, set 45, sample 65, FV *Janas*, coll. B. Fairhead & S. Smith, 21 Jan 2003; NMNZ Cr11803, 1 ovigerous female (cl 135.3 mm, pcl 108.5 mm, cw 94.2 mm), SE of Balleny Islands, 67°41.10'S, 167°08.50'E, 1307–1340 m, *San Liberatore*, Trip 1743, set 31, coll. G. Dolan, 11 Mar 2003; NMNZ Cr11802, 1 male (pcl 109.3 mm, cw 99.8 mm), area 88.1, 67°34.00–32.00'S, 167°18.00–13.00'E, 1290–1248 m, FV *San Liberatore*, trip 1743, line 34, 12 Mar 2003; NMNZ Cr11801, 1 male (cl 134.8 mm, pcl 120.5 mm, cw 113.0 mm), Ross Sea region, S of Antarctic Convergence, >70°S, Antarctic toothfish survey, 2000/2001; NIWA 38209, 1 male (pcl 116.9 mm, cw 106.1 mm), 68°06.52–07.63'S, 179°14.30–15.36'W, 855–879 m, TAN0802/211, 3 Mar 2008; NIWA 42888, 1 male (cl 170.4 mm, pcl 146.1 mm, cw 135.6 mm), area 88.1K, 75°31.20'S, 170°10.80–27.00'W, 1177–1258 m, trip 2526/92, coll. R. Coy, 5 Feb 2008.

Diagnosis. Carapace dorsal surface with thick, conical spines and scattered, widely separated granules or minute secondary spines; longest spine not exceeding 0.2 pcl. Rostrum 0.1–0.3 pcl; ventral surface smooth. Posterior orbital margin near vertical; outer orbital spine slender, reaching to but not beyond cornea. Outer spine of antennal article 2 reaching beyond article 4; scaphocerite shorter than antennal article 4. Major cheliped palm with prominent dorsal and lateral spines; dactylus with convex dorsal margin, slightly longer than dorsal margin of palm. Walking legs spinose, secondary spines absent or sparsely distributed; merus ovate in cross-section, shorter than carapace; propodus subcylindrical in cross section; dactylus distinctly longer than half propodus length.

Description. *Carapace*: Pyriform, 1.02–1.15 times longer than wide; dorsal surface armed with short, stout spines and scattered, widely separated granules or minute secondary spines; longest spine (hepatic spine) shorter than 0.2 pcl. Gastric region bearing 6 long spines forming transverse hexagon; with 2 near transverse rows of 3 and 4 smaller spines, first across centre of hexagon and second slightly anterior to hexagon. Hepatic spine strong, anterodorsally directed; margin between outer orbital spine and hepatic spine with 2 or 3 short spines. Branchial regions with 8 or 9 large dorsal spines in addition to scattered, smaller, secondary spines; margins with 13 large spines in addition to several shorter spines. Cardiac region with 4 equal spines of similar size to gastric spines and pair of shorter spines near intestinal region. Intestinal region with 2 upright spines in transverse row and 2 smaller, posterior directed spines. Pterygostomian region spinose or tuberculate, with small subdistal spine.

Rostrum 0.12–0.25 pcl; median spine straight, nearly horizontal, with pair of divergent dorsal spines and smaller pair of basal spines; ventral surface smooth. Posterior orbital near vertical in lateral view; outer orbital spine slender, reaching to but not beyond cornea (when eyes directed forwards). Anterolateral spine about half to two-thirds length of outer orbital spine.

Ocular peduncle: Longer than cornea; with scattered dorsal granules or spinules.

Antennule: Peduncle unarmed, reaching anteriorly beyond antennal peduncle by half to two-thirds length of distal antennular peduncle article.

Antenna: Basal antennal article with small anterolateral spine; outer margin of article 2 with small basal spine and long slender spine that reaches beyond end of article 4 but not to midlength of article 5; article 3 with sharp to angular inner distal tooth; scaphocerite a minute, shorter than article 4, apex sharp to blunt; article 4 unarmed, about half length of article 5.

Abdomen: Somite 2 bearing 10 long spines and several smaller spines on median plate; submedian plate with 4 or 5 long spines on posterior border and 2–4 shorter spines on surface; lateral border of marginal plates with 6–8 teeth. Somites 3–5 multispinose or nodulose in males; as calcified plates covered with stout, well-spaced spines in females. Telson spinose. Egg diameter 2.70 mm.

Pereopod 1 (chelipeds): Dimorphic but spination similar. Coxa with blunt tubercles and tufts of setae, unarmed; ischiobasis with 4 or 5 stout ventral spines. Merus inner margin smooth or granular, with stout subdistal spine; ventral margin with two rows of stout

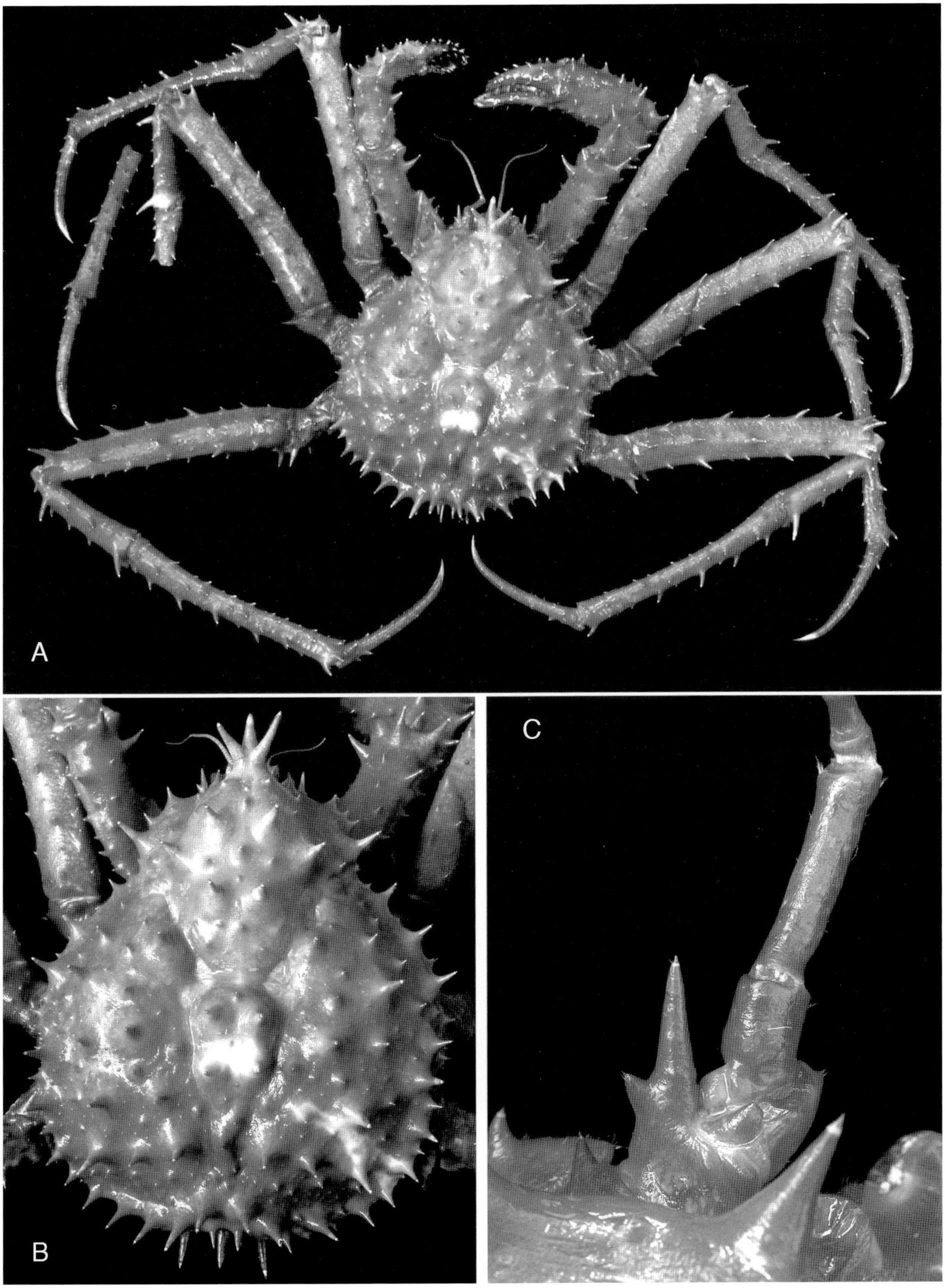

Figure 59. *Neolithodes yaldwyni* Ahyong & Dawson, 2006, male, cl 159 mm, pcl 141.5 mm, cw 124.1 mm, N of Sturge Island, Balleny Islands (NIWA 3434). A, dorsal habitus. B, carapace. C, left antenna.

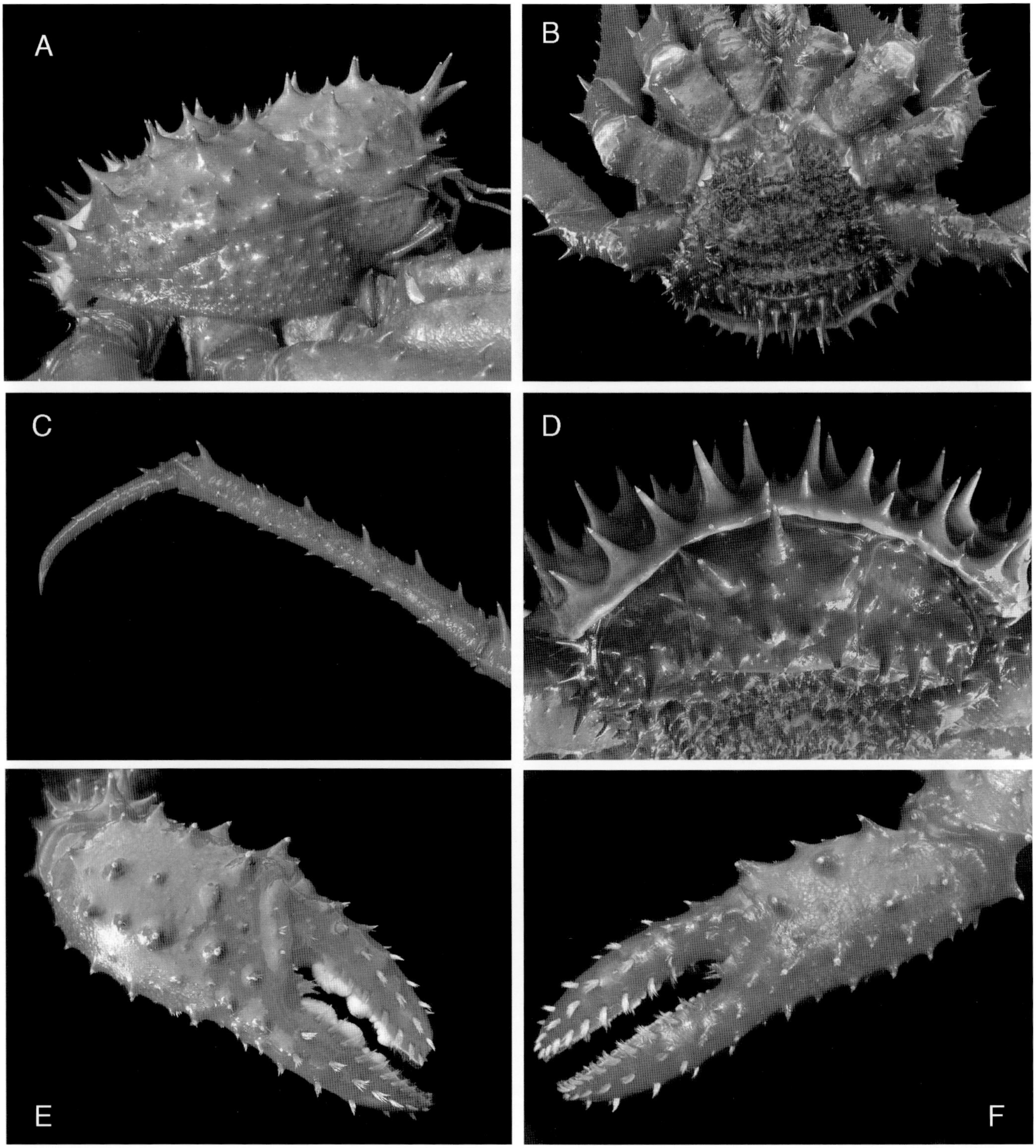

Figure 60 ***(above).*** *Neolithodes yaldwyni* Ahyong & Dawson, 2006, male, cl 159 mm, pcl 141.5 mm, cw 124.1 mm, N of Sturge Island, Balleny Islands (NIWA 3434). A, carapace, right lateral. B, ventral surface and abdomen. C, left pereopod 4 dactylus and carpus. D, posterior carapace and abdominal somite 2. E, right chela. F, left chela.

Figure 61 ***(opposite).*** *Neolithodes yaldwyni* Ahyong & Dawson, 2006, ovigerous female, pcl 108.5 mm, Ross Sea (NMNZ Cr11803). A, dorsal habitus. B, abdomen.

A
B

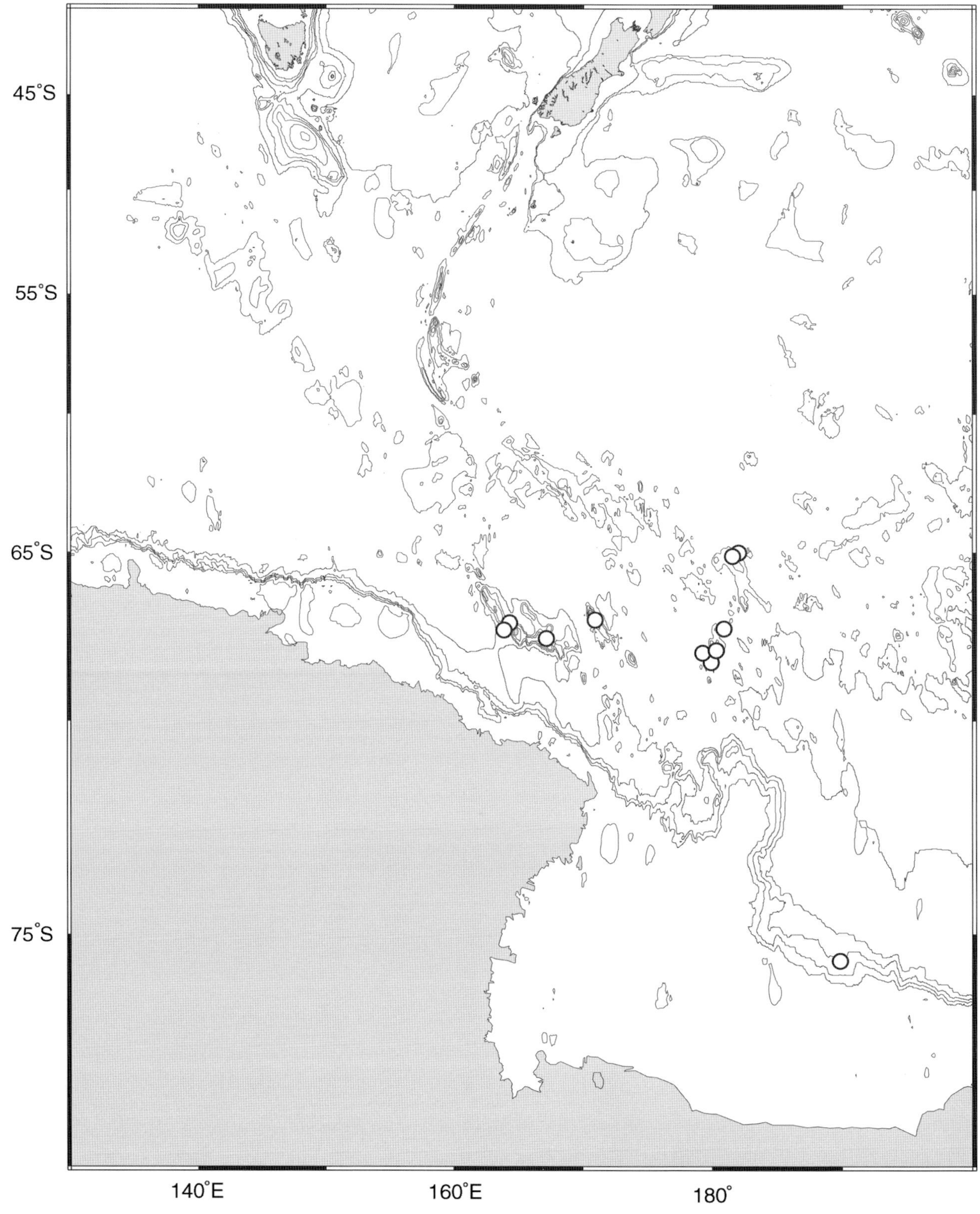

Figure 62. Geographic distribution of *Neolithodes yaldwyni* Ahyong & Dawson, 2006.

spines, mesial row with 2 spines, lateral row with 4 spines; dorsal and lateral surface spinose, spines largest distally. Carpus with prominent spines on dorsal and lateral surfaces; dorsal margin with row of 4 or 5 stout spines; lateral margin with 2 or 3 rows of 4 or 5 stout spines of similar size to dorsal row; ventral margin small, scattered, acute tubercles. Palm prominently spinose on dorsal, lateral and ventral surfaces, inner surface with acute tubercles; dorsal margin with row of about 5 prominent conical spines; midlateral surface with 2 rows of 4 or 5 spines of similar size to dorsal row; ventral surface with 2 rows of about 6 or 7 spines, smaller than lateral and dorsal spines.

Major cheliped 1.32–1.55 pcl (male), 1.15 pcl (female); upper palm length 1.13–1.29 times height (male), 1.19 (female); occlusal margins corneous for slightly less than distal half, proximally with 3 calcareous nodules; dactylus dorsal margin broadly convex, with rows of tufts of golden setae and proximal cluster of 2 or 3 small spines, 1.15–1.26 times longer than dorsal margin of palm (male), 1.27 (female).

Minor cheliped 1.25–1.40 pcl (male), 1.15 pcl (female); upper palm length 1.42 times height (male), 1.30 (female); occlusal margins corneous for slightly more than distal half, proximally crenulate; dactylus dorsal margin broadly convex, with rows of tufts of golden setae and proximal cluster of 2 or 3 small spines, 1.54–1.65 times longer than dorsal margin of palm (male), 1.57 (female).

Pereopods 2–4 (walking legs 1–3): Similar; segments spinose, surface between major spines smooth or with few, minute, sparsely distributed spines, pereopod 4 longest. Distal margins of coxae with short spines, longest on coxa 3. Ischiobasis with 3 or 4 distal spines and 2 or 3 smaller ventral spines. Merus ovate in cross section, shorter than carapace; extensor margin with 8 or 9 spines in addition to paired distal spines; dorsal surface with row of 4–6 spines of similar size to extensor spines; flexor margin with 2 rows of 5–7 spines. Carpus slightly longer than half merus length; extensor margin with 5 or 6 spines, distal and second proximal spines longest, about twice length of other spines; dorsal surface with 4 or 5 spines. Propodus ovate to subcircular in cross section; with 7–9 spines on extensor margin and 8–10 similar spines on dorsal surface; flexor margin with 6–10 smaller spines. Dactylus curved, rounded in cross section, with 4–6 small proximal spines, apex corneous; distinctly longer than half propodus length.

Pereopod 2 length 2.54–2.63 pcl (male), 1.91 pcl (female). Merus 0.85–0.92 pcl (male), 0.61 pcl (female); length:height ratio 5.15–6.27 (male), 4.19 (female). Carpus 0.50–0.54 merus length (male), 0.64 (female). Propodus 0.89–0.92 merus length (male), 0.98 (female); length:height ratio 6.49–10.90 (male), 7.15 (female). Dactylus 0.57–0.64 propodus length (male), 0.67 (female).

Pereopod 3 length 2.72–2.85 pcl (male), 2.06 pcl (female). Merus 0.94–1.01 pcl (male), 0.66 pcl (female); length:height ratio 5.45–6.58 (male), 4.81 (female). Carpus 0.49–0.52 merus length (male), 0.61 (female). Propodus 0.89–0.92 merus length (male), 0.90 (female); length:height ratio 7.44–11.94 (male), 6.91 (female). Dactylus 0.55–0.63 propodus length (male), 0.73 (female).

Pereopod 4 length 2.83–2.94 pcl (male), 2.13 pcl (female). Merus 0.92–1.01 pcl (male), 0.67 pcl (female); length:height ratio 5.68–7.12 (male), 5.06 (female). Carpus 0.52–0.55 merus length (male), 0.61 (female). Propodus 0.93–0.96 merus length (male), 7.64 (female); length:height ratio 7.67–8.44 (male), 7.64 (female). Dactylus 0.55–0.60 propodus length (male), 0.73 (female).

Colour in life. Deep-red overall (Pl 2D).

Remarks. *Neolithodes yaldwyni* Ahyong & Dawson, 2006, from the Ross Sea, Antarctica, was recently described in detail based on male specimens. The first known female of the species, a 108.5 mm pcl ovigerous specimen, is reported here. The female differs from males in having stouter, thicker carapace spines, and as with other lithodids, proportionally shorter pereopods. As in *N. capensis*, the female abdomen of *N. yaldwyni* is composed of calcified plates covered with stout, well-spaced spines. Notably, the dactyli of the walking legs are proportionally longer than in any of the males studied (0.7 versus 0.6 times propodus length).

Neolithodes yaldwyni closely resembles *N. capensis* Stebbing, 1905 (type locality: South Africa) in almost all respects, but differs in having proportionally longer pereopod 2–4 dactyli. The pereopod 2–4 dactyli of *N. capensis* are half the propodus length in adult males versus 0.6 propodus length or greater in *N. yaldwyni*. *Neolithodes capensis* has been reported also from the Kerguelen Islands (Macpherson 2004) and Bellingshausen Sea (García Raso *et al.* 2005). As remarked by Ahyong & Dawson (2006), the Bellingshausen Sea record might be referable to *N. yaldwyni*. At present, *N. yaldwyni* is known with certainty from localities in the Ross Sea region, from the Scott and Balleny Islands to the Ross Sea slope at 75°31'S. The affinities of *N. yaldwyni* were discussed in detail by Ahyong & Dawson (2006).

Distribution. Known only from the Ross Sea, Antarctica; 124–1950 m.

Paralomis White, 1856

Paralomis White, 1856: 134 [type species: *Lithodes granulosa* Hombron & Jacquinot, 1846, pl. 8: figs 15–25. Gender: feminine].

Acantholithus Stimpson, 1858: 69 [type species: *Lithodes histrix* de Haan, 1844. Gender: masculine].

Leptolithodes Benedict, 1895: 484 [type species: *Paralomis aculeata* Henderson, 1888, by subsequent designation (Ahyong *et al.*, 2010). Gender: masculine].

Pristopus Benedict, 1895: 486 [type species: *Pristopus verrilli* Benedict, 1895, by subsequent designation (Ahyong *et al.*, 2010). Gender: masculine].

Diagnosis. Carapace pentagonal or pyriform, not covering bases of walking legs; regions indicated; gastric region elevated above other regions; cardiac region triangular, separated from gastric region by deep groove; cervical groove shallow, indistinct. Rostrum comprised of median spine and at least one pair of dorsal spines near base. Abdominal somite 2 comprising a single plate. Abdominal somites 3–5 comprising well-calcified plates, without nodules and membranous areas; each somite comprising median, paired submedian and paired marginal plates (sometimes subdivided or notched); marginal and submedian plates of somites 3 (and rarely 4) fused. Sternite 5 (between pereopods 2) without deep median fissure. Scaphocerite well-developed, usually spinose. Walking legs (pereopods 2–4) similar in form, walking leg 2 usually longest; walking leg 3 always longer than pcl; dactyli of adults with corneous spines along flexor margin.

Composition

P. aculeata Henderson, 1888 [Crozet and Prince Edwards Islands, southwest Indian Ocean, Southern Ocean]

P. africana Macpherson, 1982 [Namibia, southeastern Atlantic Ocean]

P. alcockiana Hall & Thatje, 2009b [South Carolina, western Atlantic Ocean]

P. anamerae Macpherson, 1988b [Argentina, western Atlantic Ocean]

P. arae Macpherson, 2001 [Fiji and Taiwan, western Pacific Ocean]

P. arethusa Macpherson, 1994 [Barbados, western Atlantic]

P. aspera Faxon, 1893 [central eastern Pacific Ocean]

P. birsteini Macpherson, 1988a [Antarctic and Southern Ocean, Pacific sector]

P. bouvieri Hansen, 1908 [northeastern Atlantic Ocean]

P. ceres Macpherson, 1989 [western Indian Ocean]

P. chilensis Andrade, 1980 [Chile, Peru, eastern Pacific Ocean]

P. cristata Takeda & Ohta, 1979 [Japan, western Pacific Ocean]

P. cristulata Macpherson, 1988b [eastern Atlantic Ocean]

P. cubensis Chace, 1939 [Caribbean Sea, western Atlantic Ocean]

P. danida Takeda & Bussarawit, 2007 [Andaman Sea, eastern Indian Ocean]

P. dawsoni Macpherson, 2001 [southwestern Pacific Ocean]

P. debodeorum Feldmann, 1998 [extinct; Miocene, New Zealand]

P. diomedeae (Faxon, 1893) [eastern Pacific Ocean]

P. dofleini Balss, 1911 [western Pacific Ocean]

P. echidna sp. nov. [Australia, New Zealand]

P. elongata Spiridonov, Türkay, Arntz & Thatje, 2006 [Bouvet Island, Southern Ocean]

P. erinacea Macpherson, 1988b [eastern Atlantic Ocean]

P. formosa Henderson, 1888 [southwestern Atlantic Ocean]

P. gowlettholmes sp. nov. [southwestern Australia]

P. granulosa (Hombron & Jacquinot, 1846) [southwestern Atlantic Ocean, southeastern Pacific Ocean]

P. grossmani Macpherson, 1988b [eastern Atlantic Ocean]

P. haigae Eldredge, 1976 [southwestern Pacific Ocean]

P. hirtella de Saint Laurent & Macpherson, 1997 [southwestern Pacific Ocean]

P. histrix (de Haan, 1844) [Japan, western Pacific Ocean]

P. hystrixoides Sakai, 1980 [Japan, western Pacific Ocean]

P. inca Haig, 1974 [southeastern Pacific Ocean]

P. indica Alcock & Anderson, 1899 [northwestern Indian Ocean]

P. investigatoris Alcock & Anderson, 1899 [northwestern Indian Ocean]

P. japonica Balss, 1911 [Japan, western Pacific Ocean]

P. jamsteci Takeda & Hashimoto, 1990 [Japan, western Pacific Ocean]

P. kyushupalauensis Takeda, 1985 [Japan, western Pacific Ocean]

P. longidactylus Birstein & Vinogradov, 1972 [Uruguay, southwestern Atlantic Ocean]

P. longipes Faxon, 1893 [Columbia to Peru, eastern Pacific Ocean]

P. makarovi Hall & Thatje, 2009b [Bering Sea, northern Pacific Ocean]

P. manningi Williams, Smith & Baco, 2000 [California, eastern Pacific Ocean]

P. medipacifica Takeda, 1974 [western central Pacific Ocean]

P. mendagnai Macpherson, 2003 [Solomon Islands, western Pacific Ocean]

P. microps Filhol, 1884 [eastern Atlantic Ocean]

P. multispina (Benedict, 1895) [northern Pacific Ocean]

P. nivosa Hall & Thatje, 2009b [Philippines, western Pacific Ocean]

P. ochthodes Macpherson, 1988a [Indonesia, western Pacific Ocean]
P. odawarai (Sakai, 1980) [Japan, western Pacific Ocean]
P. otsuae Wilson, 1990 [Chile, southeastern Pacific Ocean]
P. pacifica Sakai, 1978 [northwestern Pacific Ocean]
P. papillata (Benedict, 1895) [eastern Pacific Ocean]
P. pectinata Macpherson, 1988b [Venezuela, western Atlantic Ocean]
P. phrixa Macpherson, 1992 [Peru, southeastern Pacific Ocean]
P. poorei sp. nov. [New Zealand, Australia]
P. roeleveldae Kensley, 1980 [South Africa, southwestern Indian Ocean]
P. seagranti Eldredge, 1976 [southwestern Pacific Ocean]
P. serrata Macpherson, 1988b [western Atlantic Ocean]
P. sonne Guzmán, 2009 [Chile, southeastern Pacific Ocean]
P. spectabilis Hansen, 1908 [northeast Atlantic Ocean]
P. spinosissima Birstein & Vinogradov, 1972 [southwest Atlantic Ocean, Antarctica]
P. staplesi sp. nov. [New Zealand, Australia]
P. stella Macpherson, 1988c [Reunion Islands, western Indian Ocean]
P. stevensi Ahyong & Dawson, 2006 [Ross Sea, Antarctica]
P. taylorae sp.nov. [southestern Australia]
P. truncatispinosa Takeda & Miyake, 1980 (= *P. heterotuberculata* Tung, Wang & Li, 1984) [East China Sea, western Pacific Ocean]
P. tuberipes Macpherson, 1988a [Chile, southwestern Pacific Ocean]
P. verrilli (Benedict, 1895) [northern Pacific Ocean]
P. webberi sp. nov. [New Zealand, southwestern Pacific Ocean]
P. zealandica Dawson & Yaldwyn, 1971 (= *P. shinkaimaruae* Takeda & Hatanaka, 1984) [New Zealand, southwestern Pacific Ocean]

Remarks. *Paralomis* is the most speciose and most morphologically diverse genus in the Lithodidae. Prior to the present study, 62 extant and one fossil species were recognised (De Grave *et al.* 2009; Hall *et al.* 2009). Eleven species of *Paralomis* are known from the study area, of which six are new species: one is known only from New Zealand (*P. webberi* sp. nov.), two exclusively from Australia (*P. gowlettholmes* sp. nov., *P. taylorae* sp. nov.), and three occur in both regions (*P. echidna* sp. nov., *P. poorei* sp. nov., *P. staplesi* sp. nov.). *Paralomis shinkaimaruae* Takeda & Hatanaka, 1984 is shown to be a junior synonym of *P. zealandica* Dawson & Yaldwyn, 1971. Thus, 67 extant species of *Paralomis* are currently recognised.

The overall monophyly of *Paralomis* is not presently in question (though further investigation is warranted into the the status of *Lopholithodes* with respect to *P. odawarai*; Macpherson, 1988b) and despite the wide range of carapace and pereopod forms, the genus is united by the undivided plate of abdominal somite 2 and the calcified, rather than variously membranous, plates of abdominal somites 3–5. Nevertheless, some potentially natural groups may be present within *Paralomis*, such as the short-legged, densely spiny forms similar to *P. histrix* and *P. echidna*, relatively less spiny, long-legged forms similar to *P. aculeata* and *P. spinosissima*, generally sparsely adorned forms similar to *P. spectabilis* and *P. birsteini*, long-legged forms with cristate carapace margins similar to *P. verrilli* and *P. cristulata*, and forms with a subhexagonal carapace densely covered by nodular tubercles, such as *P. granulosa*, *P. dawsoni*, and *P. dofleini*. Clearly, the range of forms within *Paralomis* is extensive and phylogenetic analysis is much needed to further understand morphological diversity within the genus. As discussed by Macpherson (1988b) and Ahyong *et al.* (2010), the taxonomic history of *Paralomis* has been rather unstable, and three nominal genera are presently included in its synonymy: *Acantholithus* Stimpson, 1858 (type species: *Lithodes histrix* de Haan, 1844), *Leptolithodes* Benedict, 1895, and *Pristopus* Benedict, 1895. Benedict (1895) incorrectly regarded the type species of *Paralomis*, *Lithodes granulosa* Hombron & Jacquinot, 1846, as a species of *Echinocerus* White, 1848 (a junior synonym of *Lopholithodes* Brandt, 1848), effectively sinking *Paralomis*. As such, Benedict (1895) erected two new genera to accommodate other species of *Paralomis*: *Leptolithodes* and *Pristopus*. *Leptolithodes* was erected for *Paralomis aculeata* Henderson, 1888, *P. longipes* Faxon, 1893, and two new species described therein, *P. multispina* Benedict, 1895 and *P. papillata* Benedict, 1895. *Pristopus* was erected for two species, *Pristopus verrilli* Benedict, 1895, described therein as new, and *Paralomis formosa* Henderson, 1888. Benedict (1895) did not designate type species for either *Leptolithodes* or *Pristopus*, so in anticipation of further study of *Paralomis*, Ahyong *et al.* (2010) selected *Paralomis aculeata* Henderson, 1888 and *Pristopus verrilli* Benedict, 1895 as the respective type species for the two nominal genera.

Paralomis includes the deepest living lithodid, *P. bouvieri* from the North Atlantic, which is known from as deep as 4152 m (Macpherson 1988b).

Key to adult *Paralomis* from New Zealand, Australia, and the Ross Sea

1. Carapace densely covered with slender or conical spines 2
- Carapace surface unarmed or with short, sparsely distributed dorsal spines 6
2. Meri of walking legs long, slender, pereopod 4 merus at least 0.85 pcl [southeastern Australia] .... *P. taylorae* sp. nov.
- Meri of walking legs short, pereopod 4 merus less than 0.75 pcl 3

3. Abdominal somites 4–6 with well-spaced granules and nodules [New Zealand and southeastern Australia] ....*P. poorei* sp. nov.
- Abdomen densely covered with slender spines or sharp conical tubercles in adults 4
4. Carapace covered with slender spines in juveniles but low conical tubercles in adults [New Zealand] ....*P. zealandica*
- Carapace covered with slender spines in adults 5
5. Subrostral lobe unarmed. Length of propodus of last walking leg about 3.7 times height in female [Tasman Sea] .... *P. echidna* sp. nov.
- Subrostral lobe multispinulate. Length of propodus of last walking leg less than 4.4 times height in female [Kermadec Ridge].......... *P. webberi* sp. nov.
6. Carapace surface smooth, without spines or tubercles; with tufts of stiff setae [hydrothermal vents; Fiji to Kermadec Ridge] ....*P. hirtella*
- Carapace surface rugose, with spines or tubercles 7
7. Carapace surface entirely covered with blunt, wart-like tubercles; no dorsal spines [New Zealand to Solomon Islands] .... *P. dawsoni*
- Carapace surface with scattered short spines or small tubercles; surface between spines and tubercles generally smooth...8
8. Walking leg dactyli longer than dorsal extensor margin of respective propodi 9
- Walking leg dactyli shorter than dorsal extensor margin of respective propodi 10
9. Walking leg 3 propodus length:height ratio exceeding 6.2 in males, exceeding 7.3 in females [Antarctica and Macquarie Ridge] ....*P. birsteini*
- Walking leg 3 propodus length:height ratio 5.5 or less in males, less than 5.6 in females [southeastern Australia] .... *P. gowlettholmes* sp. nov.
10. Carapace surface with scattered granules and few, low conical spines. Cardiac region with 4 low tubercles [Ross Sea, Antarctica] .... *P. stevensi*
- Carapace surface prominently multispinose. Cardiac region with 4 spines [New Zealand and Australia] .... *P. staplesi* sp. nov.

Paralomis birsteini Macpherson, 1988

(Figs 63–66, Pl. 3A, 4E)

Paralomis spectabilis. — Birstein & Vinogradov, 1967: 390, figs 1, 2. — Kirkwood, 1984: 37, fig. 45 [map only]. — Tkachuk, 1985: 1733–1734, figs A, B. — Zaklan, 2002: 773 [Ross Sea records] [not *P. spectabilis* Hansen, 1908].

Paralomis birsteini Macpherson, 1988a: 72–74, figs 4, 5a–e [type locality: Scott Island, Ross Sea, Antarctica, 67°29'S, 179°55'W, 1080 m]. — Dawson, 1989: 319. — Arana & Retamal, 1999: 101 [Bellingshausen Sea records]. — Zaklan, 2002: 770, 788. — Macpherson, 2004: 421, tab. 1. — Stevens, 2004: 6, 7 [part]. — Ahyong & Dawson, 2006: 55–58, figs 5–6, tab. 1.

Neolithodes brodiei. — Thatje & Lörz, 2005: 335 [NIWA 3432 only; not *N. brodiei* Dawson & Yaldwyn, 1970].

Type material. *Holotype*: USNM 228830, female (cl 63.3, pcl 55.0 mm, cw 53.1 mm), Ross Sea, N of Scott Island, 67°29'S, 179°55'W, 1080 m, RV *Eltanin*, Stn 27-1946, 2 Feb 1967.

Other material examined. *Macquarie Ridge*: NMV J61053, 1 male (pcl 64.8, cw 61.1 mm), N end of Gap, N of Macquarie Island, 53°37.2'S, 159°11.3–17.4'E, with *Briarosaccus callosus* Boschma, 1158–1500 m, SS01/99/122, 30 Jan 1999.

Ross Sea, Antarctica: NMNZ Cr11779, 1 male (cl 89.9 mm, pcl 78.9 mm, cw 79.2 mm), N of Ross Sea, 62°34.80'S, 173°35.40'E, 900 m, trip 1430, set 197B, FV *San Aotea II*, 22 Apr 2001; NMNZ Cr11745–11749, 5 males (pcl 83.3–89.1 mm, cw 82.1–83.2 mm), Ross Sea, area 88.1C, 64°30.50'S, 171°13.8'W, 1128–1083 m, trip 1845, set 14, FV *Janas*, 21 Dec 2003; NIWA 23842, 1 male (cl 41.4 mm, cw 40.7 mm), 1 female (cl 35.2 mm, cw 32.5 mm), both infected with *Briarosaccus callosus* Boschma, 1930 (Rhizocephala), near Scott Island, 64°31.0–30.0'S, 171°14.0–08.0'W, 1225–1617 m, from stomach of *Dissostichus mawsoni* Norman, 1937, *Avro Chieftain*, set 24, 17 Dec 2003; NIWA 41869, 1 male (cl 86.0 mm, pcl 76.4 mm, cw 72.5 mm), Ross Sea, area 88.1C, 65°07.20–08.40'S, 178°29.40–25.80'W, 1669–1950 m, trip 2530, set 11, coll. G. Higgins, 7 Dec 2007; NIWA 3432, ovigerous female (cl 85.4 mm, pcl 73.9 mm, cw 72.5 mm), Balleny Seamount, 65°28.53–28.97'S, 161°02.88–02.75'E, 760–750 m, TAN0402/269, RV *Tangaroa*, 7 Mar 2004; NMNZ 11255, 1 male (cl 89.1 mm, pcl 78.2 mm, cw 73.8 mm), 66°35'S, 177°33'W, 1549 m, trip 1728, set 48, FV *Janas*; NIWA 38668, 1 male (cl 50.0 mm, pcl 42.6 mm, cw 40.0 mm), 67°20.42–20.21'S, 179°55.91–56.53'W, 1130–1235 m, TAN0802/256, 8 Mar 2008; NIWA 24217, 1 male (cl 89.8 mm, cl 79.5 mm, cw 76.4 mm), near Scott I., 67°21.01–21.02'S, 179°52.68–52.34'W, 540–600 m, sled tow, TAN0602/394, 6 Mar 2006; NIWA 38712, 1 ovigerous female (cl 69.6 mm, pcl 61.1 mm, cw 57.1 mm), 67°21.80–21.54'S, 179°57.11–57.25'W, 540–456 m, TAN0802/258, 8 Mar 2008; NIWA 38716, 1 male (cl 97.8 mm, pcl 90.9 mm, cw 86.6 mm), 1 ovigerous female (cl 69.9 mm, pcl 61.5 mm, cw 57.3 mm), 67°21.80–21.54'S, 179°57.11–57.25'W, 540–456 m, TAN0802/258, 8 Mar 2008; NIWA 38505, 1 male (pcl 69.5 mm, cw 66.2 mm), 67°22.92–22.67'S, 179°52.33–51.81'W, 420–456 m, TAN0802/243, 7 Mar 2008; NMNZ Cr11001, 1 male (cl 79.8, cw 74.7 mm), 1 female (cl 61.7, cw 59.4 mm), both infected with *Briarosaccus callosus* Boschma, 1930 (Rhizocephala), near Scott Island, 68°23–21'S,

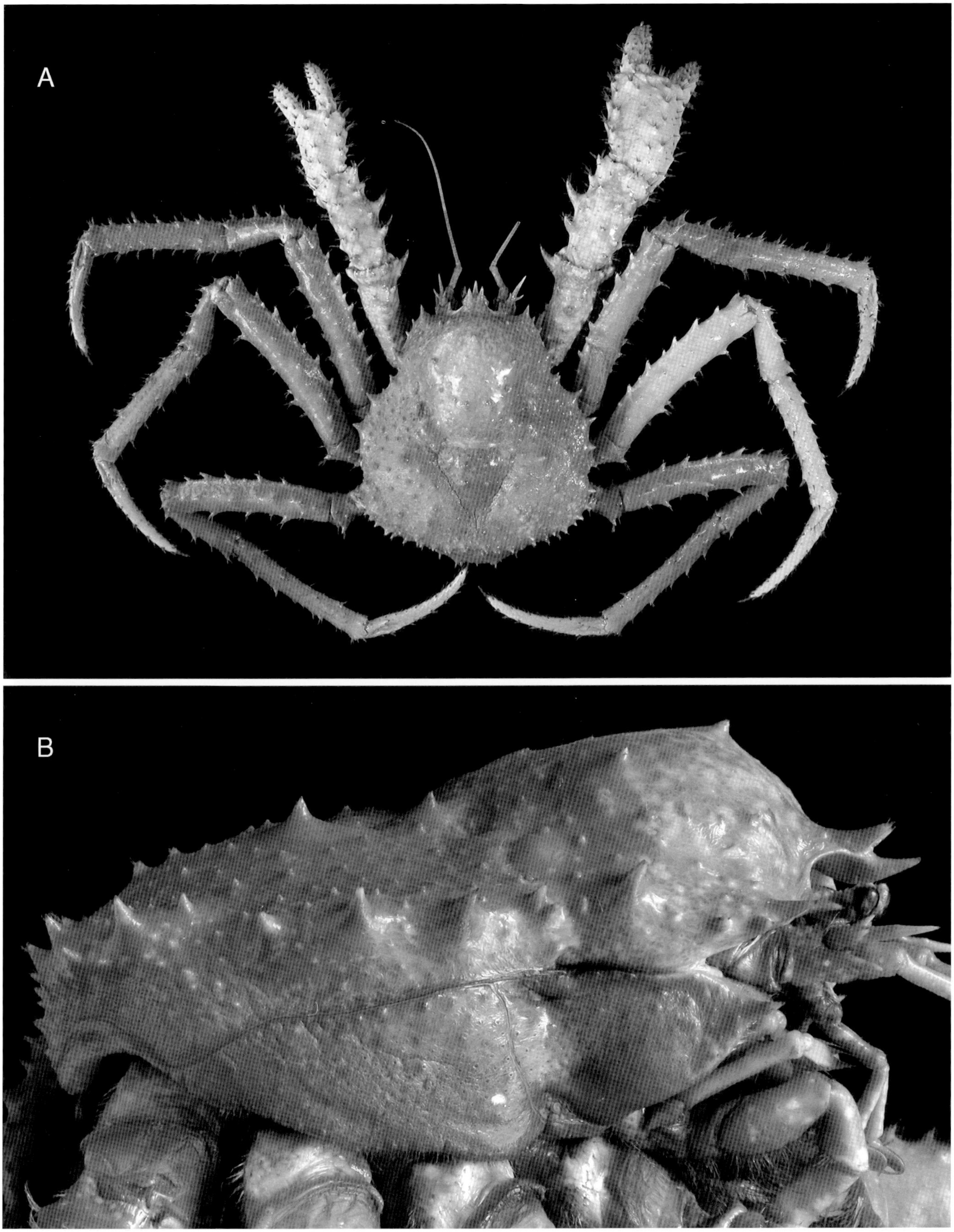

Figure 63. *Paralomis birsteini* Macpherson, 1988a, male, pcl 79.4 mm, Ross Sea (NIWA 24217). A, dorsal habitus. B, carapace, right lateral view.

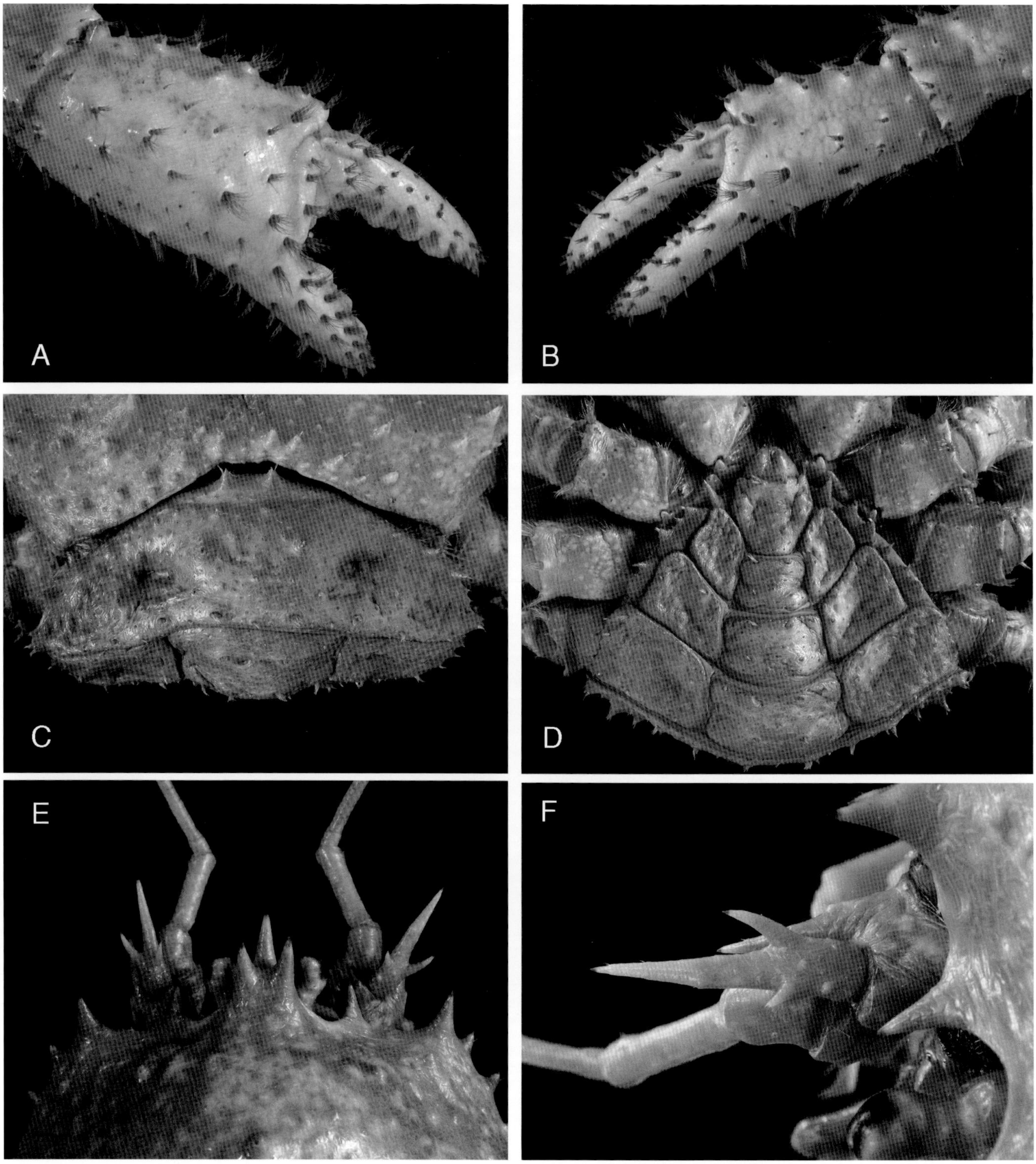

Figure 64. *Paralomis birsteini* Macpherson, 1988a, male, pcl 79.4 mm, Ross Sea (NIWA 24217). A, right chela. B, left chela. C, posterior carapace and abdominal somite 2. D, ventral surface and abdomen. E, anterior carapace, dorsal view. F, orbit and right antenna.

179°53–57′E, 1337–1050 m, RV *San Aotea II*, trip 2010, haul 137B, on hook, B.E. Winslade, 31 Jan 2005; NMNZ Cr11000, 1 male (cl 76.9, cw 73.4 mm), Ross Sea, Trip 1862, haul 147, area 88.1G, RV *San Aotea*, obs. S. Voice [note: specific station data not available].

Diagnosis. Carapace subpentagonal, slightly wider than long; surface glabrous, sparsely covered with small, scattered granules and short conical spines, including median gastric spine; lateral margins with short conical spines; outer distance between bases of anterolateral

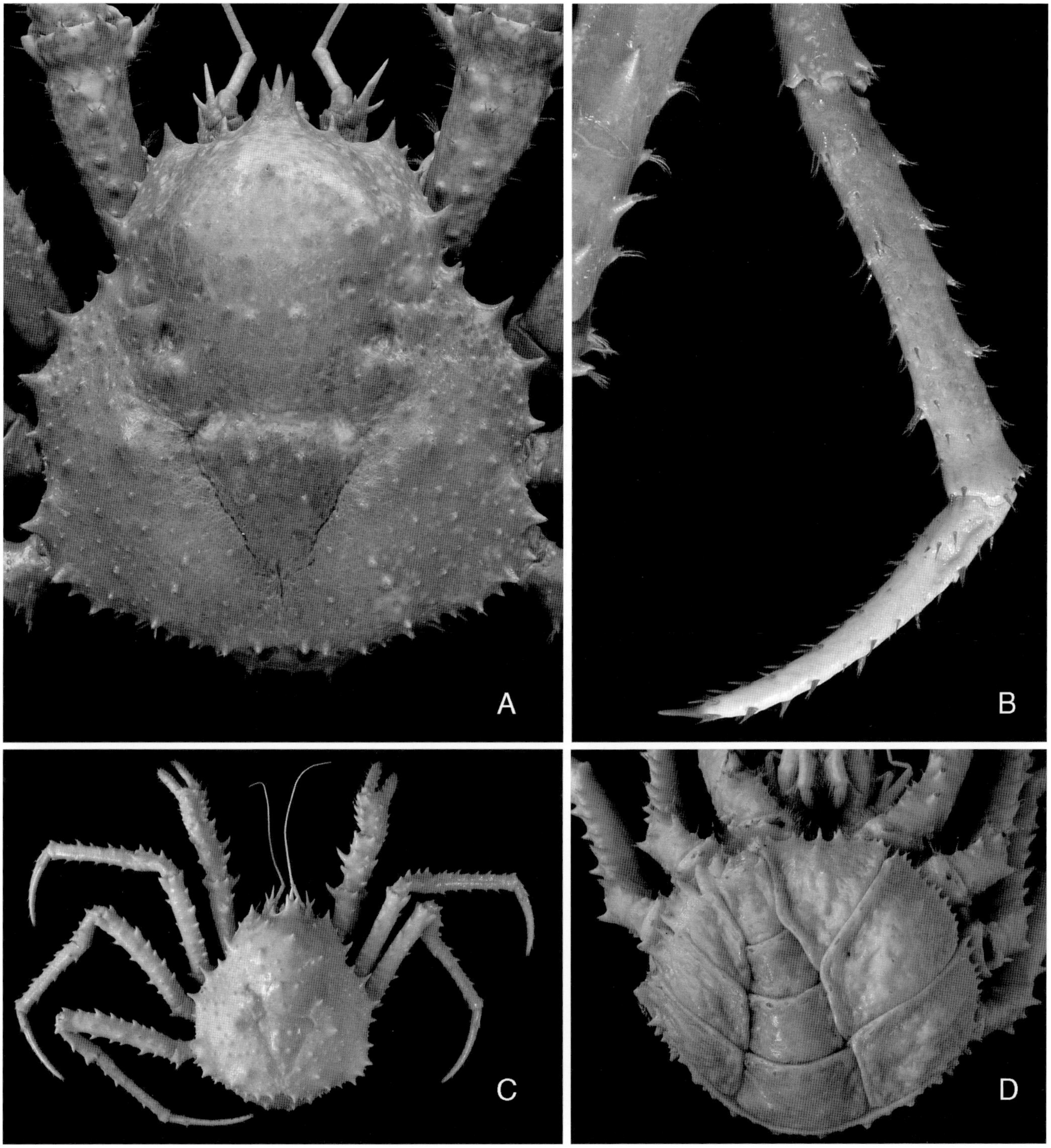

Figure 65. *Paralomis birsteini* Macpherson, 1988a. A–B, male, pcl 79.4 mm, Ross Sea (NIWA 24217). C–D, ovigerous female, pcl 73.9 mm, Ross Sea (NIWA 3432). A, carapace. B, right pereopod 4 dactylus and propodus. C, dorsal habitus. D, abdomen.

spines slightly less than half carapace width. Rostrum trispinose, broadest basally, without basal constriction. Scaphocerite with 1 or 2 short inner basal spines and 1 or 2 outer spines. Male chelipeds dimorphic, spinose; major cheliped palm height less than 1.6 times that of minor cheliped. Major cheliped palm of female 1.2–1.3 times height of minor cheliped. Walking legs elongate, spinose, 2.2–2.7 pcl (male), 2.0–2.3 pcl (female); ventral spines of propodus widely spaced; dactyli longer than extensor margin of propodi. Walking leg 3 merus 0.7–0.8 pcl (male), 0.6–0.7 pcl (female), about 4–5 times longer than high; propodus length 6.2–7.9 times height.

Description. *Carapace*: Subpentagonal, length 1.00–1.06 times width; surface glabrous, sparsely covered with small, scattered granules. Gastric region convex, elevated, more prominent than other regions, with 5 short conical spines forming pentagon with anteriormost in midline. Lateral margin of hepatic region with 3 short conical spines, anterior shorter. Branchial regions with 12–14 short marginal spines; surface with 4 or 5 low conical spines in addition to low, scattered tubercles. Cardiac region subtriangular, with 2 pairs of short conical spines forming square. Intestinal region sparsely granular. Pterygostomian region sparsely granular, with prominent anterior spine.

Rostrum 0.13–0.17 pcl; broadest basally, without constriction; median spine smooth ventrally; paired dorsal spines divergent or subparallel, directed obliquely upwards. Posterior orbital margin near concave; outer orbital spine reaching to base of cornea. Anterolateral spine as long as or shorter than outer orbital spine, with low marginal granule or small spine midway between outer orbital spine and anterolateral spine; outer distance between bases of anterolateral spines slightly less than half carapace width.

Ocular peduncle: Longer than cornea; with 2–5 dorsal granules.

Antennule: Peduncle unarmed, reaching anteriorly beyond antennal peduncle by about three-quarters length of distal antennular peduncle article.

Antenna: Basal antennal article with small anterolateral spine. Article 2 with angular to sharp inner distal margin; outer margin with small basal spine and slender spine that does not reach beyond article 4. Article 3 unarmed. Scaphocerite slender, reaching beyond midlength but not beyond distal three-quarters of article 5, with 1 or 2 short inner spines and 1 or 2 outer spines. Article 4 unarmed, about half length of article 5.

Abdomen: Ornamentation similar in both sexes. Somites sparsely granular. Somite 2 with small, widely separated spines along margins. Somite 3 median plate unarmed; submedian plates with low, irregular, triangular teeth on lateral margin; marginal plates absent, apparently undifferentiated. Somites 4–5 with unarmed median and submedian plates; marginal plates undivided, with 2–4 low angular protrusions. Somite 6 as long as or longer than wide, subquadrate, with pair of small distal teeth; marginal plates short, triangular. Telson wider than long, apex rounded.

Pereopod 1 (chelipeds): Spination similar in both sexes. Major cheliped of males inflated, 1.39–1.57 times height of minor cheliped; minor cheliped slender. Chelipeds of female unequal, major cheliped 1.24–1.25 times height of minor cheliped. Coxae smooth, unarmed; distal margins with dense tufts of setae. Ischiobasis with 4–8 stout, apically setose ventral spines. Merus with smooth mesial margins and tuberculate lateral margins; dorsal and ventral margins spinose, inner distal spine largest. Carpus with prominent spines on dorsal and lateral surfaces; dorsal margin with row of 3–5 spines; mesial margin with 2 spines, proximal largest; ventral surface with 3 spines. Palm mesial margin with 6 spines, other surfaces with prominent, apically setose tubercles or acute tubercles. Fingers with short basal tubercle and rows of tufts of golden setae.

Major cheliped 1.48–1.84 pcl (male), 1.38–1.44 (female); upper palm length 1.13–1.35 times height (male), 1.32–1.43 (female); occlusal margins of fingers corneous for distal third, proximally with 3 calcareous nodules; dactylus dorsal margin broadly convex, with rows of tufts of golden setae and blunt proximal tooth, 1.01–1.16 times longer than dorsal margin of palm (male), 1.33–1.34 (female).

Minor cheliped 1.44–1.66 pcl (male), 1.35–1.47 (female); upper palm length 1.18–1.38 times height (male), 1.35–1.50 (female); occlusal margin corneous for slightly less than distal third, proximally crenulate; dactylus dorsal margin broadly convex, with rows of tufts of golden setae and blunt proximal tooth, 1.42–1.67 times longer than dorsal margin of palm (male), 1.61–1.67 (female).

Pereopods 2–4 (walking legs 1–3): Similar, elongate, spinose. Coxae smooth, unarmed; distal margins with dense tufts of setae. Ischiobasis with 3–5 apically setose ventral spines. Merus compressed, shorter than carapace in both sexes; extensor margin with 8–11 spines in addition to paired distal spines; dorsal surface with 4–8 spines; flexor margin with rows of 5–7 spines; merus of pereopod 3 longer than that of pereopod 2 and pereopod 4. Carpus slightly longer than half merus length, subcircular in cross section; extensor margin with 5 or 6 spines; dorsal surface with 4–6 spines; flexor margin unarmed. Propodus dorsoventrally flattened; slightly shorter than merus; with 9–14 spines on extensor margin; dorsal surface with small scattered spines; flexor margin with 5–8 spines. Dactylus broadly curved; slightly longer than extensor margin of propodus; surface with tufts of setae; extensor margin with 3 or 4 apically corneous spines proximally; lateral proximal surfaces with short, distinct sulcus, flanked ventrally by 1 or 2 small, corneous spines; flexor margin lined with corneous spinules

Pereopod 2 length 2.22–2.69 pcl (male), 2.01–2.25 pcl (female). Merus 0.72–0.85 pcl (male), 0.60–0.67 pcl (female); length:height ratio 4.05–5.04 (male), 4.01–4.67 (female). Carpus 0.55–0.58 merus length (male), 0.58–0.64 (female). Propodus 0.82–0.86 merus length (male), 0.90–0.96 (female); length:height ratio 5.43–7.11 (male), 6.69–6.95 (female). Dactylus 0.93–1.01 propodus length (male), 0.88–0.96 (female).

Pereopod 3 length 2.31–2.69 pcl (male), 2.10–2.33 pcl (female). Merus 0.73–0.86 pcl (male), 0.64–0.69 pcl (female); length:height ratio 4.08–5.26 (male), 4.81–4.99

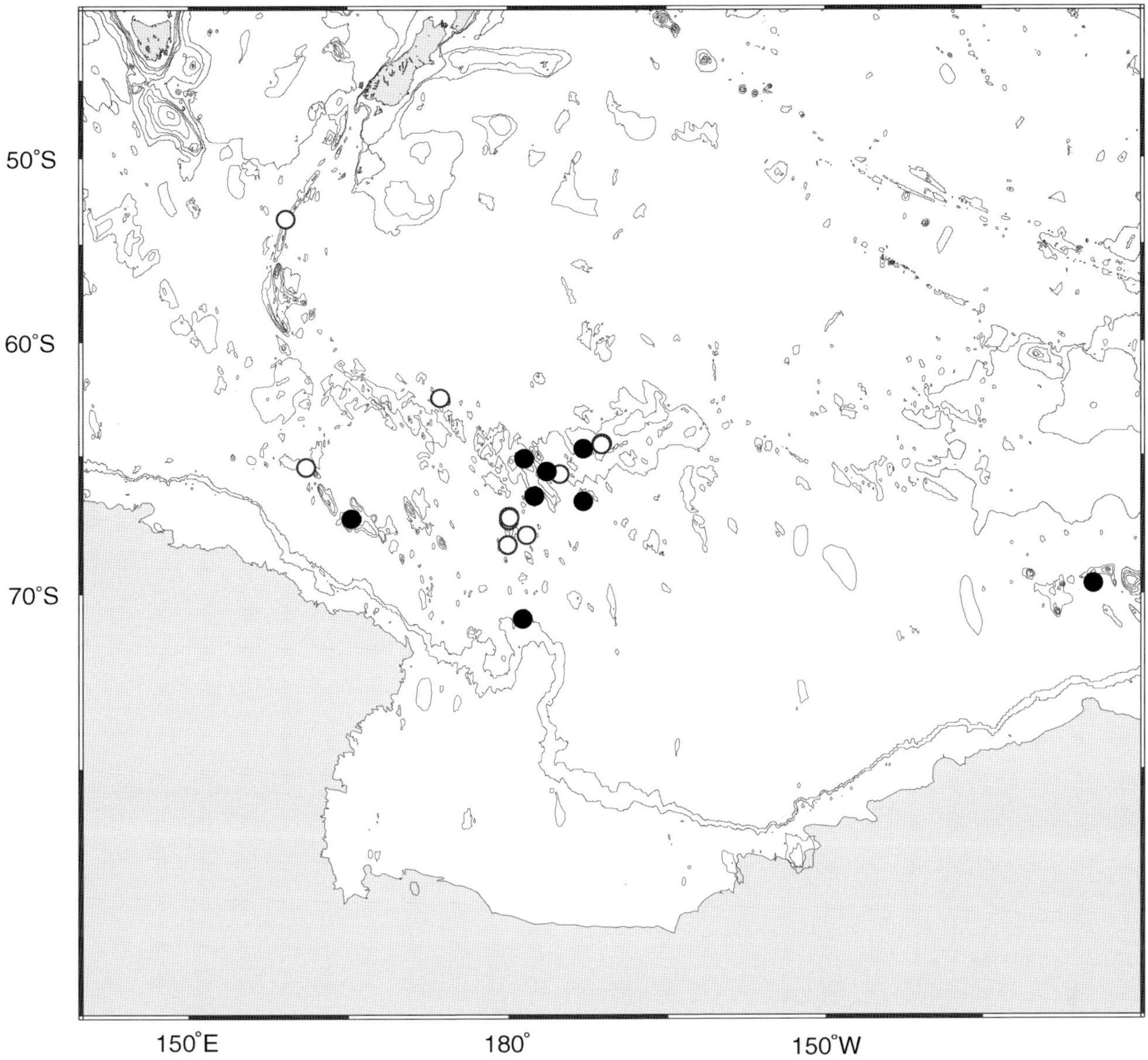

Figure 66. Geographic distributions of *Paralomis birsteini* Macpherson, 1988 (○) and *P. stevensi* Ahyong & Dawson, 2006 (●).

(female). Carpus 0.55–0.57 merus length (male), 0.58–0.64 (female). Propodus 0.85–0.87 merus length (male), 0.92–0.96 (female); length:height ratio 6.01–7.53 (male), 6.81–7.66 (female). Dactylus 0.92–0.99 propodus length (male), 0.88–0.94 (female).

Pereopod 4 length 2.27–2.72 pcl (male), 2.09–2.34 pcl (female). Merus 0.69–0.82 pcl (male), 0.60–0.68 pcl (female); length:height ratio 4.26–5.23 (male), 5.01–5.11 (female). Carpus 0.56–0.59 merus length (male), 0.58–0.64 (female). Propodus 0.91–0.94 merus length (male), 0.97–1.02 (female); length:height ratio 6.22–7.84 (male), 7.37–7.86 (female). Dactylus 0.90–0.96 propodus length (male), 0.89–0.94 (female).

Egg diameter 2.2–2.3 mm.

Colour in life. Light orange to orange-red; carapace often paler than pereopods, especially on branchial regions (Pl. 3A).

Remarks. *Paralomis birsteini* Macpherson, 1988a has been discussed extensively by Macpherson (1988a) and Ahyong & Dawson (2006). It closely resembles *P. spectabilis* Hansen, 1908, *P. stevensi* Ahyong & Dawson, 2006, and *P. gowlettholmes* sp. nov.; distinguishing features are discussed under the accounts of *P. gowlettholmes* and *P. stevensi* (Table 3).

The single specimen from Macquarie Ridge is a significant northern range extension for the species. The specimen is parasitised by *Briarosaccus callosus* and, although a male, it has feminised pereopod morphology in which the chelipeds are not significantly unequal, and the walking leg morphometrics are in the range reported here for female *P. birsteini* from the Ross Sea. The smallest ovigerous female examined is 61.1 mm pcl, and the smallest female overall (pcl 35.2 mm) has an immature abdomen in which full asymmetry has not yet been reached. The female holotype (pcl 55.0 mm) appears to be mature.

Paralomis birsteini has been attributed a near circum-Antarctic distribution, with records from the Ross Sea in the vicinity of Scott and Balleny Islands, the Crozet and Kerguelen Islands, the Bellingshausen Sea (Macpherson 1988b, 2004; Arana & Retamal 1999; García Raso *et al.* 2005; Ahyong & Dawson 2006), and now Macquarie Ridge, near Macquarie Island. Specimens from across the reported range of the species, however, have not been directly compared with topotypic material. Notably, Arana & Retamal's (1999) figure of a specimen from the Bellingshausen Sea differs from *P. birsteini* sensu stricto, most significantly in the walking leg dactyli being shorter rather than longer than the respective propodal extensor margins. A preliminary molecular study found a paraphyletic relationship between putative *P. birsteini* from the Bellingshausen Sea, Crozet and Kerguelen Islands (Thatje *et al.* 2008). Thus *P. birsteini*, as presently understood, probably represents a species complex, and is currently under investigation in collaboration with S. Hall. The present account of *P. birsteini* is based only on specimens collected at or near the type locality (i.e., Ross Sea) and will serve as a basis for future comparison with *P. birsteini* from other localities. Several so-called circum-Antarctic Crustacea, when more closely scrutinised, have also proven to be composites, including various amphipods (Lörz *et al.* 2009), isopods (see Held & Wägele 2005) and the king crab *Lithodes murrayi* (see Macpherson 1988b, c; this study).

Distribution. Known with certainty only from the Ross Sea, north to the Macquarie Ridge. Records from the Bellingshausen Sea, Crozet and Kerguelen Islands require confirmation; 420–1549 m in the Ross Sea, and 620–1947 m outside of the Ross Sea (Macpherson 2004; Thatje *et al.*, 2008).

Table 3. Selected measurements of *P. birsteini, P. gowlettholmes, P. stevensi*, and *P. webberi*. Measurements in millimetres (mm). Abbreviations: postorbital carapace length (pcl).

	Major cheliped height: minor cheliped height (Male)	Major cheliped height: minor cheliped height (Female)	Pereopod 2–4 dactylus length relative to extensor propodal margin	Pereopod 4 propodus length/ht (Male)	Pereopod 4 propodus length/ht (Female)	Maximum known pcl	Smallest ovigerous female (pcl)
P. gowlettholmes	1.65–1.82	1.35–1.46	Longer	4.96–5.53	3.67–5.59	50.1	28.3
P. birsteini	1.39–1.57	1.24–1.26	Longer	6.22–7.84	7.37–7.86	90.9	61.1
P. stevensi	1.66–1.91	1.31–1.33	Shorter	8.45–11.22	6.72–7.52	81.7	62.4
P. staplesi	1.59	?		9.81	?	50.0	?

Paralomis dawsoni Macpherson, 2001 (Figs 67–72A, 73, Pl. 3B)

Paralomis sp. — Macpherson, 1990: 225, figs. 2c, 4. — O'Shea *et al.*, 1999: 49, fig. 16. — Zaklan, 2002: 810.
Paralomis sp. nov. — Zaklan, 2002: 773, 794.
Paralomis dawsoni Macpherson, 2001: 802–805, fig. 4 [type locality: New Caledonia]. — Webber & Naylor, 2004b: 62. — Naylor *et al.*, 2005: 46. — Ahyong *et al.*, 2007: 156. — Dawson, 2008: 7, fig. 8.
Paralomis sp. ["multipustulata"]. — Dawson, 2008: fig. 6.

Type material. *Holotype*: MNHN Pg4278, ovigerous female (pcl 71.0 mm, cw 72.8 mm), New Caledonia, seaward side of reef, from trap.

Other material examined. *West Norfolk Ridge*: NMNZ Cr12028, 1 ovigerous female (cl 128.1 mm, pcl 112.6 mm, cw 114.7 mm, W of Ninety Mile Beach, 34°49.5′S, 169°50.9′E, 820–1118 m, trip 1945, tow 27, 26 May 2004.

Challenger Plateau: NIWA 48671, 1 male (cl 140.6 mm, pcl 120.8 mm, cw 126.5 mm), 37°25.03′S, 168°06.06′E, 900 m, Z9724, coll. A. Knox, 3 Mar 1999.

Bay of Plenty: NMNZ Cr11566–11569, 1 male (cl 127.9 mm, pcl 114.7 mm, cw 117.8 mm), 2 ovigerous females (cl 116.7–122.1 mm, pcl 102.3–107.4 mm, cw 104.3–109.4 mm), 1 female (cl 132.4 mm, pcl 114.7 mm, cw 121.1 mm), Bay of Plenty, FV *Savannah*, coll. B. Roughlin; NMNZ Cr11767, 1 female (pcl 142.7 mm, cw 139.8 mm), Bay of Plenty, coll. Gregg Gibbs.

Hawkes Bay: NMNZ Cr5642, 1 male (cl 129.9 mm, pcl 112.0 mm, cw 118.4 mm), Ritchie Bank, 39°28.39′S, 178°23.57′E, 400–600 m, FV *Betty T*, coll. P. Terry, 30 Sep 1986.

Louisville Ridge: NMNZ Cr11730, 1 male (cl 94.3+ mm, pcl 85.4 mm, cw 89.7 mm), 40°58′S, 165°04′W, 800

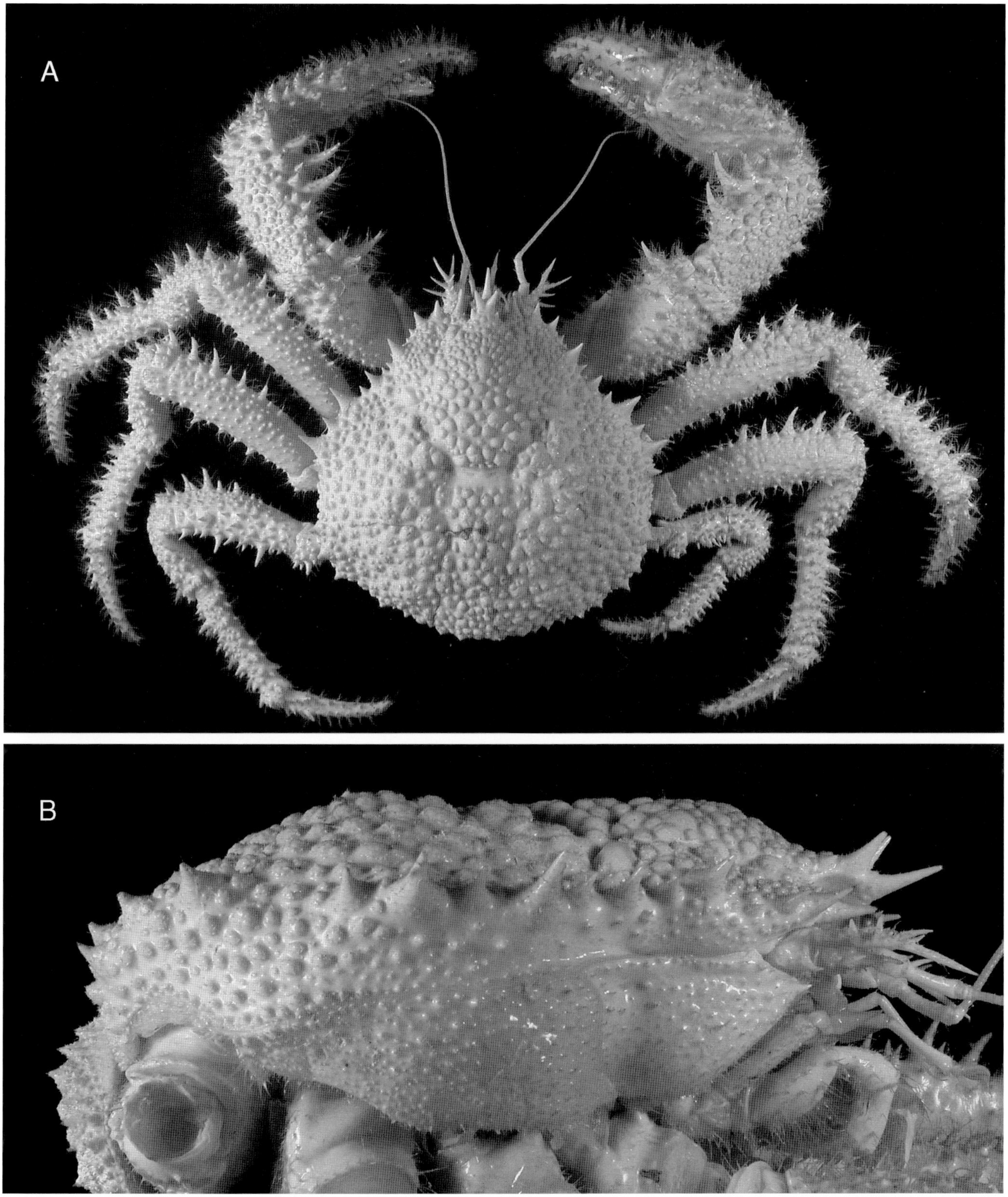

Figure 67. *Paralomis dawsoni* Macpherson, 2001, male, pcl 120.8 mm, Challenger Plateau (NIWA 48671). A, dorsal habitus. B, carapace, right lateral view.

m, FV *Peterson*, coll. M. Miranovich, 2 Mar 1995.

Hikurangi Margin: NIWA 29295, 1 ovigerous female (cl 105.8 mm, pcl 91.9 mm, cw 92.3 mm), Rock Garden cold seep, 40°02.31–02.47′S, 178°08.58–08.62′E, 730–747 m, TAN0616/06, 4 Nov 2006.

Chatham Rise: NIWA 3905, 1 female (cl 80.6+ mm,

Figure 68. *Paralomis dawsoni* Macpherson, 2001, male, pcl 120.8 mm, Challenger Plateau (NIWA 48671). A, carapace. B, anterior carapace, dorsal view.

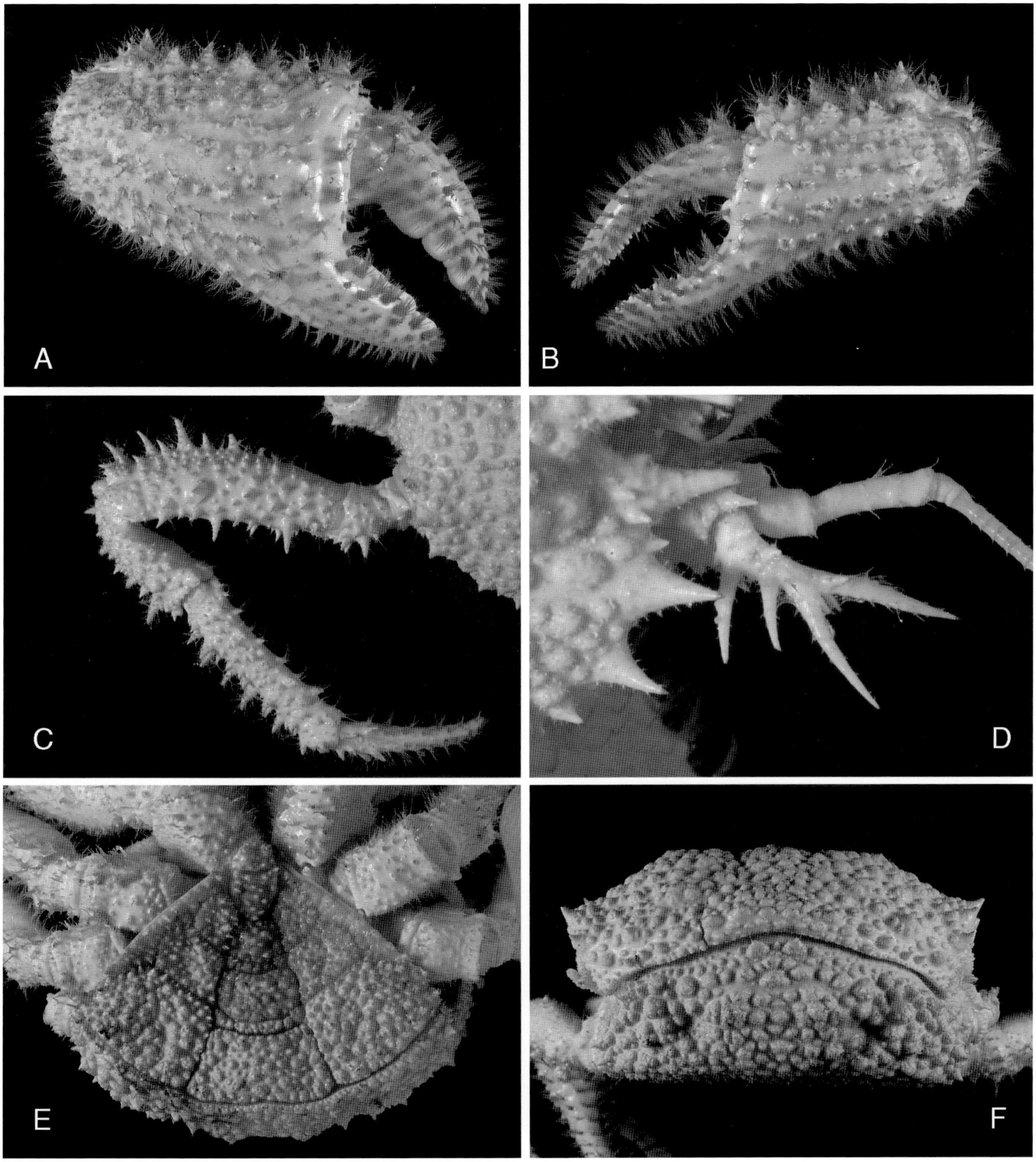

Figure 69. *Paralomis dawsoni* Macpherson, 2001, male, pcl 120.8 mm, Challenger Plateau (NIWA 48671). A, right chela. B, left chela. C, left pereopod 4. D, right antenna. E, ventral surface and abdomen. F, posterior carapace and abdominal somite 2.

pcl 72.9 mm, cw 71.8 mm), top of Morgue Seamount, 42°43.17–43.00′E, 179°57.78–58.24′E, 986 m, baited pot, TAN0104/193, Z10715, 19 Apr 2001; NMNZ Cr11969, 1 ovigerous female (cl 118.5 mm, pcl 104.9 mm, cw 108.1 mm), 42°48.2–48.4′S, 179°48.7–49.1′W, 866–970 m, FV *San Waitaki*, Trip 1702/140, coll. T. Brunning & M. Leslie, 16 Oct 2002; NIWA 54363, 1 juvenile male (cl 12.6 mm, pcl 10.3 mm, cw 9.0 mm), 44°01.67–01.87′S, 174°35.46–35.44′E, 801–823 m, TAN0905/121, 28 Jun 2009.

Figure 70. *Paralomis dawsoni* Macpherson, 2001, ovigerous female, pcl 91.9 mm, Rock Garden cold seep (NIWA 29295). A, dorsal habitus. B, abdomen.

Diagnosis. Carapace subhexagonal, about as long as wide; surface without spines, uniformly covered with nodulose, minutely setose, wart-like, microscopically setose, compound tubercles, tubercles smallest anterior to level of hepatic spines; margins microscopically setose; hepatic to anterior branchial margins lined with short but prominent conical spines; lateral branchial margins subparallel, straight to slightly concave; posterior branchial margins nodular or with short acute tubercles or spines. Rostrum trispinose, dorsal pair of spines prominent, without basal constriction. Scaphocerite multispinose. Abdomen covered with wart-like, compound tubercles. Chelipeds and walking legs covered with compound tubercles and coarse spines.

Description. *Carapace*: Subhexagonal, 0.95–1.02 times as long as wide; regions distinct, gastric region more inflated than others; surface uniformly covered with nodulose, minutely setose, wart-like compound tubercles, tubercles smallest anterior to level of hepatic spines. Hepatic and anterior branchial margins divergent (approximately to level of pereopod 3); with 9–15 conical, microscopically setose spines of varying length. Lateral branchial margins subparallel, straight or slightly convave, granular. Posterior branchial margin arcuate, tuberculate, usually unarmed, occasionally with acute tubercle anteriorly (largest male also with 2–4 short acute tubercles along margin). Pterygostomian region with small anterior spine; surface with small, coarse granules, densest and largest on posterior half.

Rostrum 0.11–0.16 pcl; broad basally, not constricted proximal to dorsal spines; median spine curved slightly dorsally, overreaching outerorbital spine; with blunt ventral tooth; paired dorsal spines short, divergent, directed obliquely upwards; with single small upright spine at base of paired dorsal spines. Posterior orbital margin concave, unarmed; outer orbital spine slender, directed anteriorly, overreaching eyes, with or without small inner basal spine. Anterolateral spine half as long as outer orbital spine; outer distance between bases of anterolateral spines distinctly less than half carapace width.

Ocular peduncle: Longer than cornea; dorsally with short, scattered granules or spines, distalmost spine longest; distal spine about half corneal diameter, slightly overreaching cornea.

Antennule: Peduncle unarmed, reaching anteriorly beyond apex of antennal peduncle by two-thirds length of distal antennular peduncle article.

Antenna: Basal antennal article with acute granules and slender, arcuate anterolateral spine. Article 2 with slender inner spine and short dorsal spine; outer margin with granules and 2 small spines proximally, and long, distal spine overreaching article 3. Article 3 unarmed. Scaphocerite multispinose; primary spine long, slender, slightly overreaching distal peduncular article; with multiple lateral spines of varying length, occasionally secondarily bifurcate; dorsomesially with 2 or 3 spines, shorter than half length of lateral spines. Article 4 unarmed, about half length of article 5.

Abdomen: Ornamentation similar in both sexes. Somites 2–3 covered with nodulose tubercles. Somites 4–6 covered with short conical spines; margins dentate to spinose. Telson spinose; semicircular.

Pereopod 1 (chelipeds): Strongly spinose and tuberculate, setose, unequal; ornamentation similar on both sides and in both sexes. Coxal margins tuberculate. Ischiobasis tuberculate laterally and ventrally. Merus outer surfaces granular; mesial surface relatively smooth, with strong inner distal spine, flanked by several small spines; dorsal surface granular proximally becoming spinose distally. Carpus surfaces tuberculate; dorsal margin with 6–8 strong, upright spines; ventral margin with short, conical spines and acute tubercles. Palm surfaces tuberculate, dorsal margin spinose; inner (mesial) surface sparsely tuberculate, apically setose.

Major cheliped 1.69–1.93 pcl (male), 1.20–1.23 (female); upper palm length 0.99–1.05 times height (male), 1.01–1.08 (female); occlusal margins of fingers corneous for distal third, proximally with 3 calcareous nodules; dactylus dorsal margin broadly convex, with rows of tufts of golden setae and blunt proximal tooth, 1.04–1.12 times longer than dorsal margin of palm (male), 1.27–1.31 (female).

Minor cheliped 1.85–1.61 pcl, (male), 1.17–1.21 (female); upper palm length 1.00–1.06 times height (male), 1.03–1.11 (female); occlusal margin corneous in distal third to half, proximally crenulate to weakly dentate; dactylus dorsal margin broadly convex, with rows of tufts of golden setae and blunt proximal tooth, 1.59–1.64 times longer than dorsal margin of palm (male), 1.85–1.91 (female).

Pereopods 2–4 (walking legs 1–3): Similar, elongate, spinose. Pereopod 2 longest. Coxae distal margins irregular; ventral surfaces tuberculate, becoming low and blunt in mature females and obsolete on pereopod 3–4. Ischiobasis spinose. Merus and carpus dorsal surfaces tuberculate (also spinose on pereopod 4), extensor and flexor margins prominently spinose; ventral surfaces unarmed, at most with a row of low tubercles or small spines near flexor and extensor margins. Propodus shorter than merus; tuberculate and spinose on all surfaces, spines longest on dorsal margins. Dactylus curved, laterally compressed; longer than carpus; with cluster of proximal spines and scattered tufts of golden setae; flexor margin lined with 9–13 obliquely inclined corneous spinules; apex corneous.

Pereopod 2 length 1.87–1.95 pcl (male), 1.34–1.46 pcl (female). Merus 0.59–0.66 pcl (male), 0.40–0.45 pcl (female); length:height ratio 3.45–3.71 (male), 2.56–2.96 (female). Carpus 0.61–0.63 merus length (male), 0.68–

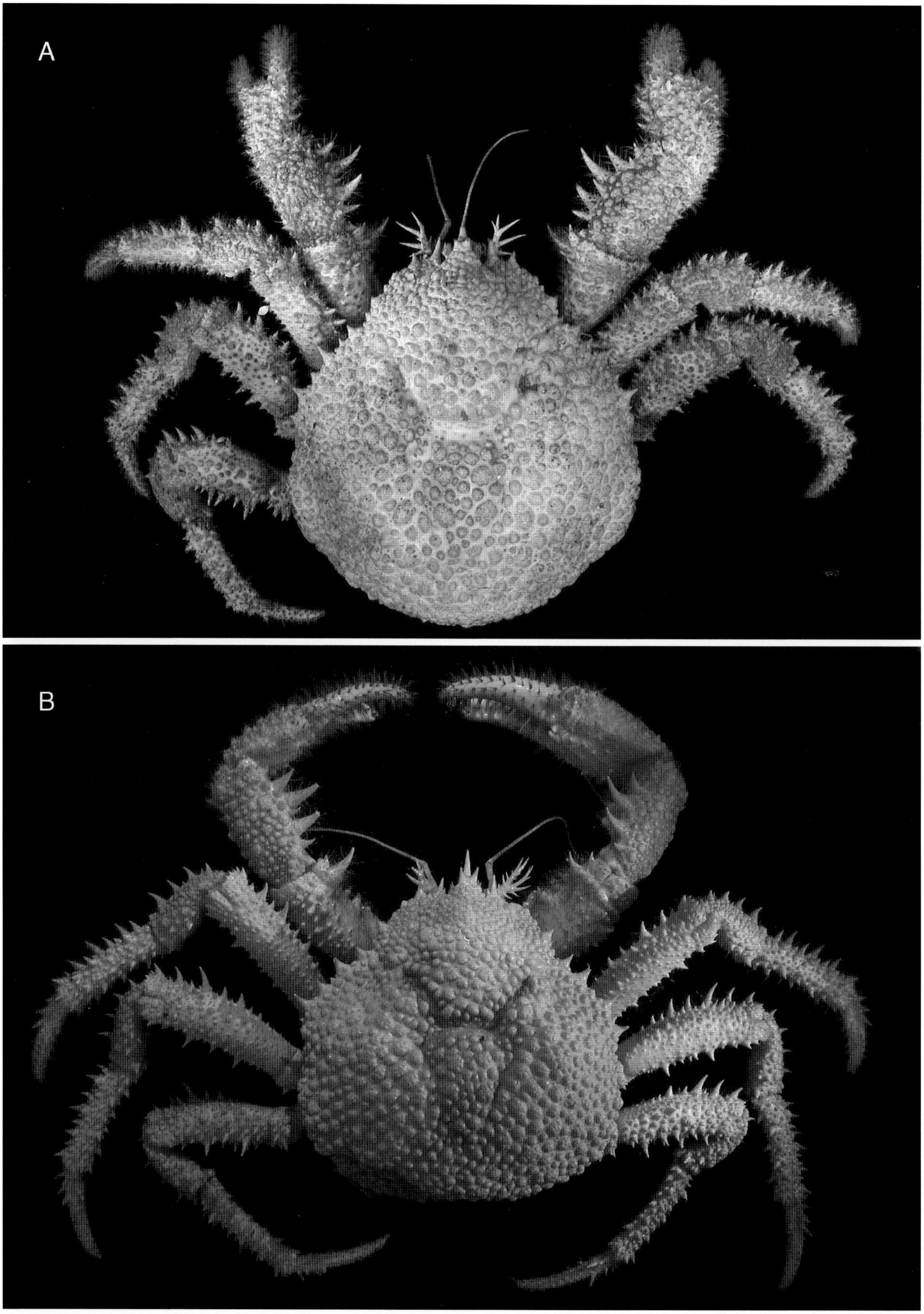

Figure 71. *Paralomis dawsoni* Macpherson, 2001. A, female, pcl 142.7 mm, Bay of Plenty (NMNZ Cr11767). B, male, pcl 112.0, Ritchie Bank (NMNZ Cr5642).

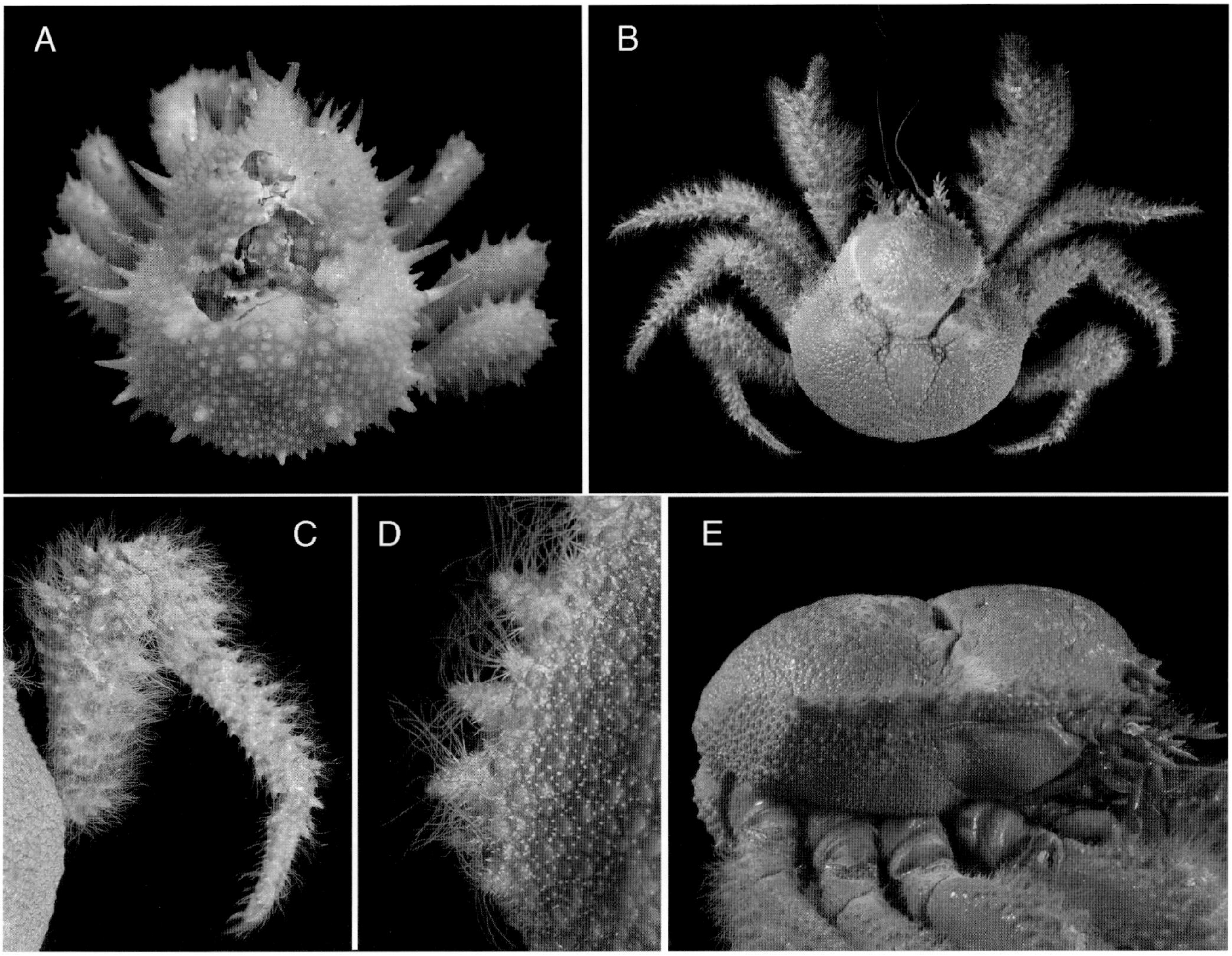

Figure 72. A, *Paralomis dawsoni* Macpherson, 2001, juvenile, pcl 10.3 mm, Chatham Rise (NIWA 54363). B–E, *Paralomis seagranti* Eldredge, 1976, male, cl 63.5 mm, pcl 58.1 mm, cw 63.9 mm, South Bank, Guam (ZRC 2000.0738). A–B, dorsal habitus. C, right pereopod 4. D, right anterior branchial margin of carapace. E, carapace, right lateral view.

0.77 (female). Propodus 0.75–0.76 merus length (male), 0.78–0.79 (female); length:height ratio 4.23–4.39 (male), 3.24–3.76 (female). Dactylus 0.93–1.01 propodus length (male), 0.98–1.05 (female).

Pereopod 3 length 1.81–1.94 pcl (male), 1.33–1.46 pcl (female). Merus 0.56–0.62 pcl (male), 0.41–0.45 pcl (female); length:height ratio 3.37–3.45 (male), 2.54–2.75 (female). Carpus 0.64–0.65 merus length (male), 0.69–0.75 (female). Propodus 0.81 merus length (male), 0.82–0.83 (female); length:height ratio 4.54–4.90 (male), 3.65–4.04 (female). Dactylus 0.90–1.02 propodus length (male), 0.92–1.04 (female).

Pereopod 4 length 1.70–1.88 pcl (male), 1.30–1.31 pcl (female). Merus 0.49–0.56 pcl (male), 0.36–0.36 pcl (female); length:height ratio 3.15–3.58 (male), 2.73–3.01 (female). Carpus 0.68–0.71 merus length (male), 0.72–0.81 (female). Propodus 0.90–0.91 merus length (male), 0.91–0.92 (female); length:height ratio 4.74–5.20 (male), 3.86–4.24 (female). Dactylus 0.91–1.03 propodus length (male), 0.94–1.00 (female).

COLOUR IN LIFE. Uniform orange-red to red-brown (Pl. 3B).

REMARKS. *Paralomis dawsoni* was first formally recorded from New Zealand by Webber & Naylor (2004b). It is the largest species of the genus in New Zealand waters and the 142.7 mm pcl female from the Bay of Plenty is the largest known specimen of the species. Allometric variation within the material at hand of *P. dawsoni* is minor in comparison to 'spiny' species of *Paralomis* such as *P. zealandica* and *P. histrix*. The primary changes are in the size and sharpness of the spines along the anterolateral carapace margins (proportionally shortest in the largest specimens) and in the degree of concavity of the branchial margin at the level of pereopods 3 and 4, being least concave in the largest specimens. The smallest specimen, a juvenile male (pcl 10.3 mm), differs from adults chiefly in the following respects: the marginal carapace spines are proportionally longer; the posterior branchial margin bears several prominent spines; the compound tubercles of the carapace and abdomen are less developed and do not fully crowd the surface; and several short conical spines are present on the carapace surface (three gastric including median; two cardiac; two branchial). The dorsal carapace spines of the early juvenile are completely lost in adults; and the posterior branchial spines are lost or reduced to acute tubercles in some adult males, lost in adult females.

As in other species of *Paralomis*, sexual dimorphism is evident in the more strongly inflated chelipeds and proportionally longer walking legs in adult males. Additionally, the fingers of the minor cheliped of the males above 110 mm pcl exhibit a distinct gape, rather than occluding for the full length as in smaller specimens.

Paralomis dawsoni resembles those species with a broadly hexagonal carapace outline with spines along the hepatic and at least the anterior branchial margin, and in which the carapace and abdomen are densely covered with blunt, rounded, compound tubercles, such as *P. alcockiana* Hall & Thatje, 2009b [South Carolina, western Atlantic], *P. cubensis* Chace, 1939 [Caribbean Sea], and *P. seagranti* Eldredge, 1976 [Guam, western Pacific]. *Paralomis dawsoni* is readily distinguished from *P. alcockiana* by the absence of the prominently enlarged median gastric tubercle on the carapace, and from *P. cubensis* by having much more prominent carapace tubercles with margins well-demarcated from the adjacent carapace surface. In *P. cubensis*, the carapace tubercles are relatively low with ill-defined margins. Of the aforementioned species, *P. dawsoni* most closely resembles *P. seagranti* (Fig. 72B–E) from Guam and Kiribati in bearing spines along the hepatic and anterior branchial margins, with margins unarmed posterior to the level of the gastro-cardiac groove. *Paralomis dawsoni* differs from *P. seagranti* in more pronounced

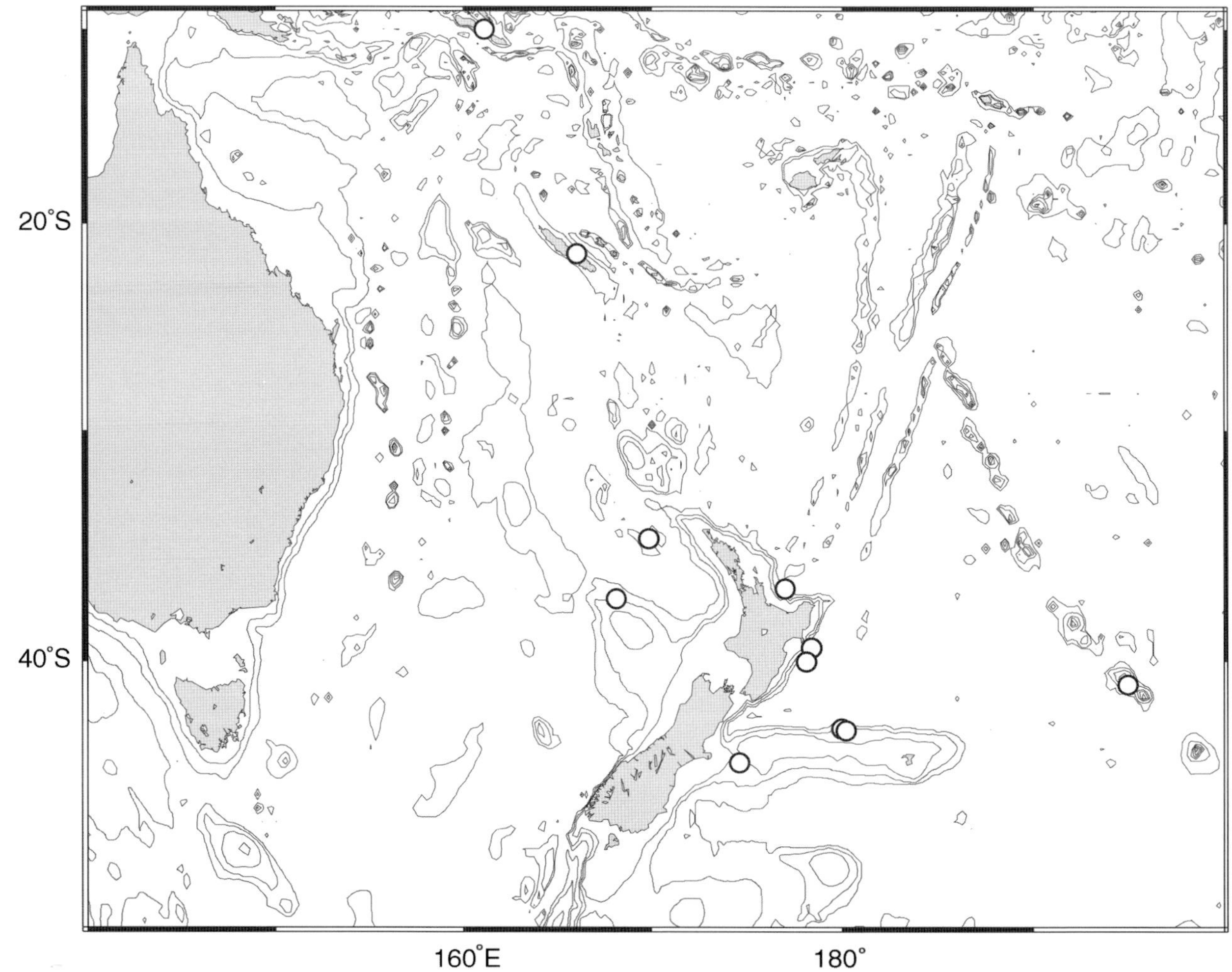

Figure 73. Geographic distribution of *Paralomis dawsoni* Macpherson, 2001.

and nodular carapace tubercles (rather than relatively flat), a lower, less strongly inflated gastric region, in the concave rather than evenly convex midlateral branchial margins, and in lacking the long, dense, golden setae covering the marginal carapace spines. The two species also differ in the degree of setation of the pereopods: sparse in *P. dawsoni* and dense in *P. seagranti*, giving the latter a distinctly hirsute appearance.

Paralomis dawsoni and *P. zealandica* are the two most common species of the genus in New Zealand. Their ranges are largely discrete, with *P. dawsoni* a northern form, and *P. zealandica* a southern form, with their ranges overlapping at the Subtropical Convergence in the vicinity of the Chatham Rise.

The ovigerous female of *P. dawsoni* from Rock Garden cold seep (NIWA 29295) is the first lithodid to be recorded from a New Zealand cold seep.

Distribution. New Caledonia, the Solomon Islands and northern New Zealand; 400–1118 m. The New Zealand range includes the West Norfolk Ridge and Challenger Plateau in the west, and in the east, the Bay of Plenty, Ritchie Bank, and the Louisville Ridge, south to the Chatham Rise.

Paralomis echidna sp. nov. (Figs 74–81)

Paralomis cf. *histrix*. — Poore, 2004: 269, fig. 76e.

Type material. *Holotype*: NMNZ Cr9285, ovigerous female (cl 41.9 mm, pcl 33.6 mm, cw 29.6 mm), Gascoyne Seamount, Tasman Sea, 36°30′S, 156°30′E, about 650 m, in baited pots, coll. J. Cave, Feb 1992.

Paratype: NMV J15122, ovigerous female (cl + 47.4 mm, pcl 38.8 mm, cw 33.5 mm), 75 km SE of Point Hicks, Victoria, 38°11.50′S, 149°56.50′E, 636 m, Engels high-lift demersal trawl, RV *Soela*, S06/84, 28 Nov 1984.

Other material examined. *Norfolk Ridge*: NMNZ Cr16862, 1 ovigerous female (pcl 32.1 mm, cw 28.5 mm), Norfolk Ridge, S of Wanganella Bank, 33°41.0′S, 167°52.6′E, 817 m, trawl, FV *Seamount Explorer*, coll. J. Houston, 13 Mar 2006.

No specific locality: NMNZ Cr11636, 1 male (cl 61.7 mm, pcl 49.7 mm, cw 46.9 mm), no data, coll. G. McInninnie.

Diagnosis. Carapace pyriform, slightly longer than wide; surface and margins densely and uniformly covered with slender spines of similar length; outer distance between bases of anterolateral spines slightly less than half carapace width. Rostrum ventral lobe rounded, unarmed. Ocular peduncle dorsally spinulate, spinules overreaching cornea. Basal antennal article with slender anterolateral spine reaching to base of scaphocerite. Scaphocerite a long, slender spine overreaching cornea spines; with lateral and dorsal spines. Abdominal somites densely covered with slender spines, becoming shorter on successive somites. Chelipeds short, stout, with outer and upper surfaces densely spinose; mesial surface of palm with low, scattered, setose tubercles, otherwise smooth; dactyli proximally spinose. Walking legs similar, short, stout, not exceeding 1.8 pcl (male), 1.4 pcl (female); segments densely spinose except for ventral surface of meri and coxae; pereopod 4 merus length less than 3.0 times height (male), 2.5 (female); propodus length about 4.3 times height (male), 3.7 (female); dactylus shorter than carpus, surface spinose for proximal half. Somite 6 length 1.08–1.15 times width.

Description. *Carapace*: Pyriform, 1.06–1.13 times longer than wide; regions distinct; surface and margins densely and uniformly covered with slender spines of similar length; surface of spines with scattered, minute, well-spaced simple setae; longest spine (on anterior branchial margin) 0.27 pcl; cardiac region with 27 (male) or 18–22 (female) spines; cervical groove distinct. Pterygostomian region with slender anterior spine; posterior two-thirds with field of spinules; surface of anterior third smooth except for scattered granules near dorsal margin and with cluster of 6 or 7 blunt spinules anteriorly.

Rostrum 0.24–0.25 pcl; broad basally, not constricted proximal to dorsal spines; median spine slender, ventral lobe triangular with blunt apex, unarmed; dorsally with 1 or 2 median spines and pair of laterally divergent spines directed obliquely upwards. Posterior orbital margin concave, unarmed; outer orbital spine slender, directed anterolaterally, reaching midlength of cornea. Anterolateral spine as long as outer orbital spine; outer distance between bases of anterolateral spines slightly less than half carapace width.

Ocular peduncle: Longer than cornea; with scattered granules and 9–12 dorsal spines, anterior 3 or 4 the longest, anteriorly directed, arranged in arcuate row above cornea; longest spine overreaching cornea by more than length of cornea.

Antennule: Peduncle unarmed, reaching anteriorly beyond antennal peduncle by length of distal antennular peduncle article.

Antenna: Basal antennal article with slender, curved, anterolateral spine, not reaching base of scaphocerite. Article 2 with inner distal spine, sometimes bearing 1 or 2 subdistal granules; outer margin with 3–5 spines, distalmost longest, reaching midlength of ultimate peduncular article. Article 3 unarmed. Scaphocerite a long, slender spine overreaching cornea spines; lateral margins with 2 or 3 long spines; dorsally with 1–4 small spines; inner margin unarmed. Article 4 unarmed, almost half length of article 5.

Figure 74. *Paralomis echidna* sp. nov., ovigerous female holotype, cl 41.9 mm, pcl 33.6 mm, cw 29.6 mm, Gascoyne Seamount (NMNZ Cr9285). A, dorsal habitus. B, ventral habitus.

Figure 75. *Paralomis echidna* sp. nov., ovigerous female holotype, cl 41.9 mm, pcl 33.6 mm, cw 29.6 mm, Gascoyne Seamount (NMNZ Cr9285). A, right chela, outer surface. B, right chela, inner surface. C, left chela, outer surface. D, carapace, right lateral view. E, right pereopod 4, dorsal view. F, right pereopod 4, ventral view.

Abdomen: Ornamentation similar in both sexes. Somites densely covered with slender spines, becoming shorter on successive somites. Spines on somite 2 similar to dorsal carapace spines. Spines of somites 3–6 less than half-length of dorsal carapace spines. Somite 6 length 1.08–1.15 times width. Telson subtriangular with rounded apex, with 5–7 blunt, apically setose tubercles.

Figure 76. *Paralomis echidna* sp. nov., ovigerous female holotype, cl 41.9 mm, pcl 33.6 mm, cw 29.6 mm, Gascoyne Seamount (NMNZ Cr9285). A, anterior carapace, dorsal view. B, anterior carapace, right lateral view. C, carapace, anterior view. D, posterior carapace and abdomen. E, abdomen, posterior somites and telson. F, abdomen.

Pereopod 1 (chelipeds): Densely spinose, unequal. Major cheliped 1.65 times height of minor cheliped (male), 1.48–1.56 (female). Coxae unarmed, setose; distal margins with dense tufts of setae. Ischiobasis spinose laterally, ventrally with clusters of golden setae. Merus with smooth mesial surface, other surfaces spinose. Carpal surfaces densely spinose except for mesial surface with clusters of golden setae. Palm

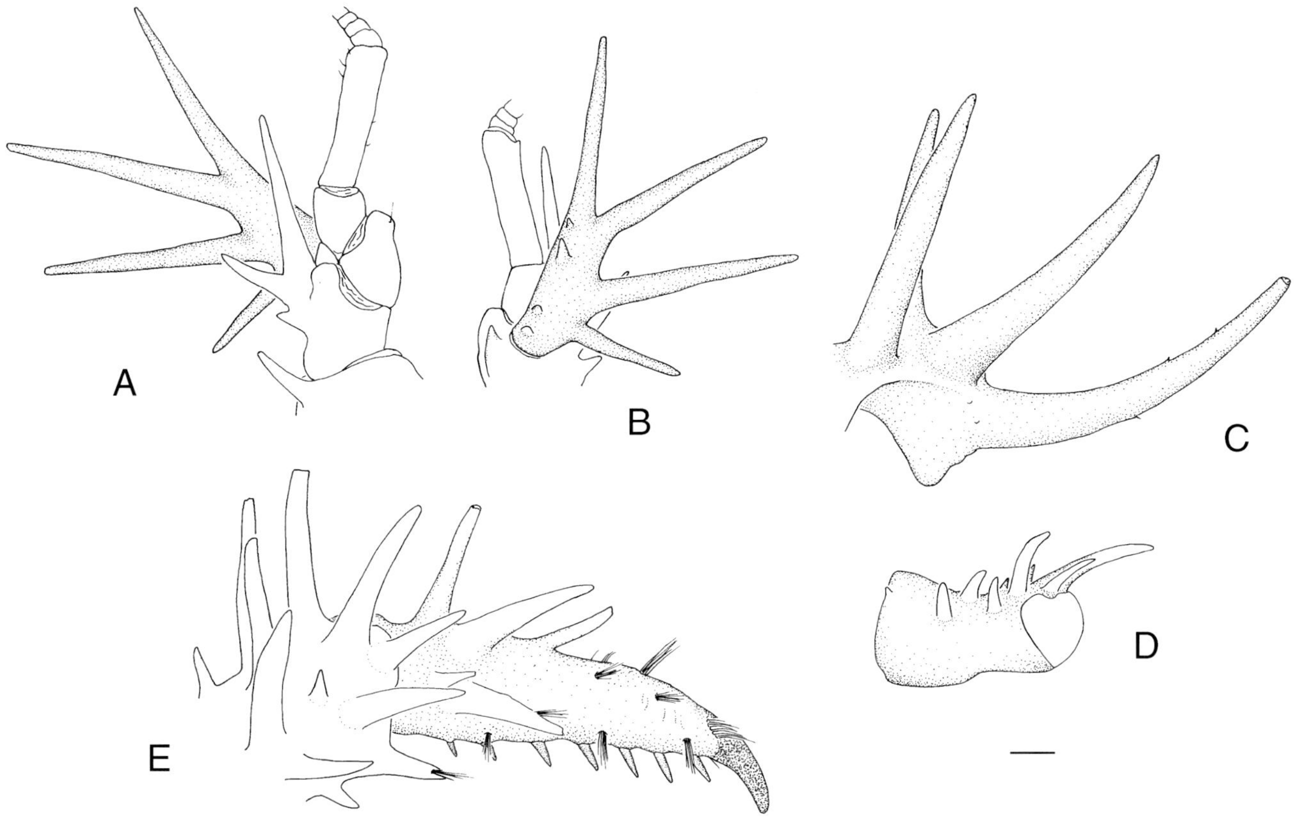

Figure 77. *Paralomis echidna* sp. nov., ovigerous female holotype, cl 41.9 mm, pcl 33.6 mm, cw 29.6 mm, Gascoyne Seamount (NMNZ Cr9285). A, right antenna, ventral view. B, right antenna, dorsal view. C, rostrum, right lateral view. D, right eye, lateral view. E, right pereopod 4 dactylus. Scale = 1.0 mm.

surfaces densely spinose except for mesial surface with low scattered tubercles bearing tufts of golden setae. Fingers with tufts of golden setae; dactyli with proximal cluster of 2–4 spines.

Major cheliped 1.73 pcl (male), 1.21–1.35 (female); upper palm length 1.02 times height (male), 0.94–1.08 (female); occlusal margins corneous for distal quarter, proximally with 4 low calcareous prominences; dactylus dorsal margin broadly convex, with rows of tufts of golden setae and proximal cluster of 2–4 spines, 1.07 times longer than dorsal margin of palm (male), 1.30–1.34 (female).

Minor cheliped 1.64 pcl (male), 1.21–1.26 (female); upper palm length 1.06 times height (male), 1.13–1.18 (female); occlusal margins corneous for distal third to half, proximally crenulate; dactylus dorsal margin broadly convex, with rows of tufts of golden setae and proximal cluster of 2–4 spines, 1.53 times longer than dorsal margin of palm (male), 1.58–1.67 (female).

Pereopods 2–4 (walking legs 1–3): Similar, short, stout, densely spinose; pereopod 2 longest. Coxae, unarmed, surface setose, distal margins crenulate. Ischiobasis spinose. Merus ovate in cross section, extensor and flexor surfaces spinose, longest extensor spines at about distal third; ventral surface smooth, unarmed, at most with row of 2–5 small setose tubercles near flexor margin. Carpus spinose, longest spines as high as carpus. Propodus dorsoventrally flattened; all surfaces spinose, longest spines twice propodal height. Dactylus curved; shorter than carpus; surface spinose for proximal half; distal half with tufts of golden setae; flexor margin lined with 7–9 obliquely inclined corneous spinules; apex corneous.

Pereopod 2 length 1.79 pcl (male), 1.33–1.38 pcl (female). Merus 0.58 pcl (male), 0.41–0.43 pcl (female); length:height ratio 2.88 (male), 2.45–2.51 (female). Carpus 0.71 merus length (male), 0.76–0.78 (female). Propodus 0.85 merus length (male), 0.86–0.91 (female); length:height ratio 4.60 (male), 3.84–3.96 (female). Dactylus 0.74 propodus length (male), 0.75–0.83 (female).

Pereopod 3 length 1.74 pcl (male), 1.30–1.34 pcl (female). Merus 0.58 pcl (male), 0.39–0.40 pcl (female); length:height ratio 2.95 (male), 2.36–2.39 (female). Carpus 0.72 merus length (male), 0.78–0.84 (female). Propodus 0.85 merus length (male), 0.91–0.98 (female); length:height ratio 4.64 (male), 3.52–3.83 (female). Dactylus 0.72 propodus length (male), 0.74–0.84 (female).

Pereopod 4 length 1.56 pcl (male), 1.20–1.28 pcl (female). Merus 0.48 pcl (male), 0.35–0.37 pcl (female); length:height ratio 2.91 (male), 2.43–2.52 (female).

Figure 78. *Paralomis echidna* sp. nov., male, cl 61.7 mm, pcl 49.7 mm, cw 46.9 mm, no locality (NMNZ Cr11636). A, dorsal habitus. B, ventral habitus.

Figure 79. *Paralomis echidna* sp. nov., male, cl 61.7 mm, pcl 49.7 mm, cw 46.9 mm, no locality (NMNZ Cr11636). A, right chela. B, left chela. C, anterior carapace, left lateral view. D, left pereopod 4. E, carapace, right lateral view. F, posterior carapace and abdominal somite 2.

Carpus 0.80 merus length (male), 0.84–0.88 (female). Propodus 0.91 merus length (male), 1.01–1.07 (female); length:height ratio 4.34 (male), 3.68–3.73 (female). Dactylus 0.74 propodus length (male), 0.74–0.81 (female).

Egg diameter: 2.0–2.3 mm at 'black eye-spot' stage.

Colour in life. Not known.

Figure 80. *Paralomis echidna* sp. nov., ovigerous female, pcl 32.1 mm, cw 28.5 mm, Norfolk Ridge (NMNZ Cr16862). A, dorsal habitus. B, ventral habitus.

ETYMOLOGY. Name *echidna*, the common name of the spiny Australian and Papua New Guinean monotremes; used as a noun in apposition.

REMARKS. *Paralomis echidna* sp. nov. most closely resembles *P. histrix* (de Haan, 1844) from Japan, and *P. webberi* sp. nov. from the Kermadec Ridge, which together are readily distinguished from other congeners by the densely spinose carapace, abdomen, chelipeds and walking legs, and the relatively short walking legs in which the meri do not exceed 0.6 pcl in males or 0.5 pcl in females. *Paralomis echidna* differs from *P. histrix* in the pattern of carapace and abdominal spination in which most of the adjacent spines are of approximately equal length, and in having a comparatively wider frontal region, with the distance between the bases of the anterolateral spines only slightly rather than distinctly less than half carapace width. In contrast, the carapace and abdominal ornamentation of *P. histrix* is composed of numerous long spines interspersed with numerous short, apically rounded, anteriorly recurved spines of approximately half the length of the longer spines (Fig. 81; see also Ikeda 1998: pl. 73). Also, the carapace spines of *P. echidna* (and *P. webberi*) are minutely and sparsely setose rather than glabrous as in *P. histrix*. Features distinguishing *P. echidna* from *P. webberi* are outlined under the account of the latter. Differences between the male and female *P. echidna* are chiefly sexually dimorphic and allometric: the walking legs are more elongate and the right cheliped is stouter in the male. The male further differs from the females in bearing 30 instead of 18–22 cardiac spines. *Paralomis echidna* and *P. webberi* also appear to mature at a smaller size than *P. histrix*. Both *P. echidna* and *P. webberi* are ovigerous at less than 35 mm pcl, whereas a Japanese specimen of *P. histrix* examined here (ZLKU 13247; Fig. 82) appears to have just reached maturity at 68.6 mm pcl.

Unfortunately, the specific provenance of the only known male of *P. echidna*, collected by a New Zealand commercial fisherman based in Tauranga, is not known (R. Webber, pers. comm.), though it was probably collected from northern New Zealand waters. The female from the Norfolk Ridge and the male differ from southern Tasman Sea specimens of *P. echidna* in bearing a single spine in the midline of the rostrum, rather than two, and in having proportionally shorter dorsal carapace spines. The shorter carapace spines in the single known male might be attributable to its larger body size, but the Norfolk Ridge female is of similar size to the southern Tasman Sea specimens. The Norfolk Ridge female also differs from southern Tasman Sea specimens in having 22 instead of 18 or 19 cardiac spines. The aforementioned differences are presently regarded as intraspecific variation, although further study might reveal these differences to be significant. For this reason, the type series is restricted to the southern Tasman Sea specimens.

Poore (2004) reported *P. echidna* as *P.* cf. *histrix*. The only species presently known from Australia that *P.*

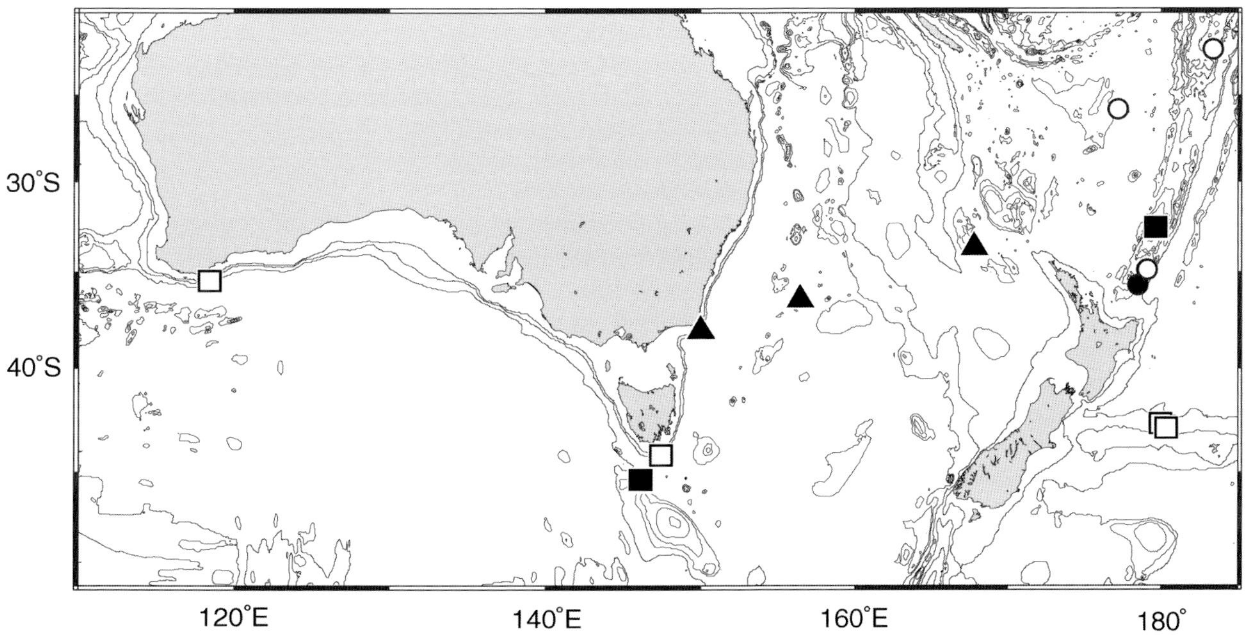

Figure 81. Geographic distributions of *Paralomis echidna* sp. nov. (▲), *P. hirtella* de Saint Laurent & Macpherson, 1997 (○), *P. poorei* sp. nov. (□), *P. staplesi* sp. nov. (■), and *P. webberi* sp. nov. (●).

Figure 82. *Paralomis histrix* (de Haan, 1844), female, cl 85.0 mm, pcl 68.6 mm, cw 66.7 mm, Kochi Prefecture, Japan (ZLKU 13247). A, carapace. B, anterior carapace, dorsal view. C, dorsal habitus. E, right posterior carapace surface.

echidna is likely to be confused with is *P. poorei* (also known from New Zealand), a species also densely covered with spines. *Paralomis echidna* can be immediately distinguished from *P. poorei* by its spinose rather than tuberculate abdominal somites 4–6.

Distribution. Presently known only from the Tasman Sea between Victoria, Australia, the Gascoyne Seamount, and the southern Norfolk Ridge; 636–817 m.

Paralomis gowlettholmes sp. nov.

(Figs 83–87, Pl. 3C–D)

Paralomis sp. MoV2716. — Poore *et al.*, 1998: 66, 72.
Paralomis cf. *birsteini*. — Poore, 2004: 269, fig. 76b.

TYPE MATERIAL. (All Tasmania). *Holotype*: SAM C6859, male (pcl 50.1 mm, cw 49.1 mm), ovigerous female (cl 52.5 mm, pcl 45.42 mm, cw 43.64 mm), approx. 425 nautical miles [787 km] SE of South East Cape, 44°09.2′S, 147°31.8′E, 970–1120 m, trawled, FV *Belinda*, coll. K. Gowlett-Holmes, 1 Feb 1992.

Paratypes: SAM C6860, ovigerous female (cl 52.5 mm, pcl 45.4 mm, cw 43.6 mm), approx. 425 nautical miles [787 km] SE of South East Cape, 44°09.2′S, 147°31.8′E, 970–1120 m, trawled, FV *Belinda*, coll. K. Gowlett-Holmes, 1 Feb 1992; NMV J44018, 2 ovigerous females (cl 34.1–45.3 mm, pcl 28.78–39.0 mm, cw 28.13–37.42 mm), J1 seamount, 82.6 km SSE of South East Cape, 44°14.4′S, 147°21.6′E, 1200 m, RV *Southern Surveyor*, SS01/97 stn 40, epibenthic sled, coll. T.N. Stranks *et al.*, 27 Jan 1997; SAM C6861, 1 male (cl 44.9 mm, pcl 40.7 mm, cw 37.7 mm), 1 female (cl 41.5 mm, pcl 34.5 mm, cw 33.6 mm), approximately 46 nautical miles [85 km] SE of South East Cape, 44°14.8′S, 147°27.5′E, 1081–1130 m, trawled, coll. K. Gowlett-Holmes, 12 Feb 1992; SAM C6862, 1 female (cl 46.3 mm, pcl 39.9 mm, cw 38.0 mm), approximately 44 nautical miles [81 km] SSE of South East Cape, 44°15.2′S, 147°21.2′E, 1050–1120 m, trawled, FV *Belinda*, coll. K. Gowlett-Holmes, 12 Feb 1992; NMV J44013, 1 juvenile female (cl 10.2 mm, pcl 8.5 mm, cw 7.8 mm), 1 crushed specimen (cl approx. 19 mm), Sister 1 seamount, 82.9 km SSE of South East Cape, 44°16.2′S, 147°17.4′E, 1100 m, RV *Southern Surveyor*, SS01/97 stn 15, epibenthic sled, coll. T.N. Stranks *et al.*, 23 Jan 1997; TM G3583, 1 female (cl 28.9 mm, pcl 23.2 mm, cw 22.1 mm), Hill K1, approx. 88 km SSE of Southeast Cape, 44°17′S, 147°25′E, 1600 m, epibenthic sled, RV *Southern Surveyor*, SS0197/17, 23 Jan 1997; NMV J44016, 1 female (cl 33.5 mm, pcl 28.3 mm, cw 26.8 mm), Dory Hill seamount, 83.2 km SSE of South East Cape, 44°19.2′S, 147°07.2′E, 1280 m, RV *Southern Surveyor*, SS01/97 stn 49, epibenthic sled, coll. T.N. Stranks *et al.*, 29 Jan 1997; NMV J61054, 1 juvenile female (cl 19.5 mm, pcl 15.9 mm, cw 14.5 mm), 44°19.78′S, 146°53.23′E, 1305 m, FV *Tommy Thompson*, ROV *Jason 2*, TN228-J2-382-016-004, 17 Dec 2008; TM G3607, 1 ovigerous female (cl 35.8 mm, pcl 30.5 mm, cw 28.6 mm), hill U, approximately 85 km SSE of South East Cape, 44°20′S, 147°10′E, 1083 m, RV *Southern Surveyor*, SS0197/34, CSIRO, 27 Jan 1997; NMV J44019, 1 ovigerous female (cl 48.7 mm, pcl 41.4 mm, cw 40.27 mm), U seamount, 86.4 km SSE of South East Cape, 44°20.4′S, 147°09.6′E, 1250 m, RV *Southern Surveyor*, SS01/97 stn 35, epibenthic sled, coll. T.N. Stranks *et al.*, 27 Jan 1997; SAM C6863, 1 male (cl 49.0 mm, pcl 38.4 mm, cw 37.9 mm), 1 ovigerous female (cl 39.2 mm, pcl 34.5 mm, cw 33.3 mm), approximately 46.5 nautical miles [86 km] SSE of South East Cape, 44°22.7′S, 147°07.3′E, 1050–1170 m, trawled, coll. K. Gowlett-Holmes, 12 Feb 1992; NMV J44020, 1 juvenile male (cl 11.5 mm, pcl 9.91 mm, cw 8.21 mm), D1 seamount, 97.4 km SSE of South East Cape, 44°24.0′S, 147°19.2′E, 1900 m, RV *Southern Surveyor*, SS01/97 stn 22, epibenthic sled, coll. T.N. Stranks *et al.*, 25 Jan 1997.

DIAGNOSIS. Carapace subpentagonal, slightly wider than long; surface glabrous, sparsely covered with small, scattered granules and short conical spines, including median gastric spine; lateral margins with short conical spines; outer distance between bases of anterolateral spines slightly exceeding half carapace width. Rostrum trispinose, broadest basally, without basal constriction. Scaphocerite with 1 or 2 short inner basal spines and 1 or 2 outer spines. Male chelipeds dimorphic, spinose; major cheliped palm height 1.7–1.8 times that of minor cheliped. Major cheliped palm of female 1.4–1.5 times height of minor cheliped. Walking legs elongate, spinose, length 2.0–2.6 pcl; ventral spines of propodus widely spaced; dactyli longer than extensor margin of propodi. Walking leg 3 merus 0.7–0.8 pcl (male), 0.5–0.6 pcl (female), less than 5 times longer than high; propodus length 3.7–5.6 times height.

DESCRIPTION. *Carapace*: Subpentagonal, 1.01–1.09 times longer than wider; surface glabrous, sparsely covered with small, scattered granules. Gastric region convex, elevated, more prominent than other regions, with 5 short conical spines forming pentagon with anteriormost in midline. Lateral margin of hepatic region with 3 short conical spines, anterior shorter. Branchial regions with 8 or 9 short marginal spines; surface with 4 low conical spines in addition to low, scattered tubercles. Cardiac region subtriangular, with 2 pairs of short conical spines forming square. Intestinal region sparsely granular. Pterygostomian region sparsely granular, with prominent anterior spine.

Rostrum 0.1–0.2 pcl; broadest basally, without constriction; median spine smooth ventrally, occasionally with small proximal dorsal spine or tubercle; paired dorsal spines divergent, directed obliquely upwards. Posterior orbital margin near concave; outer orbital spine reaching to base of cornea. Anterolateral spine as long as outer orbital spine, often with low marginal granule midway between outer orbital spine and anterolateral spine; outer distance between bases of anterolateral spines slightly exceeding half carapace width.

Ocular peduncle: Longer than cornea; with 2–5 dorsal granules.

Antennule: Peduncle unarmed, reaching anteriorly beyond antennal peduncle by about three-quarters length of distal antennular peduncle article.

Figure 83. *Paralomis gowlettholmes* sp. nov., male holotype, pcl 50.1 mm, cw 49.1 mm, SE of South East Cape, Tasmania (SAM). A, dorsal habitus. B, carapace. E, right pereopod 4.

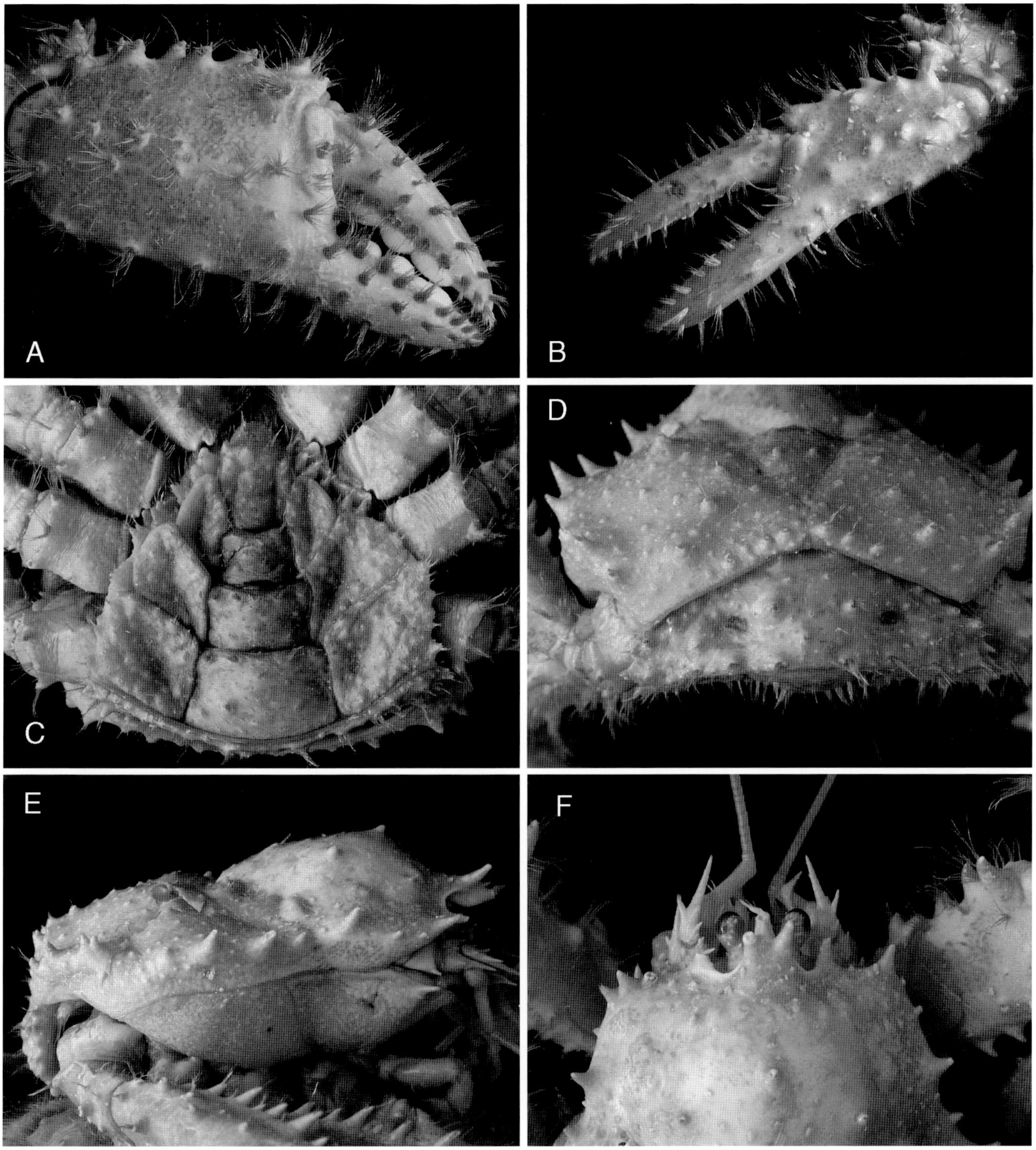

Figure 84. *Paralomis gowlettholmes* sp. nov., male holotype, pcl 50.1 mm, cw 49.1 mm, SE of South East Cape, Tasmania (SAM). A, right chela. B, left chela. C, ventral surface and abdomen. D, posterior carapace and abdominal somite 2. E, carapace, right lateral view. F, anterior carapace, dorsal view.

Antenna: Basal antennal article with small anterolateral spine. Article 2 with angular to sharp inner distal margin; outer margin with small basal spine and slender spine not reaching beyond article 4. Article 3 unarmed. Scaphocerite slender, reaching beyond midlength of article 5, with 1 or 2 short inner spines and 1–3 outer spines. Article 4 unarmed, about half length of article 5.

Abdomen: Ornamentation similar in both sexes. Somites sparsely granular. Somite 2 with small, widely separated spines along margins. Somite 3 median plate unarmed; submedian plates with low, irregular, trian-

Figure 85. *Paralomis gowlettholmes* sp. nov., female paratype, cl 33.5 mm, pcl 28.3 mm, cw 26.8 mm, Dory Hill seamount, Tasmania (NMV J44016). A, dorsal habitus. B, ventral surface and abdomen.

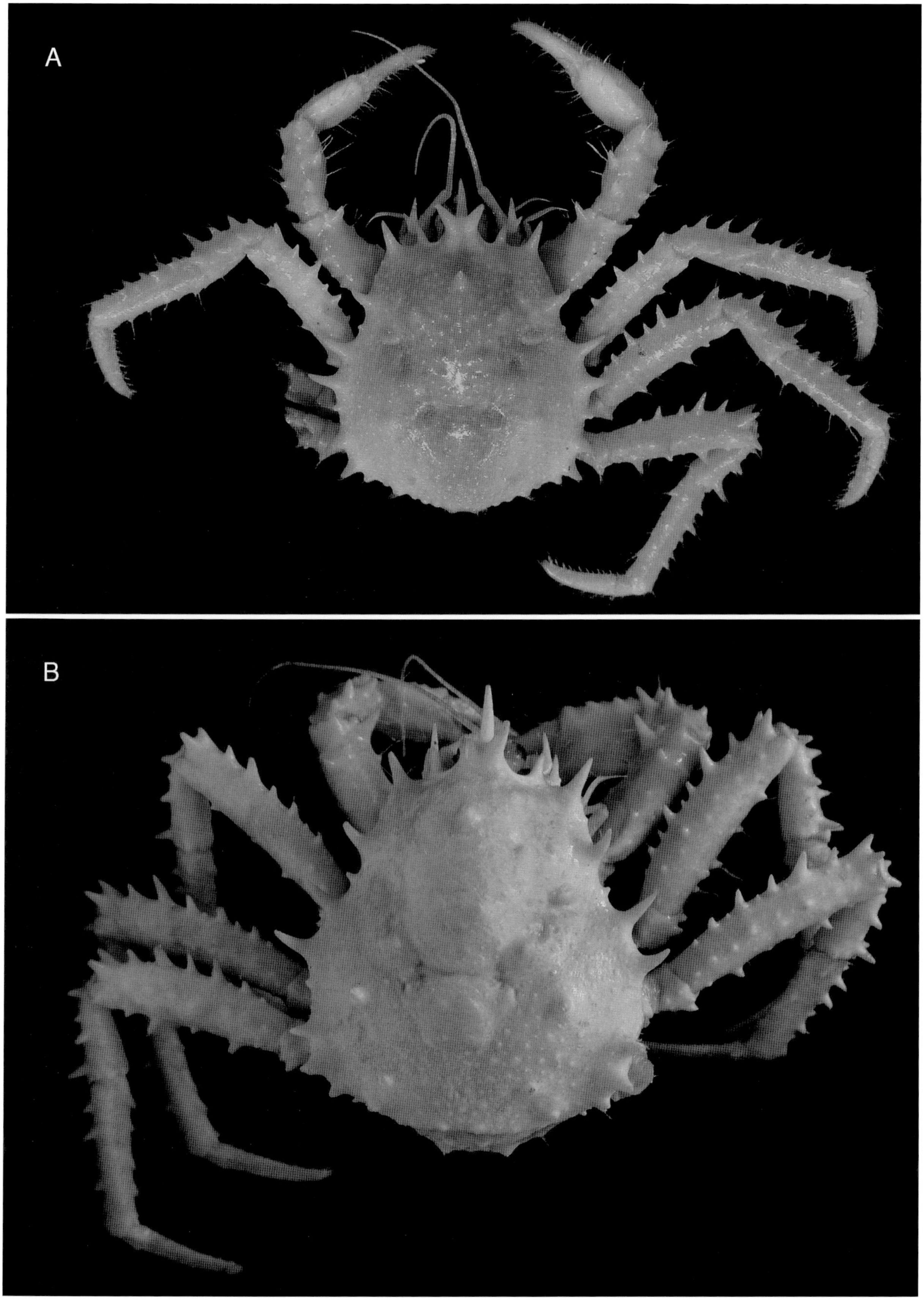

Figure 86. *Paralomis gowlettholmes* sp. nov. A, juvenile female paratype (cl 19.5 mm, pcl 15.9 mm, cw 14.5 mm (NMV J61054). B, juvenile female paratype (pcl 23.2, cw 22.1 mm), Hill K1 (TM G3583).

gular teeth on lateral margin; marginal plates absent, apparently undifferentiated. Somites 4–5 with unarmed median and submedian plates; marginal plates undivided, with irregular triangular teeth. Somite 6 slightly longer than wide, subquadrate, with pair of small distal teeth; marginal plates short, triangular. Telson wider than long, apex rounded.

Pereopod 1 (chelipeds): Spination similar in both sexes. Major cheliped of males strongly inflated, 1.7–1.8 times height of minor cheliped; minor cheliped slender. Chelipeds of female unequal, major cheliped 1.4–1.5 times height of minor cheliped. Coxae smooth, unarmed; distal margins with dense tufts of setae. Ischiobasis with 4–8 stout, apically setose ventral spines. Merus with smooth mesial margins and tuberculate lateral margins; dorsal and ventral margins spinose, inner distal spine largest. Carpus with prominent spines on dorsal and lateral surfaces; dorsal margin with row of 3–5 spines; mesial margin with 2 spines, proximal largest; ventral surface with 3 spines. Palm mesial margin with 6 spines, other surfaces with prominent, apically setose tubercles or acute tubercles. Fingers with short basal tubercle and rows of tufts of golden setae.

Major cheliped 1.77–1.96 pcl (male), 1.30–1.55 (female); upper palm length 1.11–1.14 times height (male), 1.23–1.40 (female); occlusal margins of fingers corneous for distal third, proximally with 3 calcareous nodules; dactylus dorsal margin broadly convex, with rows of tufts of golden setae and blunt proximal tooth, 1.09–1.17 times longer than dorsal margin of palm (male), 1.15–1.25 (female).

Minor cheliped 1.47–1.75 pcl (male), 1.16–1.59 (female); upper palm length 1.22–1.30 times height (male), 1.38–1.54 (female); occlusal margin corneous for slightly less than distal third, proximally crenulate; dactylus dorsal margin broadly convex, with rows of tufts of golden setae and blunt proximal tooth, 1.49–1.62 times longer than dorsal margin of palm (male), 1.42–1.68 (female).

Pereopods 2–4 (walking legs 1–3): Similar, elongate, spinose. Coxae smooth, unarmed; distal margins with dense tufts of setae. Ischiobasis with 3–5 apically setose ventral spines. Merus compressed, shorter than carapace in both sexes; extensor margin with 8–11 spines in addition to paired distal spines; dorsal surface with 4–8 spines; flexor margin with rows of 5–7 spines; merus of pereopod 3 slightly longer than that of pereopod 2 and pereopod 4. Carpus slightly longer than half merus length, subcircular in cross section; extensor margin with 5 or 6 spines; dorsal surface with 4–6 spines; flexor margin unarmed. Propodus dorsoventrally flattened; slightly shorter than merus; with 9–14 spines on extensor margin; dorsal surface with small scattered spines; flexor margin with 5–8 spines. Dactylus broadly curved; slightly longer than extensor margin of propodus; surface with tufts of setae; extensor margin with 3 or 4 apically corneous spines proximally; lateral proximal surfaces with short, distinct sulcus, flanked ventrally by 1 or 2 small, corneous spines; flexor margin lined with corneous spinules

Pereopod 2 length 2.12–2.62 pcl (male), 1.78–1.96 pcl (female). Merus 0.70–0.82 pcl (male), 0.58–0.63 pcl (female); length:height ratio 3.84–4.48 (male), 3.42–4.47 (female). Carpus 0.55–0.59 merus length (male), 0.61–0.66 (female). Propodus 0.80–0.84 merus length (male), 0.86–0.93 (female); length:height ratio 4.29–5.62 (male), 3.98–5.81 (female). Dactylus 0.94–1.00 propodus length (male), 0.85–0.99 (female).

Pereopod 3 length 2.07–2.64 pcl (male), 1.87–1.99 pcl (female). Merus 0.72–0.81 pcl (male), 0.59–0.65 pcl (female); length:height ratio 3.96–4.74 (male), 3.41–4.50 (female). Carpus 0.54–0.59 merus length (male), 0.57–0.65 (female). Propodus 0.82–0.89 merus length (male), 0.89–0.99 (female); length:height ratio 4.86–5.87 (male), 4.28–6.06 (female). Dactylus 0.92–0.94 propodus length (male), 0.83–0.96 (female).

Pereopod 4 length 2.02–2.38 pcl (male), 1.81–2.08 pcl (female). Merus 0.66–0.79 pcl (male), 0.49–0.60 pcl (female); length:height ratio 3.73–4.38 (male), 2.69–4.76 (female). Carpus 0.57–0.61 merus length (male), 0.63–0.68 (female). Propodus 0.89–0.91 merus length (male), 0.96–1.03 (female); length:height ratio 4.96–5.53 (male), 3.67–5.59 (female). Dactylus 0.93–0.95 propodus length (male), 0.81–0.96 (female).

Colour in life. Pinkish-red (Pl. 3C–D).

Etymology. Named for Karen Gowlett-Holmes, CSIRO Marine Research, who collected the holotype of the new species, in addition to many other specimens studied herein.

Remarks. *Paralomis gowlettholmes* sp. nov. closely resembles *P. spectabilis* Hansen, 1908 (type locality: off Iceland), *P. birsteini* Macpherson, 1988a, and *P. stevensi* Ahyong & Dawson, 2006 (both with type locality: Ross Sea) in carapace ornamentation in which the primary spines are low and simple with the surface between the primary spines smooth or sparsely granulate. The new species is readily distinguished from *P. spectabilis* and *P. stevensi* by the length of the dactyli of the walking legs—slightly longer, rather than shorter than the extensor margin of the propodi in *P. gowlettholmes*. *Paralomis gowlettholmes* further differs from *P. stevensi* in the having much shorter walking legs in males, and further differs from *P. spectabilis* in the armed inner margin of the scaphocerite.

Paralomis gowlettholmes is most similar to *P. birsteini* in the proportional lengths of the walking leg dactyli, but differs chiefly in the proportional size of the major chela and the less slender walking legs. As in *P. stevensi*, the right chela of male *P. gowlettholmes* is

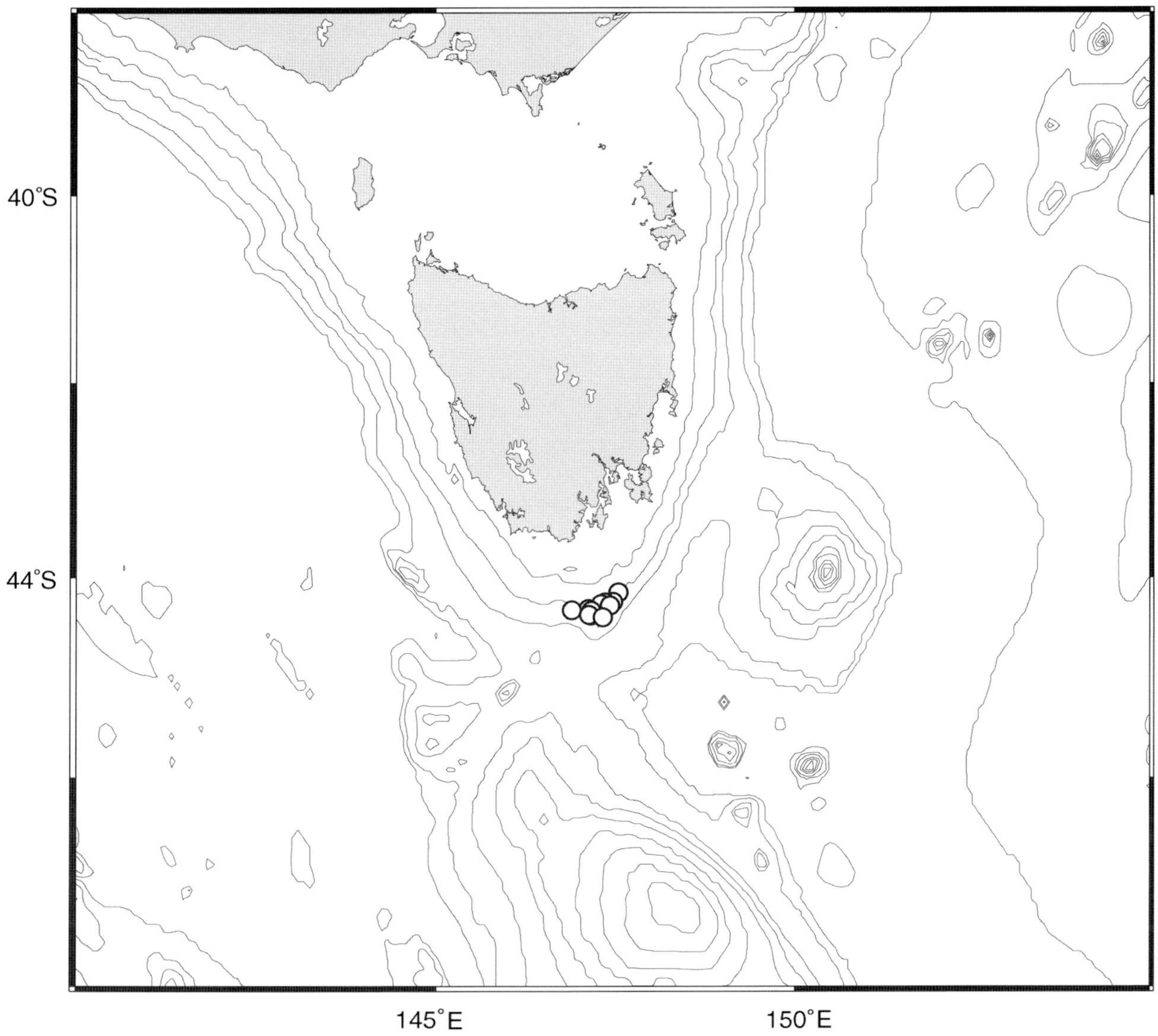

Figure 87. Geographic distribution of *Paralomis gowlettholmes* sp. nov.

considerably larger than the left, with the right palm height about 1.7–1.8 versus less than 1.6 times or less that of the left in *P. birsteini*. The right chela of females is also comparatively larger in *P. gowlettholmes*, being about 1.4–1.5 times higher than the left versus 1.1–1.2 in *P. birsteini*. The walking legs of *P. gowlettholmes* are noticeably stouter than those of *P. birsteini,* quantified in the length:height ratio of the propodi. In *P. gowlettholmes*, the length:height ratio of the walking leg 3 propodus does not exceed 5.6 but is greater than 6.2 in *P. birsteini* (see Table 3). The difference in the length: height ratio between *P. gowlettholmes* and *P. birsteini* is even more pronounced between females, <5.6 and >7.3 respectively.

Paralomis gowlettholmes appears to mature earlier and attain a smaller size than either *P. birsteini* or *P. stevensi*, both of which are known to exceed 80 mm pcl compared to 50.1 mm pcl for the largest specimen of the new species. The smallest ovigerous female of *P. gowlettholmes* is 28.3 mm pcl, less than half the size of the smallest ovigerous *P. birsteini* (pcl 61.1 mm) and *P. stevensi* (pcl 62.4 mm). Moreover, the abdomen of the smallest female *P. birsteini* examined (pcl 35.2 mm), although parasitised, has not acquired full asymmetry suggesting that females are immature at that size.

The similar morphology and Southern Ocean distribution of *Paralomis birsteini*, *P. gowlettholmes* and *P. stevensi* points to a close relationship between the three species.

Distribution. Presently known only from southern Tasmania; 970–1900 m.

Paralomis hirtella de Saint Laurent & Macpherson, 1997 (Figs 81, 88–91, Pl. 4B)

Paralomis hirtella de Saint Laurent & Macpherson, 1997: 722–726, figs 1–4 [type locality: Lau Basin, 22°13′S, 176°38′W, 1750 m]. — Macpherson & Segonzac *in* Desbruyères *et al.* 2006: 436. — Dawson, 2008: 4–8, figs 1–5.

Paralomis aff. *jamsteci.* — Webber & Naylor, 2004b: 63.

TYPE MATERIAL. *Holotype*: MNHN Pg4658, male (pcl 34.9 mm, cw 32.0 mm), Lau Basin, Valufa Rise, Hine Hina site, 22°13′S, 176°38′W, 1750 m, 14–18 May 1989.

OTHER MATERIAL EXAMINED. *New Zealand, Brothers Seamount:* NIWA 18518, 1 juvenile male (cl 10.2 mm, pcl 8.7 mm, cw 8.8 mm), 1 juvenile female (cl 16.2 mm, pcl 14.3 mm, cw 15.2 mm), Brothers Seamount, 34°51.64′S, 179°03.47′E, 1572 m, KOK0507/21, 14 May 2005; NIWA 34926, 1 male (cl 62.7 mm, pcl 58.4 mm, cw 61.0 mm), 34°51.66′S, 179°03.42′E, 1603 m, DPH0501/72, crab pot, CRB 11; NIWA 4078, 1 male (cl 36.9 mm, pcl 32.8 mm, cw 34.8 mm), 1 indeterminate juvenile (cl 9.7 mm, pcl 8.3 mm, cw 8.4 mm), 34°51.756′S, 179°03.476′E, 1649 m, DSV *Shinkai* 6500, 6K#852, RK6-B-71, 27 Oct 2004; NIWA 4079, 1 male (cl 54.5 mm, pcl 50.2 mm, cw 51.3 mm), 34°52.723′S, 179°04.304′E, 1336 m, 6K#854, RK6-B-71, 1 Nov 2004; NIWA 34925, 7 males (cl 40.7–57.6 mm, pcl 36.5–51.8 mm, cw 38.3–56.0 mm), 8 ovigerous females (cl 41.1–57.9 mm, pcl 36.8–52.6 mm, cw 39.6–57.0 mm), 34°52.9′S, 179°04.13′E, 1208 m, DPH0501/104; NIWA 34932, 1 male (cl 59.2 mm, pcl 53.4 mm, cw 55.1 mm), 1 ovigerous female (cl 37.9 mm, pcl 34.3 mm, cw 35.9 mm), 34°52.9′S, 179°04.13′E, 1208 m, DPH0501/104; NIWA 3918, 2 males (cl 18.1–29.3 mm, pcl 15.7–25.8 mm, cw 16.2–27.4 mm), 34°52.96–52.69′S, 179°04.02–04.93′E, 1197–1538 m, TAN0107/141, 22 May2001; NIWA 3914, 2 females, larger ovigerous (cl 30.5–38.8 mm, pcl 27.2–35.4 mm, cw 28.9–38.2 mm), 34°53.12–53.35′S, 179°04.49–05.93′E, 1197–1526 m, TAN0107/136, 21 May 2001.

Monowai Caldera: NIWA 18002, 1 male (cl 47.7 mm, pcl 42.8 mm, cw 45.3 mm; with rhizocephalan *Briarosaccus callosus*), 1 ovigerous female (cl 33.6 mm, pcl 30.7 mm, cw 32.0 mm), 25°48.24′S, 177°10.07′W, 1168 m, KOK0505/8, 8 Apr 2005; NIWA 18003, 1 juvenile male (cl 8.1 mm, pcl 6.7 mm, cw 6.5 mm), 25°48.33′S, 177°10.25′W, 1050 m, KOK0505/3, 7 Apr 2005; NIWA 18004, 1 male (cl 66.5 mm, pcl 61.1 mm, cw 66.8 mm; with rhizocephalan *Briarosaccus callosus*), 25°48.25′S, 177°10.11′E, 1143 m, KOK0505/14, 10 Apr 2005; NIWA 18009, 3 males (cl 16.6–41.6 mm, pcl 14.8–36.9 mm, cw 30.7–36.9 mm), 2 females (cl 33.7–53.2 mm, pcl 30.8–49.3 mm, cw 33.0–51.2 mm; largest with rhizocephalan *Briarosaccus callosus*), 25°48.29–48.59′S, 177°10.19–09.82′E, 1140–1054 m, TAN0411/06, 3 Oct 2004; NIWA 18010, 1 male (cl 62.9 mm, pcl 58.3 mm, cw 64.6 mm; with rhizocephalan *Briarosaccus callosus*), 25°48.29–48.59′S, 177°10.19–09.82′E, 1140–1054 m, TAN0411/06, 3 Oct 2004; NIWA 32233, 1 male (cl 50.8 mm, pcl 46.6 mm, cw 49.6 mm; with rhizocephalan *Briarosaccus callosus*), 2 juveniles (cl 7.7–7.9 mm), 25°48.33′S, 177°10.18′W, 1102 m, SO192-2/18, 3 May 2007; NIWA 32224, 4 males (cl 28.9–46.7 mm, pcl 22.3–42.7 mm, cw 23.9–45.8 mm), 5 females (cl 22.0–46.0 mm, pcl 19.9–41.0 mm, cw 20.7–42.7 mm), 25°48.33′S, 177°10.18′W, 1041 m, RV *Sonne*, SO192-2/20, 3 May 2007; NIWA 32223, 1 male (cl 34.5 mm, pcl 31.5 mm, cw 33.1 mm), 25°48.21′S, 177°10.08′W, 1197 m, RV *Sonne*, SO192-2/21, 3 May 2007.

DIAGNOSIS. Carapace subpentagonal, as long as or slightly wider than long; lateral margins with 12–14 short, slender, well-spaced anterolaterally directed spines; surface devoid of spines, smooth, unarmed, bearing numerous tufts of golden setae. Rostrum broad basally, without constriction, extending to or slightly beyond cornea; ventrally unarmed; paired dorsal spines short, directed anteriorly, occasionally reduced to low setose nodules. Scaphocerite with small inner and 1 or 2 outer spines. Abdominal somites unarmed, with scattered tufts of setae. Cheliped palms unarmed except for low teeth on dorsal margin. Walking legs unarmed except for row of blunt spines on meral extensor margins.

DESCRIPTION. *Carapace*: Subpentagonal, about as long as wide, length 0.90–1.04 width; regions distinct, gastric region more inflated than others; surface smooth, unarmed, bearing numerous tufts of golden setae. Anterior margin almost transverse, usually with small spine between outerorbital and anterolateral spine and occasionally with spine at inner base of anterolateral spine. Lateral margins with 12–14 short, slender, well-spaced anterolaterally directed spines. Posterior margin relatively straight, slightly convex to slightly concave. Pterygostomian region with small anterior spine; surface irregular, unarmed.

Rostrum short, 0.07–0.20 pcl; broad basally, not constricted proximal to dorsal spines; median spine horizontal or ventrally deflexed, overreaching outer-orbital spine, extending to or slightly beyond cornea; ventrally unarmed; paired dorsal spines short, directed anteriorly, occasionally reduced to low setose nodules. Posterior orbital margin concave, unarmed; outer orbital spine slender, directed anteriorly, not overreaching eyes. Anterolateral spine subequal to outer orbital spine; outer distance between bases of anterolateral spines slightly exceeding half carapace width.

Ocular peduncle: Longer than cornea; dorsally with tufts of golden setae, unarmed.

Antennule: Peduncle unarmed, reaching anteriorly beyond apex of antennal peduncle by half length of

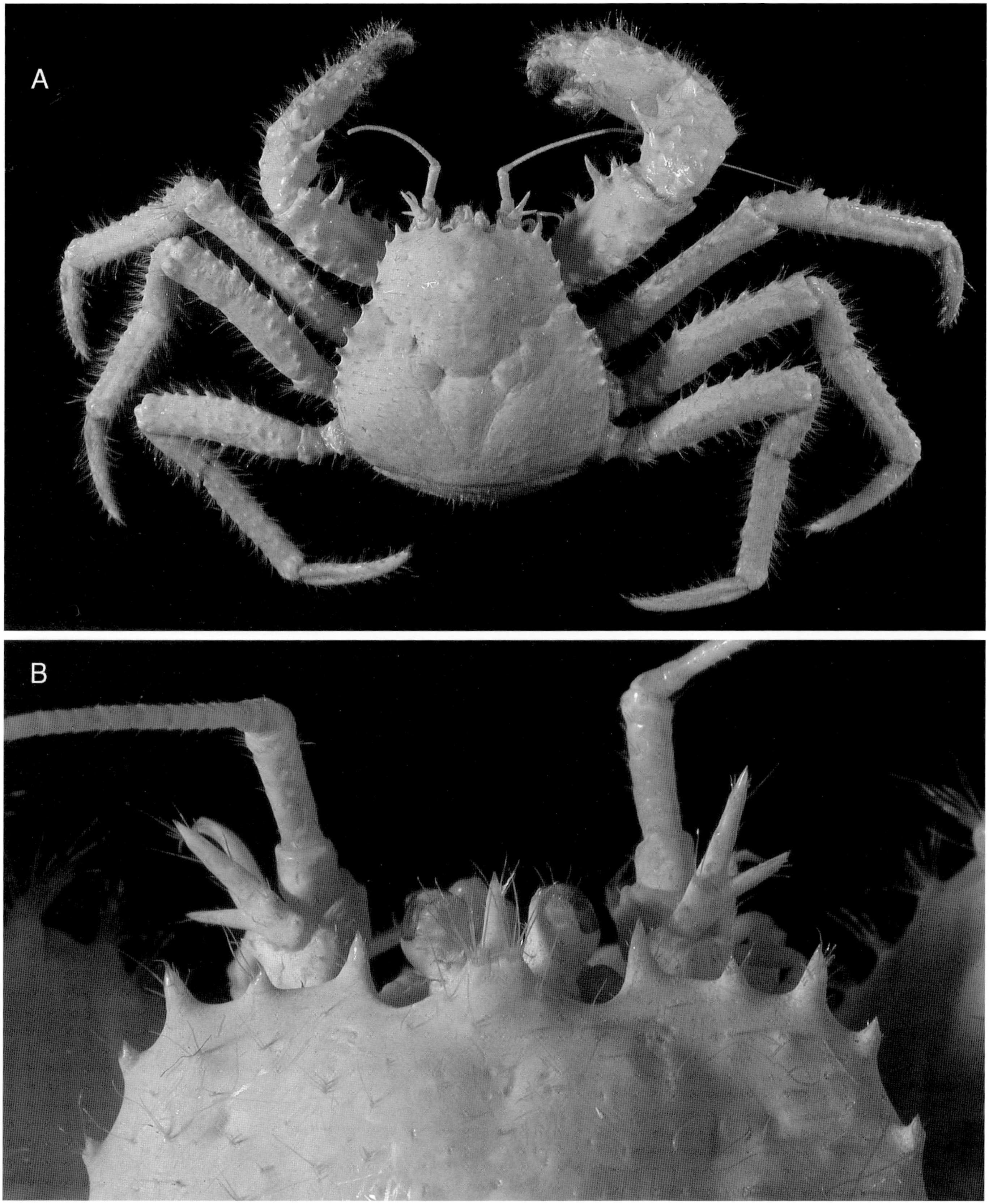

Figure 88. *Paralomis hirtella* de Saint Laurent & Macpherson, 1997, male, pcl 53.4 mm, Brothers Seamount (NIWA 34932). A, dorsal habitus. B, anterior carapace, dorsal view.

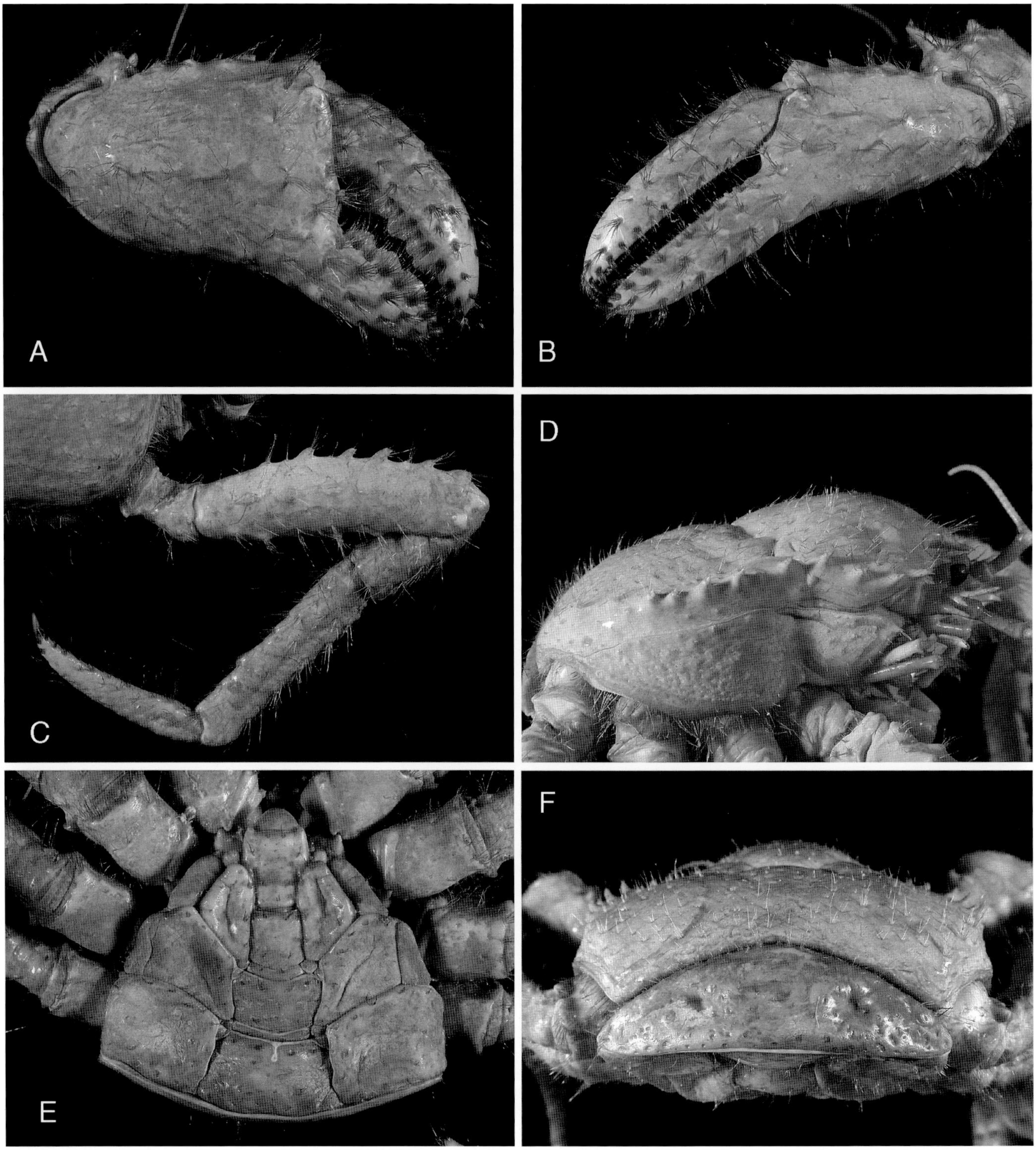

Figure 89. *Paralomis hirtella* de Saint Laurent & Macpherson, 1997, male, pcl 53.4 mm, Brothers Seamount (NIWA 34932). A, right chela. B, left chela. C, right pereopod 4. D, right carapace, lateral view. D, ventral surface and abdomen. E, posterior carapace and abdominal somite 2.

distal antennular peduncle article.

Antenna: Basal antennal article unarmed. Article 2 unarmed on inner and dorsal surface; outer margin with 1 or 2 acute granules proximally, and long, overreaching end of article 3 but not overreaching article 4. Article 3 with a few acute granules. Scaphocerite reaching to or beyond midlength of article 5; with small inner denticle and 1 or 2 outer spines, distal longer. Article 4 unarmed, about half length of article 5.

Abdomen: Ornamentation similar in both sexes. Surface smooth, covered in tufts of golden setae; margins smooth, unarmed. Telson spinose; semicircular.

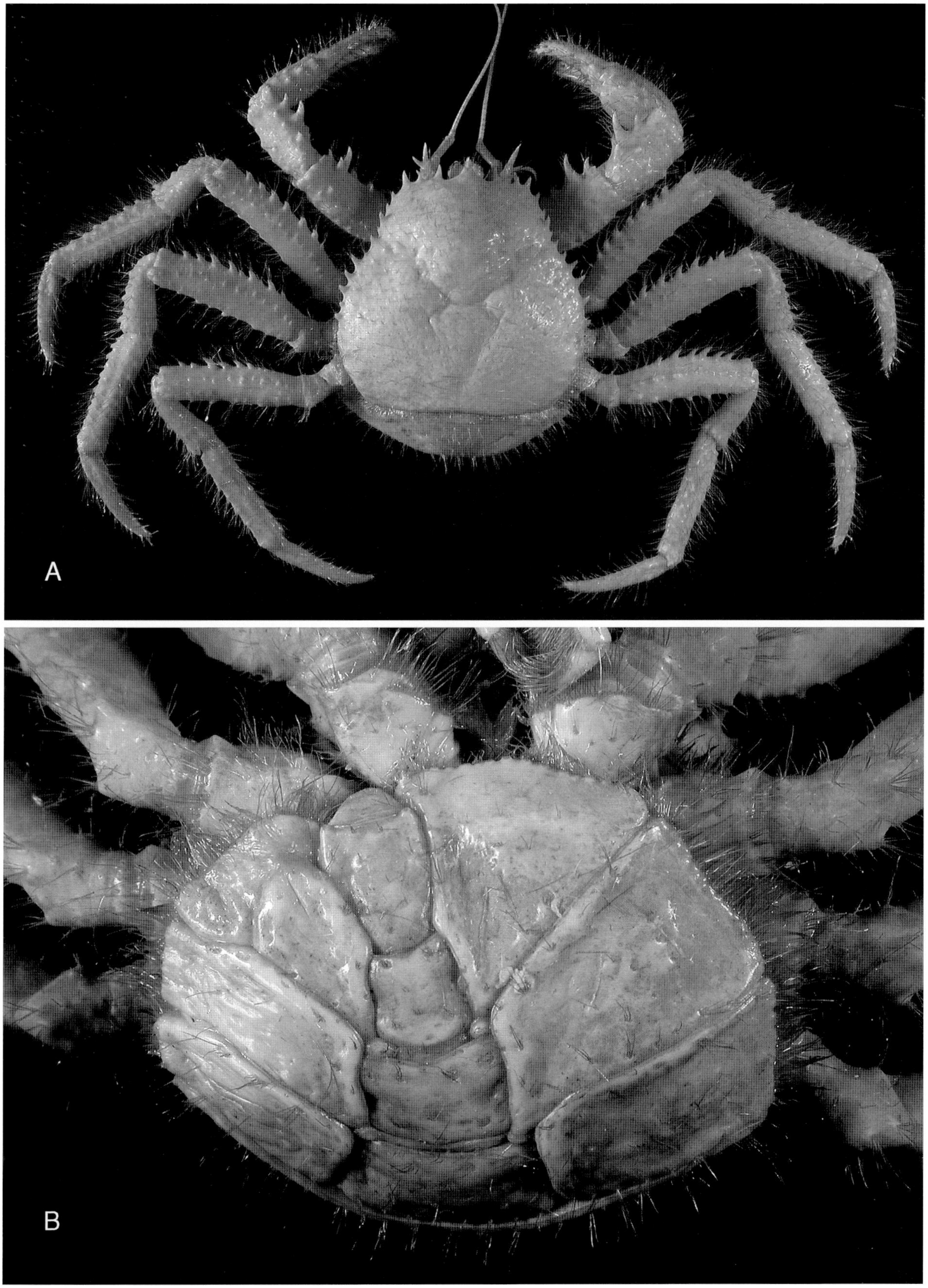

Figure 90. *Paralomis hirtella* de Saint Laurent & Macpherson, 1997, female, pcl 34.3 mm, Brothers Seamount (NIWA 34932). A, dorsal habitus. B, ventral surface and abdomen.

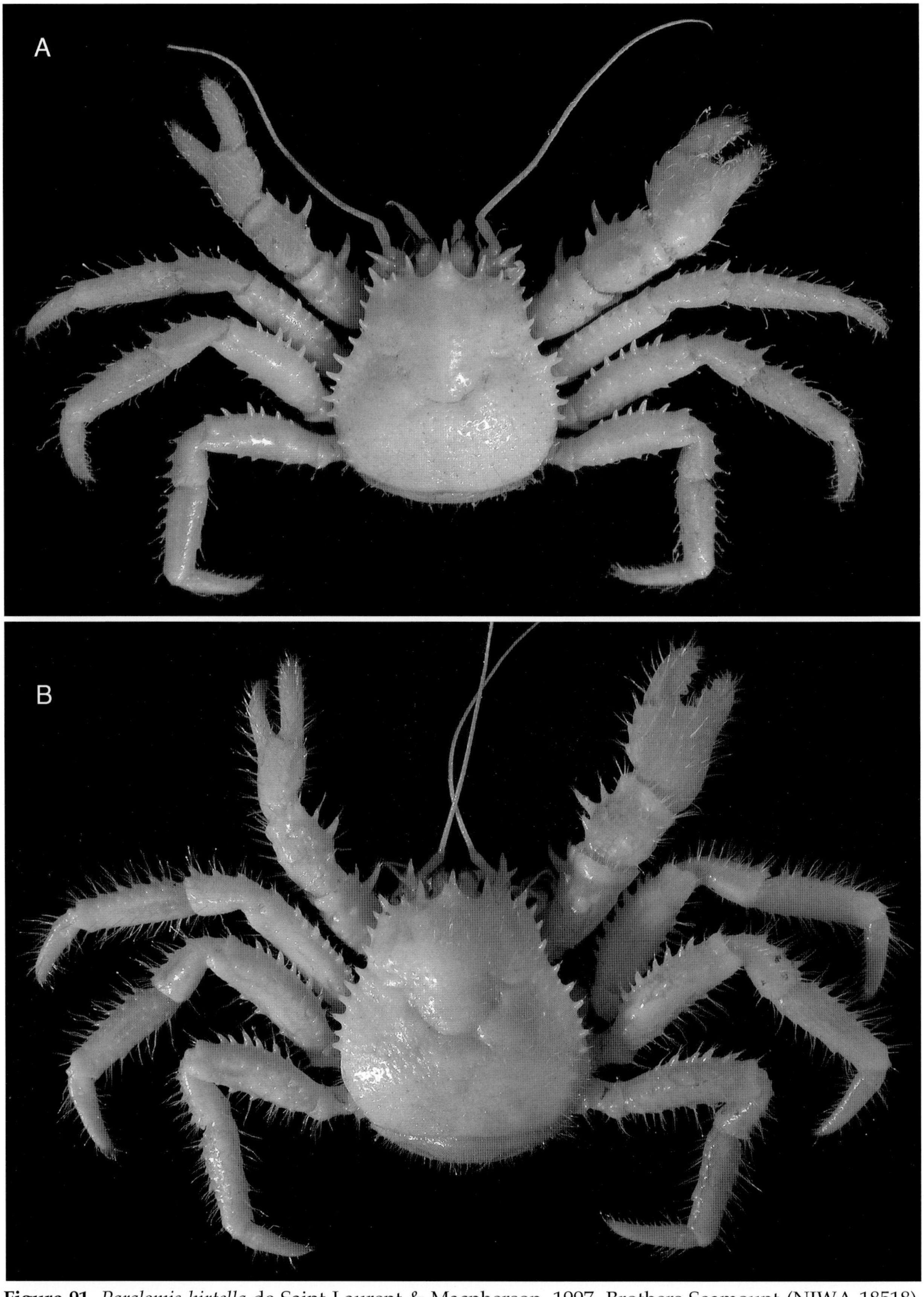

Figure 91. *Paralomis hirtella* de Saint Laurent & Macpherson, 1997, Brothers Seamount (NIWA 18518). A, male juvenile, pcl 8.7 mm. B, female juvenile, pcl 14.3 mm.

Pereopod 1 (chelipeds): Unequal; surfaces covered with low nodules bearing tufts of golden setae; ornamentation similar on both sides and in both sexes. Coxa and ischiobasis unarmed. Merus outer surfaces granular; mesial surface relatively smooth, with short ventral spines and strong inner distal spine; dorsal surface with scattered tubercles, distally with several stout spines. Carpus surfaces sparsely tuberculate; dorsal margin with 3 or 4 strong spines. Palm dorsal margin with 3–5 blunt teeth or tubercles, setose; other surfaces smooth apart from setal tufts.

Major cheliped 1.73–2.05 [1.68–1.83] pcl (male), 1.47–1.63 [1.46–1.51] (female); upper palm length 0.86–1.01 [0.90–1.02] times height (male), 0.87–0.97 [0.92–0.94] (female); occlusal margins of fingers corneous for distal third, proximally with 3 calcareous nodules; dactylus dorsal margin broadly convex, with rows of tufts of golden setae and blunt proximal tooth, 1.23–1.38 [1.22–1.36] times longer than dorsal margin of palm (male), 1.40–1.59 [1.42–1.58] (female).

Minor cheliped 1.66–1.82 [1.57–1.72] pcl (male), 1.45–1.61 [1.44–1.49] (female); upper palm length 0.92–1.28 [0.99–1.04] times height (male), 0.94–1.06 [1.01–1.07] (female); occlusal margin corneous in distal third to half, proximally crenulate to weakly dentate; dactylus dorsal margin broadly convex, with rows of tufts of golden setae and blunt proximal tooth, 1.51–1.86 [1.72–1.85] times longer than dorsal margin of palm (male), 1.74–1.86 [1.69–1.79] (female).

Pereopods 2–4 (walking legs 1–3): Similar, elongate, covered with low nodules bearing tufts of golden setae. Pereopod 3 longest. Coxae distal margins smooth, unarmed; ventral surfaces smooth, with scattered setal tufts. Ischiobasis with low setose nodules, unarmed. Merus extensor margins with 5–8 spines; flexor margins with 3–5 spinose; other surfaces with setal tufts but unarmed. Carpus with 2–5 extensor spines. Propodus shorter than merus; with 2–8 short spines on extensor margin. Dactylus curved, laterally compressed; longer than carpus, shorter than propodus; proximally unarmed; flexor margin lined with 8–12 obliquely inclined corneous spinules; apex corneous.

Pereopod 2 length 2.37–2.61 [2.17–2.29] pcl (male), 2.05–2.34 pcl [1.97–2.14] (female). Merus 0.75–0.85 [0.72–0.74] pcl (male), 0.61–0.72 [0.58–0.63] pcl (female); length:height ratio 3.28–3.72 [3.13–3.28] (male), 2.94–3.46 [2.63–3.00] (female). Carpus 0.59–0.62 merus length [0.60–0.77] (male), 0.65–0.73 [0.66–0.71] (female). Propodus 0.82–0.87 [0.81–0.83] merus length (male), 0.88–0.99 [0.86–0.91] (female); length:height ratio 3.87–4.74 [3.44–3.76] (male), 3.86–4.20 [3.07–3.41] (female). Dactylus 0.85–0.97 [0.74–0.89] propodus length (male), 0.87–0.97 [0.86–0.93] (female).

Pereopod 3 length 2.41–2.63 [2.17–2.31] pcl (male), 2.07–2.39 [2.02–2.17] pcl (female). Merus 0.75–0.84 [0.71–0.76] pcl (male), 0.62–0.72 [0.60–0.63] pcl (female); length:height ratio 3.27–4.09 [3.12–3.29] (male), 2.88–3.55 [2.59–2.80] (female). Carpus 0.57–0.61 [0.58–0.61] merus length (male), 0.64–0.71 [0.64–0.67] (female). Propodus 0.84–0.88 [0.81–0.84] merus length (male), 0.90–1.00 [0.89–0.92] (female); length:height ratio 4.13–5.45 [3.53–3.66] (male), 3.84–4.39 [3.25–3.53] (female). Dactylus 0.86–0.99 [0.86–0.92] propodus length (male), 0.88–1.00 [0.84–0.95] (female).

Pereopod 4 length 2.27–2.49 [2.02–2.21] pcl (male), 2.02–2.26 [1.92–2.03] pcl (female). Merus 0.68–0.76 [0.65–0.67] pcl (male), 0.56–0.65 [0.55–0.59] pcl (female); length:height ratio 3.23–3.81 [2.97–3.30] (male), 2.99–3.39 [2.72–3.07] (female). Carpus 0.60–0.63 [0.61–0.62] merus length (male), 0.66–0.72 [0.67–0.71] (female). Propodus 0.89–0.94 [0.89–0.91] merus length (male), 0.96–1.20 [0.92–0.97] (female); length:height ratio 4.17–4.91 [3.68–3.97] (male), 3.97–5.50 [3.29–3.70] (female). Dactylus 0.89–0.95 [0.88–0.93] propodus length (male), 0.89–1.00 [0.90–0.93] (female).

Egg diameter 1.70–2.52 mm.

Colour in life. Ivory-white, but often with light brown iron precipitates on external cuticle.

Remarks. *Paralomis hirtella* de Saint Laurent & Macpherson, 1997 was first recorded from New Zealand waters as *P.* aff. *jamsteci* based on specimens from Brothers Seamount (~35°S) (Webber & Naylor 2004b). Subsequently, Lörz *et al.* (2008) and Dawson (2008) reported *P. hirtella* from Monowai Caldera (~26°S), just outside of New Zealand waters, including specimens parasitised by *Briarosaccus callosus* Boschma. *Paralomis hirtella* differs from all other New Zealand lithodids by the smooth, unarmed carapace surface, bearing tufts of golden setae and at most a few, low, scattered granules near the anterolateral margins. The characteristic carapace ornamentation is consistent across the full size range examined (pcl 6.7–61.1 mm).

The specimens from Brothers Seamount agree well with northern specimens (Monowai Caldera material and the holotype from the Lau Basin, 22°13'S) in almost all respects, but differ subtly in having proportionally longer chelipeds and walking legs in adults. The smallest ovigerous female from Brothers Seamount is 35.4 mm pcl, and 29.9 mm pcl from Monowai. In the description above, morphometric measurements of the New Zealand specimens are given, followed by those of the northern specimens in parentheses. Both Brothers Seamount and northern specimens exhibit typical lithodid sexual dimorphism in which males have proportionally longer pereopods than respective females, differing by about 10% in *P. hirtella*. Proportional pereopod lengths in northern males, however, are similar to those of females from Brothers Seamount. The differences are subtle, and whether they represent phenotypic plasticity or reflect speciation remains to be determined. Note that the pereopod length differences between Brothers Seamount and northern specimens were determined independent of the parasitised specimens, and thus are not an artefact of parasitic feminisation.

Six of 23 Monowai Caldera specimens (26%) are infected with the rhizocephalan *Briarosaccus callosus*. Such an infection rate seems disproportionately high in comparison to other New Zealand lithodids. For instance, of the 83 specimens of *Lithodes aotearoa* examined, only four specimens (5%) were obviously infected (i.e., externae visible). The infected males of *P. hirtella* have been feminised, their pereopod proportions being similar to those of sympatric females. Saint Laurent & Macpherson (1997) also recorded two parasitised males among the 10 specimens examined. It is noteworthy that no southern specimens exhibited any evidence of parasitism, either in visible rhizocephalan externae or morphometric anomalies, despite the vents at Brothers Seamount being within the known range latitudinal and bathymetric range of *Briarosaccus callosus*.

Paralomis hirtella is a hydrothermal vent associate, and appears to be common around vents on Brothers Seamount, where other vent decapods have also been recorded: *Alvinocaris alexander* Ahyong, 2009, *A. niwa* Webber, 2004, *A. longirostris* Kikuchi & Ohta, 1995, *Nautilocaris saintlaurentae* Komai & Segonzac, 2004, *Lebbeus wera* Ahyong, 2009, and *Gandalfus puia* McLay, 2007 (McLay 2007; Ahyong 2009).

Distribution. North Fiji Back-Arc Basin (White Lady vent field); Lau Basin Back-Arc Basin (Hina Hina vent field), Monowai Caldera, and from Brothers Seamount, New Zealand; 1041–1750 m.

Paralomis poorei sp. nov.

(Figs 81, 92–97, Pl. 3E)

Paralomis cf. *phrixa*. — Koslow & Gowlett-Holmes, 1998: 43. — Poore, 2004: 269, fig. 76c. — Poore *et al.*, 2008: 27.
Paralomis sp. MoV2717. — Poore *et al.*, 1998: 66, 72.

Type material. (All Chatham Rise, New Zealand). *Holotype*: NIWA 61161, male (cl 24.6 mm, pcl 18.9 mm, cw 17.3 mm), Diabolical Seamount, 42°47.17–46.96′S, 179°59.11–59.01′W, 993–900 m, TAN0104/48, 16 Apr 2001.

Paratypes: NIWA 53273, 1 juvenile male (cl 11.8 mm, pcl 9.1 mm, cw 7.2 mm), 42°38.65–38.67′S, 179°52.88–53.05′E, 1052–1080 m, TAN0905/48, 18 Jun 2009; NIWA 29526, 1 juvenile male (cl 14.9 mm, pcl 11.3 mm, cw 10.1 mm), Pyre Seamount, 42°43.06′S, 179°54.29′W, 1025–1156 m, TAN0604/102, 4 Jun 2006; NIWA 53134, 1 male (pcl 27.3 mm, cw 25.4 mm), 42°44.68–44.90′S, 179°55.41–55.43′E, 1020–1125 m, TAN0905/41, 17 Jun 2009; NIWA 3917, 1 juvenile male (cl 9.0 mm, pcl 6.7 mm, cw 5.8 mm), Pyre Seamount, 42°43.09–43.18′S, 179°54.57–54.87′W, 1075–1008 m, TAN0104/333, 20 Apr 2001; NIWA 29642, 1 male (cl 24.0 mm, pcl 19.9 mm, cw 17.8 mm), Zombie Seamount, 42°45.76′S, 179°55.51′W, 1019–1081 m, TAN0604/09, 28 May 2006; NIWA 3916, 1 juvenile female (cl 10.0 mm, pcl 7.8 mm, cw 6.8 mm), Diabolical Seamount, 42°47.56–47.70′S, 179°58.86–58.60′W, 950–900 m, TAN0104/47, 16 Apr 2001.

Other material examined. *Western Australia*: NMV J54984, 1 juvenile female (cl 19.0 mm, pcl 14.4.mm, cw 12.5 mm), off Albany, 35°26.05′S, 118°21.00′E, 900–915 m, RV *Southern Surveyor*, SS1005/44/040, 25 Nov 2005.

Tasmania: NMV J44012, ovigerous female (cl 32.4 mm, pcl 27.3 mm, cw 23.3 mm), 83.8 km SSE of Southeast Cape, "J1" seamount, 44°16.2′S, 147°19.6′E, 987 m, epibenthic sled, RV *Southern Surveyor*, SS01/97/36, coll. T.N. Stranks *et al.*, 27 Jan 1997.

Diagnosis. Carapace pyriform, slightly longer than wide, margins and surface covered with upright spines, surfaces minutely setose; outer distance between bases of anterolateral spines exceeding half carapace width. Rostrum with 2 pairs of dorsal spines; subrostral lobe rounded. Dorsal eyespines not extending anteriorly beyond cornea. Scaphocerite multispinose. Abdominal somite 2 and 3 spinose; somites 4–6 with tubercles. Chelipeds and walking legs covered with spines, most elongate on dorsal and extensor margins; spines on dactylar extensor margin restricted to proximal half of margin.

Description. *Carapace*: Pyriform, 1.07–1.17 times longer than wide; regions distinct; surface and margins uniformly covered with slender, straight conical spines of varying length, 7 gastric, 4 cardiac and 4 branchial longer than other spines; surface of spines with short, well-spaced simple setae; longest spine (on anterior branchial margin) about 0.2 pcl; cardiac region with 20–23 spines; cervical groove distinct. Pterygostomian region sparsely granulose; with prominent, anterior, submarginal spine; spinulose on posterior two-thirds.

Rostrum 0.19–0.32 pcl; broad basally, not constricted proximal to dorsal spines; median spine slender, ventral lobe angular, unarmed; proximally with 2 pairs of laterally divergent spines directed obliquely upwards and single upright spine between proximal pair of dorsal spines. Posterior orbital margin concave, unarmed; outer orbital spine slender, directed anteriorly, not reaching beyond apex of cornea (when eyes directed anteriorly). Anterolateral spine as long as outer orbital spine; outer distance between bases of anterolateral spines exceeding half carapace width.

Figure 92. *Paralomis poorei* sp. nov., male holotype, cl 24.6 mm, pcl 18.9 mm, cw 17.3 mm, Diabolical Seamount (NIWA 61161). A, dorsal habitus. B, carapace, right lateral view.

Figure 93. *Paralomis poorei* sp. nov., male holotype, cl 24.6 mm, pcl 18.9 mm, cw 17.3 mm, Diabolical Seamount (NIWA 61161). A, right chela, outer surface. B–C, left chela, outer and inner surfaces. D–E, right pereopod 4, dorsal and ventral view. F, anterior carapace, right oblique view.

Ocular peduncle: Longer than cornea; dorsally with 5 or 6 acute granules and single short distal spine; distal spine upright, about half corneal diameter but not overreaching cornea.

Antennule: Peduncle unarmed, reaching anteriorly beyond apex of antennal peduncle by half to quarter length of distal antennular peduncle article.

Antenna: Basal antennal article with short anterolateral spine. Article 2 unarmed on inner margin; outer margin with acute granule proximally and long distal spine reaching slightly beyond apex of article 3. Article 3 unarmed. Scaphocerite a long, slender spine

Figure 94. *Paralomis poorei* sp. nov., male holotype, cl 24.6 mm, pcl 18.9 mm, cw 17.3 mm, Diabolical Seamount (NIWA 61161). A, carapace. B, anterior carapace, dorsal view. C, abdomen. D, posterior carapace and abdominal somite 2.

overreaching distal peduncular article; with 2 lateral spines, first minute, second about two-thirds length of main spine; dorsally unarmed; mesially with 2 small spines or acute granules proximally. Article 4 unarmed, almost half length of article 5.

Abdomen: Ornamentation of both sexes similar. Somite 2 covered with spines of similar length to dorsal carapace spines. Somite 3 less spinose, with scattered tubercles and some spines of about half length of spines on preceding somite. Somites 4–5 sparsely granulate, margins irregularly granulate. Somite 6 with 4 blunt, apically setose tubercles on distal margin. Telson semicircular, unarmed.

Pereopod 1 (chelipeds): Strongly spinose, unequal. Coxae unarmed, setose; distal margins with dense tufts of setae. Ischiobasis spinose laterally, ventrally with

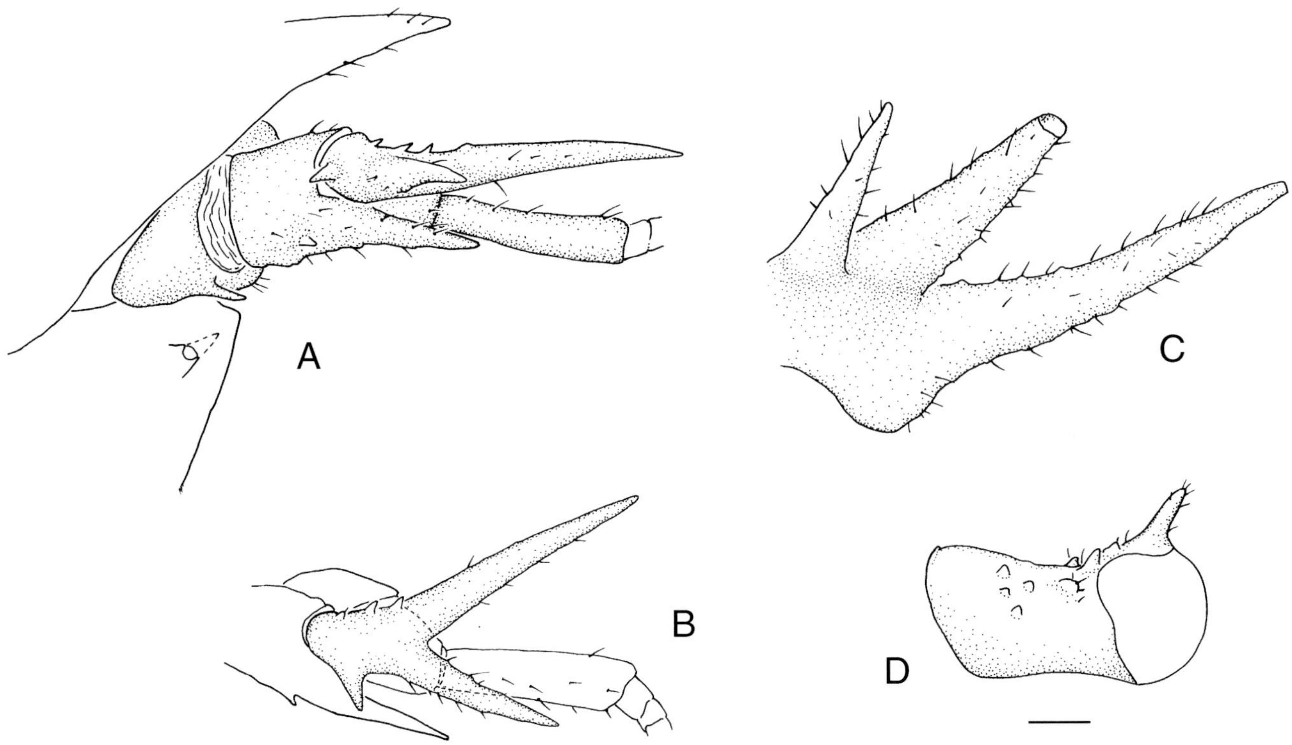

Figure 95. *Paralomis poorei* sp. nov., male holotype, cl 24.6 mm, pcl 18.9 mm, cw 17.3 mm, Diabolical Seamount (NIWA 61161). A, right antenna, lateral view. B, right antenna, dorsal view. C, rostrum, right lateral view. D, right eye, lateral view. Scale = 1.0 mm.

clusters of golden setae. Merus with smooth mesial surface, others surfaces spinose, long prominent spine dorsodistally. Carpus prominently spinose except for mesial surface, with clusters of golden setae. Palm outer surface with 2 or 3 uneven rows of 4–6 spines; upper margin with 2 uneven rows of 4–7 spines; ventral surface irregularly spinose or tuberculate; mesial surface with blunt scattered tubercles bearing tufts of golden setae.

Major cheliped 1.61–2.10 pcl (male), 1.49 pcl (female); upper palm length 1.19–1.29 times height (male), 1.31 (female); occlusal margins of fingers corneous for distal quarter to fifth, proximally with 3 calcareous nodules; dactylus dorsal margin broadly convex, with rows of tufts of golden setae and small proximal spine, 1.06–1.12 times longer than dorsal margin of palm (male), 1.09 (female).

Minor cheliped 1.54–1.93 pcl (male), 1.49 pcl (female); upper palm length 1.33–1.48 times height (male), 1.39 (female); occlusal margin corneous in distal third to half, proximally crenulate; dactylus dorsal margin broadly convex, with rows of tufts of golden setae and small proximal spine, 1.62–1.65 times longer than dorsal margin of palm (male), 1.59 (female).

Pereopods 2–4 (walking legs 1–3): Similar, elongate, spinose. Pereopod 2 longest. Coxae unarmed, or with a few low granules, distal margins crenulate to dentate. Ischiobasis spinose. Merus ovate in cross section; extensor and flexor margins each with 2 relatively even rows of 7–11 and 5–7 spines respectively; dorsal surface with smaller scattered spines; ventral surface unarmed. Carpus spinose dorsally and laterally, longest spine as high as carpus. Propodus dorsoventrally flattened; extensor margin with 2 rows of 6–8 spines; flexor margin with row of 7–9 spines; dorsal surface multispinose; ventral spines sparse and short, not exceeding 1.5 length of dorsal spines. Dactylus curved, laterally compressed; longer than carpus; with 5–7 spines on proximal half and scattered tufts of golden setae; flexor margin lined with 13–18 obliquely inclined corneous spinules; apex corneous.

Pereopod 2 length 1.90–2.31 pcl (male), 1.69–1.81 pcl (female). Merus 0.62–0.74 pcl (male), 0.51–0.56 pcl (female); length:height ratio 4.22–4.61 (male), 3.60–3.97 (female). Carpus 0.60–0.61 merus length (male), 0.64–0.70 (female). Propodus 0.85–0.87 merus length (male), 0.94–0.98 (female); length:height ratio 5.38–6.82 (male), 5.11–6.30 (female). Dactylus 0.76–0.77 propodus length (male), 0.75–0.79 (female).

Pereopod 3 length 1.91–2.26 pcl (male), 1.72–1.79 pcl (female). Merus 0.59–0.69 pcl (male), 0.54 pcl (female); length:height ratio 4.00–4.65 (male), 3.59–3.90 (female). Carpus 0.61 merus length (male), 0.66–0.68 (female). Propodus 0.93–0.94 merus length (male), 0.90–0.99

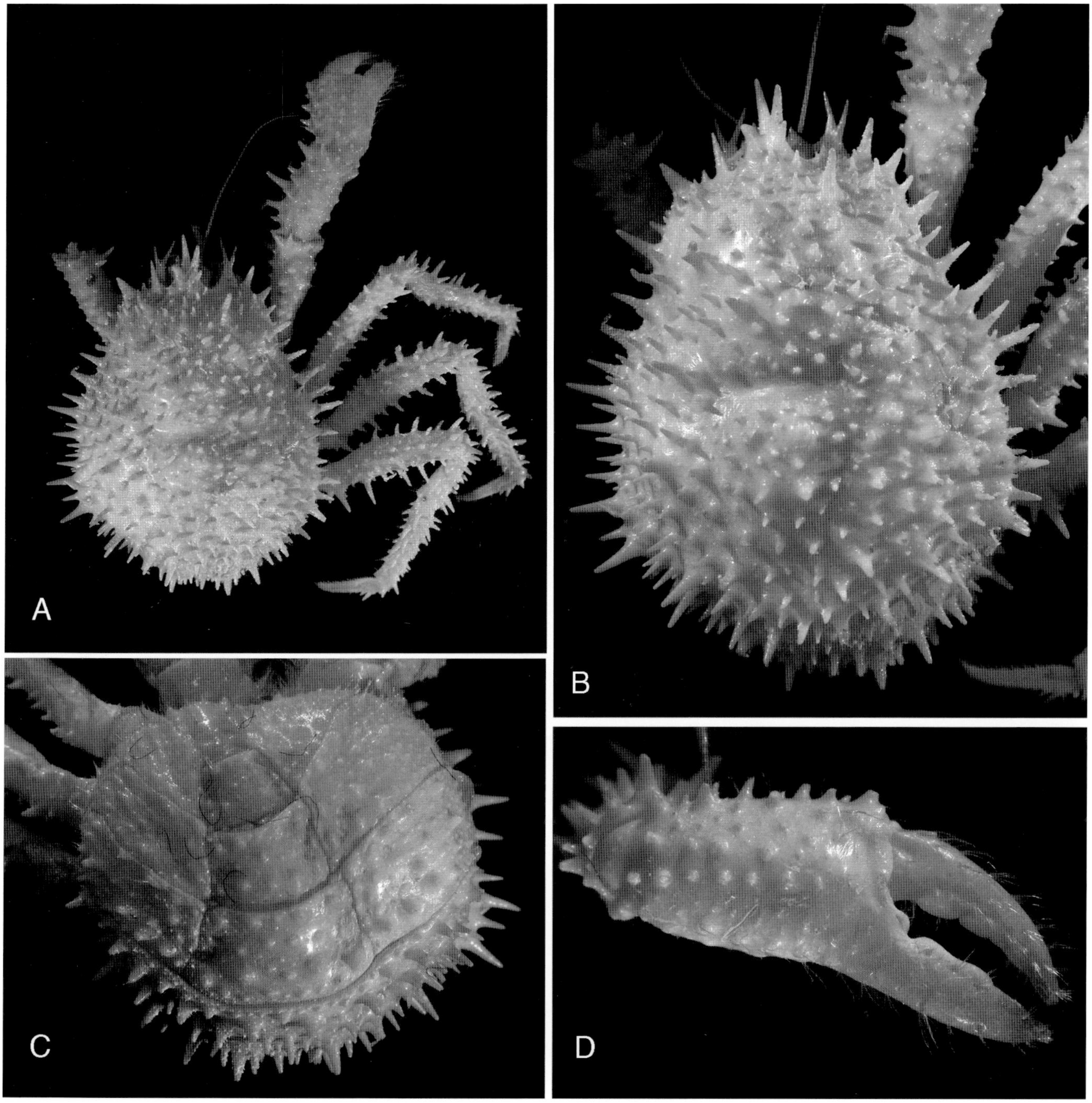

Figure 96. *Paralomis poorei* sp. nov., ovigerous female, cl 32.4 mm, pcl 27.3 mm, cw 23.3 mm, J1 seamount (NMV J44012). A, dorsal habitus. B, carapace. C, abdomen. E, right chela.

(female); length:height ratio 5.88–7.15 (male), 5.01–6.55 (female). Dactylus 0.75–0.81 propodus length (male), 0.79–0.82 (female).

Pereopod 4 length 1.81–2.17 pcl (male), 1.60–1.80 pcl (female). Merus 0.53–0.64 pcl (male), 0.46–0.49 pcl (female); length:height ratio 3.71–4.37 (male), 3.32–3.88 (female). Carpus 0.66–0.68 merus length (male), 0.72–0.74 (female). Propodus 1.00–1.04 merus length (male), 1.06–1.07 (female); length:height ratio 5.77–7.00 (male), 5.38–6.60 (female). Dactylus 0.73–0.80 propodus length (male), 0.73–0.79 (female).

Colour in life. Translucent pink-orange overall (Pl. 3E).

Etymology. Named after Gary Poore, who first reported the presence of the species from southeastern Australia.

Figure 97. *Paralomis poorei* sp. nov. A–B, male paratype, pcl 27.3 mm, cw 25.4 mm (NIWA 53134), dorsal habitus and chelipeds. C, juvenile male, pcl 11.3 mm (NIWA 29526). D, juvenile male paratype, pcl 6.7 mm (NIWA 3917).

REMARKS. *Paralomis poorei* sp. nov. most closely resembles *P. bouvieri* Hansen, 1908 [North Atlantic] and *P. makarovi* Hall & Thatje, 2009b [Bering Sea, North Pacific] in carapace physiognomy and spination, and in the similar armature and length of the walking legs. Each is apparently a small species, with *P. poorei* and *P. bouvieri* maturing at less than 30 mm pcl (size at maturity not known for *P. makarovi*). *Paralomis poorei* differs from *P. bouvieri* in the presence of a subrostral lobe; the distal eyespine does not overreach the cornea; and the ventral surfaces of the meri of the walking legs are smooth or sparsely granulate rather than spinose. *Paralomis poorei* is readily distinguished from *P. makarovi* in the presence of the subrostral lobe and in the length of the setae on the surfaces of the carapace spines — markedly shorter, rather than markedly longer than the basal width of the spines.

Of the New Zealand and Australian *Paralomis* species, *P. poorei* could be confused with *P. webberi* sp. nov. (New Zealand), juvenile *P. zealandica* Dawson & Yaldwyn, 1971 (New Zealand), *P. echidna* sp. nov. (Australia), and *P. taylorae* sp. nov. (Australia) owing to their dense dorsal spination. The tuberculate rather than spinulate abdominal somites will distinguish *P. poorei* from each of the aforementioned species. *Paralomis poorei* further differs from *P. webberi*, *P. echidna*, and size-matched *P. zealandica* in having more elongate, more slender walking legs as measured by pereopod 4 length and pereopod 4 meral proportions, and from *P. taylorae* in having shorter, less slender walking legs (see Tables 4, 5). The setation of the carapace spines also differs among species, being at most apically setose in *P. zealandica* and glabrous in *P. taylorae*, rather than sparsely scattered over the spine surface as in *P. webberi*, *P. echidna*, and *P. poorei*.

The juveniles agree with adults in having similar abdominal and pereopodal ornamentation, and distinctly unequal chelipeds (in males). The carapace ornamention is largely consistent with adults except in the smallest juveniles (pcl 6.7–9.1 mm) in which the spination of the pereopods and carapace surface is less extensive. The carapace surface of the small juveniles is armed with long dorsal spines (seven gastric, four cardiac, and four branchial) and numerous minute spines or acute granules which, in adults, become almost as long as the initial 'long' spines. These long spines are homologous with the short dorsal spines of comparatively less ornamented species such as *P. birsteini* Macpherson, 1988a, *P. gowlettholmes* sp. nov., and *P. stevensi* Ahyong & Dawson, 2006.

The Australian and New Zealand specimens of *P. poorei* agree well. Differences, namely in proportional limb lengths, are consistent with sexual dimorphism — the New Zealand specimens are all males and both Australian specimens are female. Therefore, detailed comparison of both sexes from both regions is not yet possible, but differences between New Zealand and Australian populations might become evident when more material becomes available for study.

Paralomis poorei is the smallest known lithodid in New Zealand and Australian waters (females ovigerous at less than 30 mm pcl) and is one of only four lithodid species to be shared by both regions (see also *P. echidna* sp. nov., *P. staplesi* sp. nov., and *Lithodes macquariae* sp. nov.). Species previously thought to occur on both sides of the Tasman are, as a result of the present study, now known to be composites (see *Lithodes aotearoa* sp. nov., *L. australiensis* sp. nov., *Neolithodes brodiei* Dawson & Yaldwyn, 1970, *N. flindersi* Ahyong, 2010).

Distribution. Presently known from seamounts on the Chatham Rise, New Zealand, from seamounts off southeastern Tasmania, and southwestern Australia; 900–1156 m.

Table 4. Selected measurements of male *P. poorei*, *P. echidna*, *P. webberi*, *P. taylorae*, and juvenile *P. zealandica*. Size range of *P. zealandica* selected to correspond to known size range of *P. poorei*. Abbreviations: postorbital carapace length (pcl); pereopod 4 (P4); length (L); height (H).

	Size range (pcl)	P4 L/pcl (Male)	P4 merus L/pcl (Male)	P4 merus L/H (Male)	P4 propodus L/H (Male)
P. poorei	11.3–27.3	1.81–2.17	0.53–0.64	3.71–4.37	5.40–7.00
P. echidna	49.7	1.56	0.48	2.91	4.34
P. webberi	?	?	?	?	?
P. zealandica	19.0–25.7	1.44–1.53	0.43–0.48	2.74–3.01	3.61–4.31
P. taylorae	?	?	?	?	?

Table 5. Selected measurements of female *P. poorei*, *P. echidna*, *P. webberi*, *P. taylorae*, and juvenile *P. zealandica*. Size range of *P. zealandica* selected to correspond to known size range of *P. poorei*. Abbreviations: postorbital carapace length (pcl); pereopod 4 (P4); length (L); height (H).

	Size range (pcl)	P4 L/pcl (Female)	P4 merus L/pcl (Female)	P4 merus L/H (Female)	P4 propodus L/H (Female)
P. poorei	14.4–27.3	1.60–1.80	0.46–0.51	3.32–3.88	5.38–6.60
P. echidna	33.6–38.6	1.20–1.28	0.35–0.37	2.43–2.52	3.68–3.73
P. webberi	33.6	1.23	0.36	2.47	4.67
P. zealandica	18.9	1.50	0.44	2.68	3.82
P. taylorae	39.8	2.74	0.85	4.35	10.6

Paralomis staplesi sp. nov.

(Figs 81, 98–100, Pl. 3H)

TYPE MATERIAL. *Holotype*: NMV J61052, male (cl 59.1 mm, pcl 50.0 mm, cw 49.0 mm), Tasman Fracture Zone, Tasmania, 45°18′01′S, 146°07′14″E, 2213 m, FV *Tommy Thompson*, TN228-J2-392-012-005, ROV *Jason 2*, coll. D. Staples, 11 Jun 2009.

OTHER MATERIAL EXAMINED. NIWA 3920, 1 badly fragmented male, Kermadec Ridge, 32°32.334–33.551′S, 179°39.616–38.977′W, 2312–1958 m, TAN0205/82, 25 Apr 2002.

DIAGNOSIS. Carapace subpentagonal, length and width subequal; surface glabrous, covered with small, scattered granules and numerous, small, well-spaced conical spines in addition to larger, longer spines (including 5 gastric, 4 cardiac, 4 branchial); lateral margins lined with well-spaced spines; outer distance between bases of anterolateral spines half carapace width. Rostrum trispinose, broadest basally. Scaphocerite with inner and outer spines. Chelipeds and walking legs strongly spinose. Walking legs elongate, about 3 times pcl; dactyli shorter than extensor margin of propodi. Walking leg 3 merus about as long as pcl, length exceeding 6.7 times height (males); propodus length exceeding 9 times height.

DESCRIPTION OF HOLOTYPE. *Carapace*: Subpentagonal, 1.02 times longer than wider; surface covered with small, scattered granules and numerous, well-spaced conical spines, glabrous. Gastric region convex, elevated, more prominent than other regions; in addition to scattered spines, with 5 longer conical spines forming pentagon with anteriormost in midline. Lateral margin of hepatic region with 3 conical spines, anterior 2 shorter, posterior long, curved. Branchial regions with 13 or 14 prominent conical marginal spines; surface with 4 spines distinctly longer than other surface spines. Cardiac region subtriangular, with 2 pairs of short conical spines forming square. Intestinal region sparsely spinose. Pterygostomian region sparsely granular, with prominent anterior spine.

Rostrum 0.18 pcl; broadest basally, without constriction; median spine smooth ventrally; paired dorsal spines divergent, directed obliquely upwards. Posterior orbital margin slightly concave; outer orbital spine reaching to end of cornea. Anterolateral spine as long as outer orbital spine, with short marginal spinule midway between outer orbital spine and anterolateral spine; outer distance between bases of anterolateral spines half carapace width.

Ocular peduncle: Longer than cornea; with 3 acute dorsal granules and distal spine slightly overhanging cornea.

Antennule: Peduncle unarmed, reaching anteriorly beyond antennal peduncle by half-length of distal antennular peduncle article.

Antenna: Basal antennal article with small anterolateral spine. Article 2 with inner distal spine; outer margin with minute basal spine and slender distal spine reaching slightly beyond article 4. Article 3 unarmed. Scaphocerite slender, reaching beyond midlength of article 5, with 1 or 2 short inner spines and 2 outer spines. Article 4 unarmed, about half length of article 5.

Abdomen of male: Somite 2 with small, widely separated spines along margins and on surface. Surface of somites 3–5 sparsely granular or tuberculate. Somite 3 median plate unarmed; submedian plates with low, irregular, triangular teeth on lateral margin; marginal plates absent, apparently undifferentiated. Somites 4–5 with unarmed median and submedian plates; marginal plates undivided, with irregular triangular teeth. Somite 6 slightly wider than long, subpentagonal, with pair of small distal teeth; marginal plates short, triangular. Telson wider than long, rounded.

Pereopod 1 (chelipeds): Spination similar in both sexes. Major cheliped of males inflated, 1.59 times height of minor cheliped; minor cheliped slender. Coxa smooth, unarmed; distal margins with dense tufts of setae. Ischiobasis with 8 stout, apically setose ventral spines. Merus with smooth mesial margins and tuberculate lateral margins; dorsal and ventral margins spinose, inner distal spine largest. Carpus with prominent spines on dorsal and lateral surfaces; dorsal margin with rows of 3–5 spines; mesial margin with 2 spines, proximal largest; ventral surface with 3 short spines. Palm mesial margin with 6 spines, other surfaces with prominent, apically setose tubercles or acute tubercles. Fingers with short basal tubercle and rows of tufts of golden setae.

Major cheliped 1.74 pcl (male); upper palm length 1.17 times height (male); occlusal margins of fingers corneous for distal third, proximally with 3 calcareous nodules; dactylus dorsal margin broadly convex, with rows of tufts of golden setae and blunt proximal tooth, 1.21 times longer than dorsal margin of palm (male).

Minor cheliped 1.41 pcl (male); upper palm length 1.27 times height (male); occlusal margin corneous for slightly less than distal third, proximally crenulate; dactylus dorsal margin broadly convex, with rows of tufts of golden setae and blunt proximal tooth, 1.81 times longer than dorsal margin of palm (male).

Pereopods 2–4 (walking legs 1–3): Similar, elongate, spinose. Coxae sparsely granular; distal margins with 3 blunt apically setose teeth. Ischiobasis with 3–5 apically setose ventral spines. Merus slightly compressed, shorter than carapace (ischiomerus longer than carapace); extensor margin with 11–13 spines in addition to paired distal spines; dorsal surface with 10–14 spines; flexor margin with 2 uneven rows of 5–7 spines; ventral

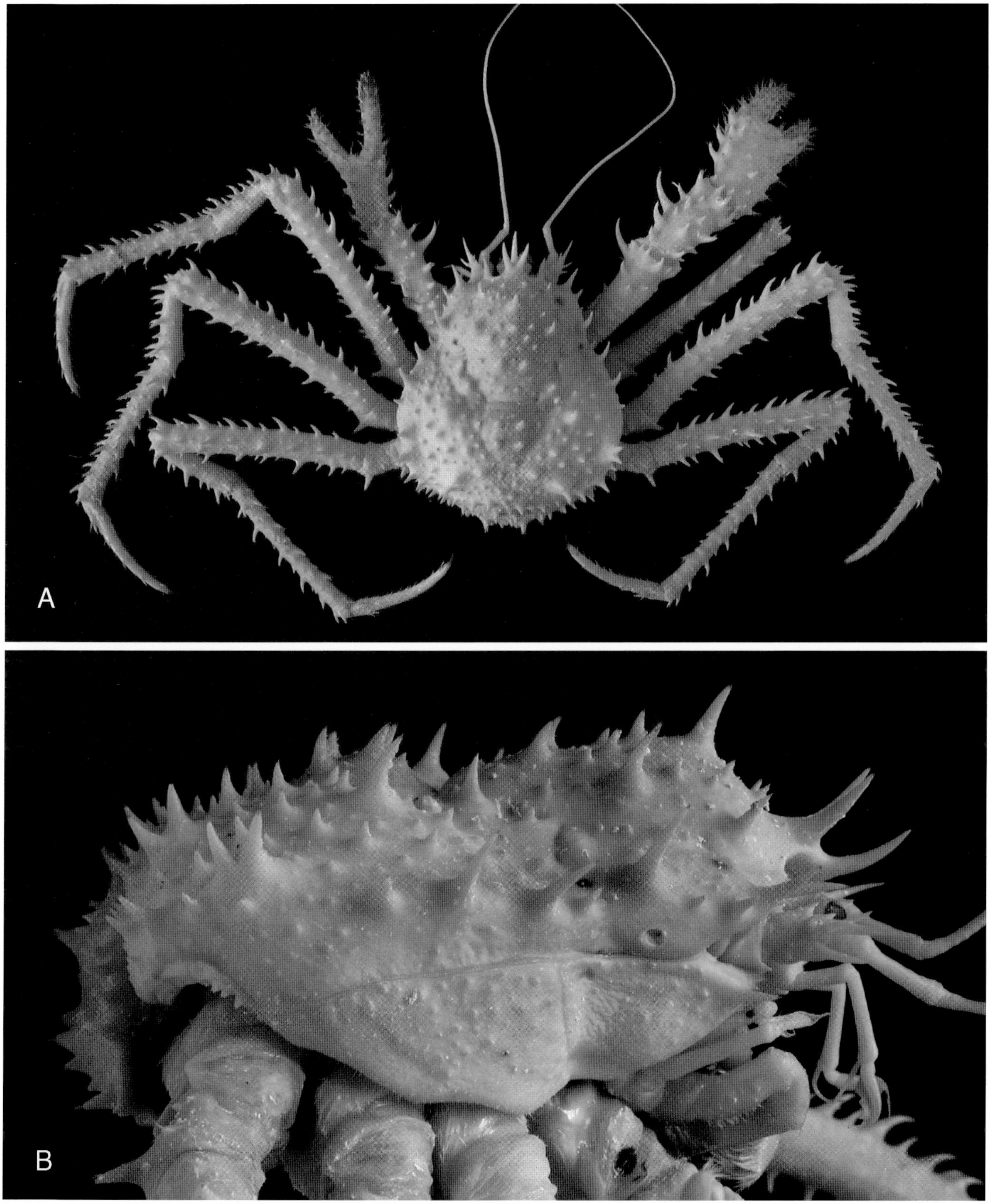

Figure 98. *Paralomis staplesi* sp. nov., male holotype, cl 59.1 mm, pcl 50.0 mm, cw 49.0 mm, Tasman Fracture Zone, Tasmania (NMV J61052). A, dorsal habitus. B, carapace, right lateral view.

Figure 99. *Paralomis staplesi* sp. nov., male holotype, cl 59.1 mm, pcl 50.0 mm, cw 49.0 mm, Tasman Fracture Zone, Tasmania (NMV J61052). A, carapace, B, right pereopod 4.

surface with scattered granules; merus of pereopod 3 slightly longer than that of pereopod 2 and pereopod 4. Carpus half merus length, subcircular in cross section; extensor margin with 9 spines of which 3 or 4 markedly longest; dorsal surface with 6–8 spines; ventral surface with 3–5 small spines; flexor margin unarmed. Propodus dorsoventrally flattened; slightly shorter than merus; dorsal surface, extensor and flexor margins prominently multispinose. Dactylus broadly curved; slightly shorter than extensor margin of propodus; surface with tufts of setae; extensor margin with 3 apically corneous spines proximally; lateral proximal surfaces with short, distinct sulcus, flanked ventrally by small, corneous spine; flexor margin lined with corneous spinules

Pereopod 2 length 2.89 pcl (male). Merus 0.98 pcl (male); length:height ratio 6.90 (male). Carpus 0.49 merus length (male). Propodus 0.81 merus length (male); length:height ratio 9.02 (male). Dactylus 0.84 propodus length (male).

Pereopod 3 length 3.03 pcl (male). Merus 1.02 pcl (male); length:height ratio 7.39 (male). Carpus 0.48 merus length (male). Propodus 0.83 merus length (male); length:height ratio 9.84 (male). Dactylus 0.82 propodus length (male).

Pereopod 4 length 2.96 pcl (male). Merus 0.95 pcl (male); length:height ratio 6.75 (male). Carpus 0.51 merus length (male). Propodus 0.89 merus length (male); length:height ratio 9.81 (male). Dactylus 0.80 propodus length (male).

COLOUR IN LIFE. Overall purplish-red (Pl. 3H).

ETYMOLOGY. Named for David Staples, Museum Victoria, who collected the holotype.

REMARKS. *Paralomis staplesi* sp. nov. closely resembles *P. birsteini* Macpherson, 1988a, *P. stevensi* Ahyong & Dawson, 2006, and *P. gowlettholmes* sp. nov. in overall carapace shape, structure, and ornamentation of the pereopods. The new species differs primarily in having more prominent and more numerous carapace and pereopodal spination (compare Figs 63, 83, 102, 106). In *P. birsteini*, *P. stevensi*, and *P. gowlettholmes*, the carapace is sparsely tuberculate or granulate, with several small conical spines on the gastric, hepatic, and branchial regions. The dorsal carapace tubercles or granules and small spines of the aforementioned species are developed in *P. staplesi* as small but prominent spines and large, prominent, conical spines, respectively.

Figure 100. *Paralomis staplesi* sp. nov., male holotype, cl 59.1 mm, pcl 50.0 mm, cw 49.0 mm, Tasman Fracture Zone, Tasmania (NMV J61052). A, right chela. B, left chela. C, posterior carapace and abdominal somite 2. D, ventral surface and abdomen. E, anterior carapace, dorsal view. F, right orbit and antenna.

Similarly, the spines of the walking legs of *P. staplesi* are more numerous and proportionally longer than in *P. birsteini*, *P. stevensi*, and *P. gowlettholmes*. In addition, *Paralomis staplesi* resembles *P. stevensi* and differs from *P. gowlettholmes* and *P. birsteini* in its more slender, more elongate walking legs in which the pereopod 4 meri are as long as or longer than the carapace (shorter in *P. birsteini* and *P. gowlettholmes*), and in which the walking leg dactyli are shorter than the propodal extensor margins.

The male specimen from the Kermadec Ridge is badly fragmented: the abdomen, chelipeds and most walking legs are largely intact but the carapace is crushed with only portions of the right branchial margin, the left anterolateral margin and the rostral/orbital regions remaining. In all details that can be compared, the damaged specimen agrees well with the holotype. A complete specimen may reveal specific differences, but until further material becomes available, it is tentatively assigned to *P. staplesi*. *Paramonis staplesi*, *P. echidna* sp. nov., and *P. poorei* sp. nov. are the only species of *Paralomis* known from both New Zealand and Australian waters.

Distribution. Presently known only from Tasmania and the Kermadec Ridge, New Zealand; 1958– 2312 m.

Paralomis stevensi Ahyong & Dawson, 2006 (Figs 66, 101–104, Pl. 3G)

Paralomis birsteini. — Stevens, 2004: 6, 7 [not *P. birsteini* Macpherson, 1988].
Neolithodes brodiei. — Thatje & Lörz, 2005: 335 [NIWA 3435 only; not *N. brodiei* Dawson & Yaldwyn, 1970].
Paralomis stevensi Ahyong & Dawson, 2006: 59–66, figs 7–10 [type locality: near Scott Island, Ross Sea, 65°40.0–36.0′S, 176°21.0–17.0′W, 1688–1924 m].

Type material. *Holotype*: NIWA 23843, male (cl 84.0 mm, pcl 74.6 mm, cw 75.7 mm), near Scott Island, 65°40.0–36.0'S, 176°21.0–17.0'W, 1688–1924 m, from stomach of *Dissostichus mawsoni* Norman, RV *Avro Chieftain*, set 3, coll. D. Stevens, 2 Dec 2003.

Paratypes: NIWA 3435, 1 ovigerous female (badly damaged, pcl about 61.2 mm), E of Sturge Island, Balleny Islands, 67°25.37'S, 165°15.82'E, 1389 m, TAN0402/214, 3 Mar 2004; MNZ Cr10856, 1 male (cl 73.4 mm, pcl 61.7 mm, cw 62.0 mm), Ross Sea, 66°46.90–44.90'S, 172°56.4–173°01.70'W, 1846–1420 m, found on hook, RV *Avro Chieftain*, trip 1742, set 51, coll. J. Buirski, 14 Mar 2003.

Other material examined. *Ross Sea, Antarctica*: NMNZ Cr11743–11744, 1 male (cl 82.4 mm, pcl 71.0 mm, cw 70.2 mm), 1 ovigerous female (cl 87.8 mm, pcl 70.4 mm, cw 67.9 mm), 64°41.00–44.7'S, 172°59.7–55.2'W, 1327–1163 m, Trip 1742, set 84, FV *Avro Chieftain*, coll. G. Anderson, 7 Apr 2003; NMNZ Cr11254, 1 parasitised male (cl 88.5 mm, pcl 75.0 mm, cw 76.0 mm), 66°35'S, 177°33'W, 1549 m, trip 1728, set 40, FV *Janas*; NIWA 38499, 1 walking leg, 67°24.48–24.33'S, 179°48.55–48.28'W, 1520–1560 m, TAN0802/237, 7 Mar 2008; NIWA 27835, 1 male (cl 96.7 mm, pcl 81.7 mm, cw 84.6 mm), 70°49'S, 178°41'W, trip 2333, set 47, FV *San Aspiring*, coll. Jano van Heerden, 17 Jan 2007.

Amundsen Sea, Antarctica: NMNZ Cr11804, 1 male (cl 68.4 mm, pcl 57.2 mm, cw 58.2 mm), area 88.2E, 69°41.00–40.00'S, 124°31.00–38.00'W, 1432–1560 m, FV *San Liberatore*, trip 1743, set 78, sample 12, camera 1, photo 18, coll. G. Dolan, 11 Apr 2003.

Diagnosis. Carapace subpentagonal, length and width subequal; surface glabrous, sparsely covered with small, scattered granules and short conical spines including median gastric spine; lateral margins with short conical spines; outer distance between bases of anterolateral spines half carapace width. Rostrum trispinose, broadest basally, without constriction. Scaphocerite with inner and outer spines. Mature male chelipeds strongly dimorphic, spinose; major cheliped palm 1.7–1.9 times height of minor cheliped. Major cheliped palm of female about 1.3 times height of minor cheliped. Walking legs elongate, 2–3 times pcl, spinose; ventral spines of propodi widely spaced; dactyli shorter than extensor margin of propodi. Walking leg 3 merus as long as or longer than pcl (male), about 0.6 pcl (female), length exceeding 6.5 times height (males), 4.7 (female); propodus length exceeding 8.4 times height (male), 6.7 (females).

Description. *Carapace*: Subpentagonal, length 0.97–1.04 times width; surface glabrous, sparsely covered with small, scattered granules. Gastric region convex, elevated, more prominent than other regions, with 5 short conical spines forming pentagon with anteriormost in midline. Lateral margin of hepatic region with 3 short conical spines, anterior shorter. Branchial regions with 12–14 short marginal spines; surface with 4 or 5 low conical spines in addition to low, scattered tubercles. Cardiac region subtriangular, with 2 pairs of short conical spines forming square. Intestinal region sparsely granular. Pterygostomian region sparsely granular, with prominent anterior spine.

Rostrum 0.13–0.25 pcl; broadest basally, without constriction; median spine smooth ventrally; paired dorsal spines divergent or subparallel, directed obliquely upwards. Posterior orbital margin near concave; outer orbital spine reaching to base of cornea. Anterolateral spine as long as or shorter than outer orbital spine, with low marginal granule or small spine midway between outer orbital spine and anterolateral spine; outer distance between bases of anterolateral spines half carapace width.

Ocular peduncle: Longer than cornea; with 2–5 dorsal granules.

Antennule: Peduncle unarmed, reaching anteriorly beyond antennal peduncle by about three-quarters length of distal antennular peduncle article.

Antenna: Basal antennal article with small anterolateral spine. Article 2 with angular to sharp inner distal margin; outer margin with small basal spine and slender spine not reaching beyond article 4. Article 3

Figure 101. *Paralomis stevensi* Ahyong & Dawson, 2006, male, pcl 71.0 mm, Ross Sea, Antarctica (MNMZ Cr11743). A, dorsal habitus. B, carapace. C, ventral surface and abdomen.

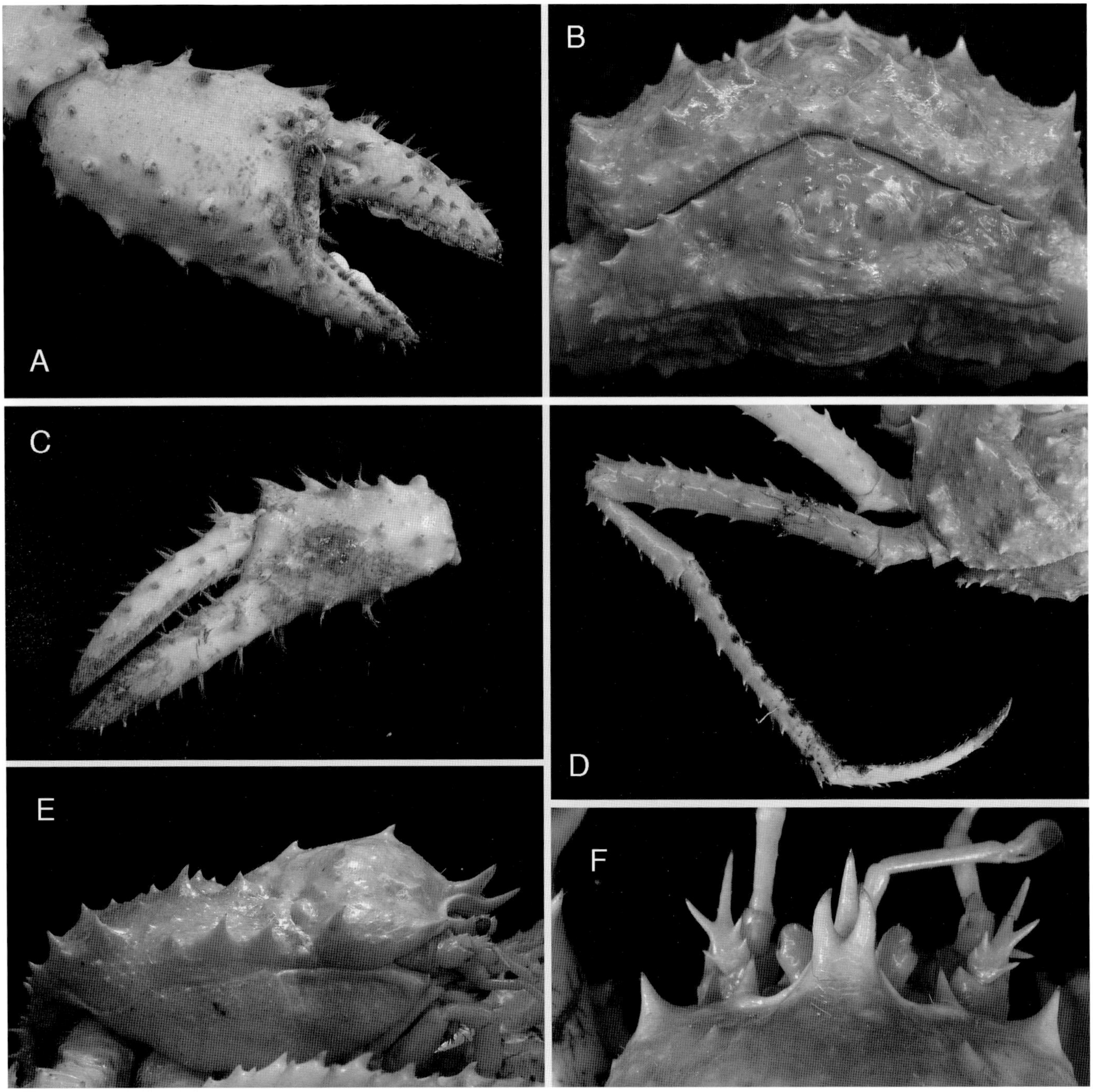

Figure 102. *Paralomis stevensi* Ahyong & Dawson, 2006, male, pcl 71.0 mm, Ross Sea, Antarctica (MNMZ Cr11743). A, right chela. B, posterior carapace and abdominal somite 2. C, left chela. D, right pereopod 4. E, carapace, right lateral view. F, anterior carapace, dorsal view.

unarmed. Scaphocerite slender, reaching beyond midlength but not beyond distal three-quarters of article 5, with 1 or 2 short inner spines and 1 or 2 outer spines. Article 4 unarmed, about half length of article 5.

Abdomen: Ornamentation similar in both sexes. Somites sparsely granular. Somite 2 with small, widely separated spines along margins. Somite 3 median plate unarmed; submedian plates with low, irregular, triangular teeth on lateral margin; marginal plates absent, apparently undifferentiated. Somites 4–5 with unarmed median and submedian plates; marginal plates undivided, with 2–4 low angular protrusions. Somite 6 as long as or longer than wide, subquadrate, with pair of small distal teeth; marginal plates short, triangular. Telson wider than long, apex rounded.

Pereopod 1 (chelipeds): Spination similar in both sexes. Major cheliped of males strongly inflated, 1.66–1.91 times height of minor cheliped; minor cheliped slender. Chelipeds of female unequal, major cheliped 1.31–1.33 times height of minor cheliped. Coxae smooth, un-

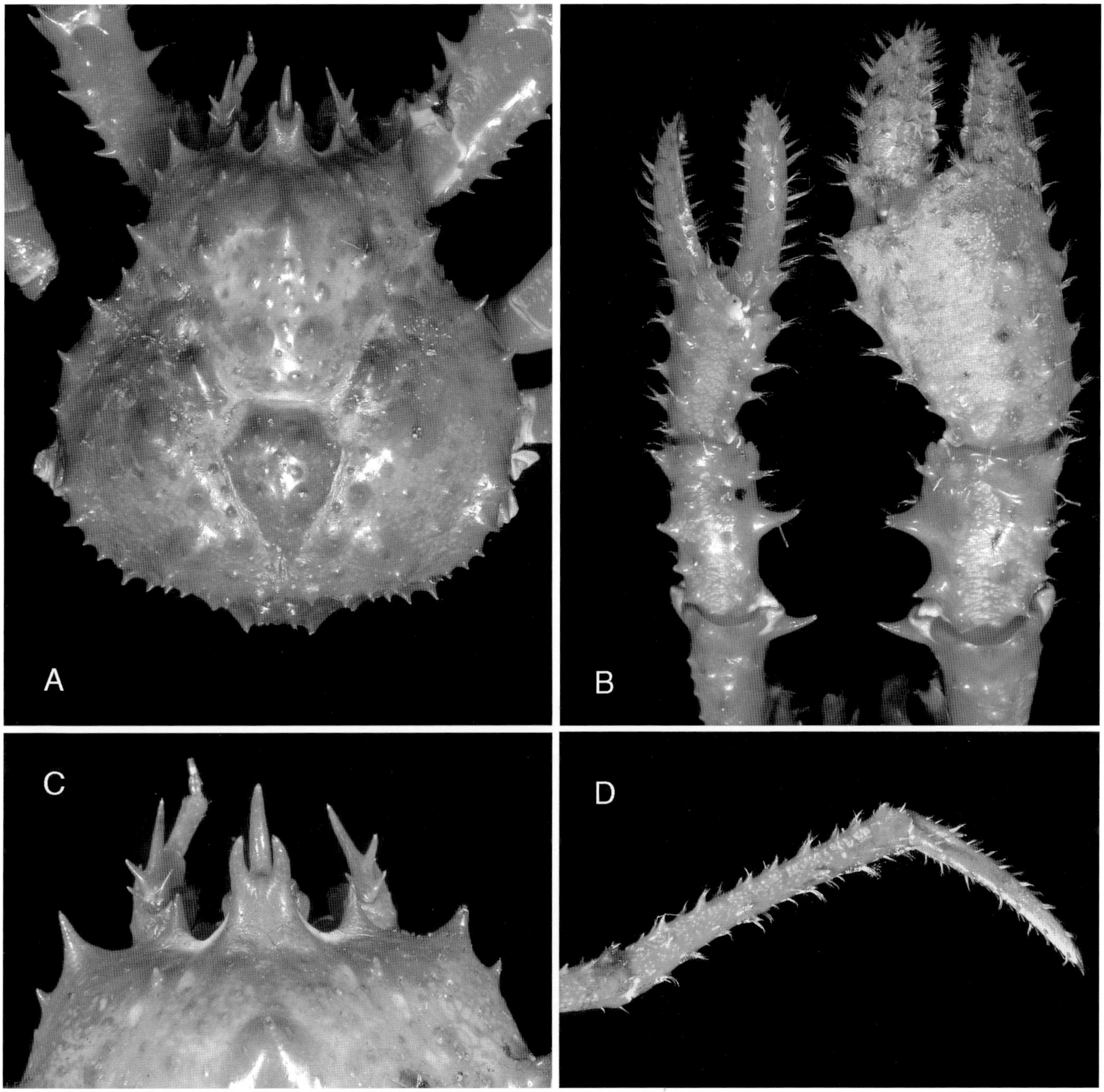

Figure 103. *Paralomis stevensi* Ahyong & Dawson, 2006, male, pcl 81.7 mm, Ross Sea, (NIWA 27835). A, carapace. B, chelipeds. C, anterior carapace, dorsal view. D, right pereopod 4 dactylus and propodus.

armed; distal margins with dense tufts of setae. Ischiobasis with 4–8 stout, apically setose ventral spines. Merus with smooth mesial margins and tuberculate lateral margins; dorsal and ventral margins spinose, inner distal spine largest. Carpus with prominent spines on dorsal and lateral surfaces; dorsal margin with row of 3–5 spines; mesial margin with 2 spines, proximal largest; ventral surface with 3 spines. Palm mesial margin with 6 spines, other surfaces with prominent, apically setose tubercles or acute tubercles. Fingers with short basal tubercle and rows of tufts of golden setae.

Major cheliped 1.62–2.22 pcl (male), 1.32–1.36 (female); upper palm length 1.03–1.15 times height (male), 1.19–1.24 (female); occlusal margins of fingers corneous for distal third, proximally with 3 calcareous nodules; dactylus dorsal margin broadly convex, with rows of tufts of golden setae and blunt proximal tooth, 1.04–1.15 times longer than dorsal margin of palm (male), 1.26–1.28 (female).

Minor cheliped 1.53–1.86 pcl (male), 1.31–1.39 (female); upper palm length 1.15–1.31 times height (male), 1.28–1.32 (female); occlusal margin corneous

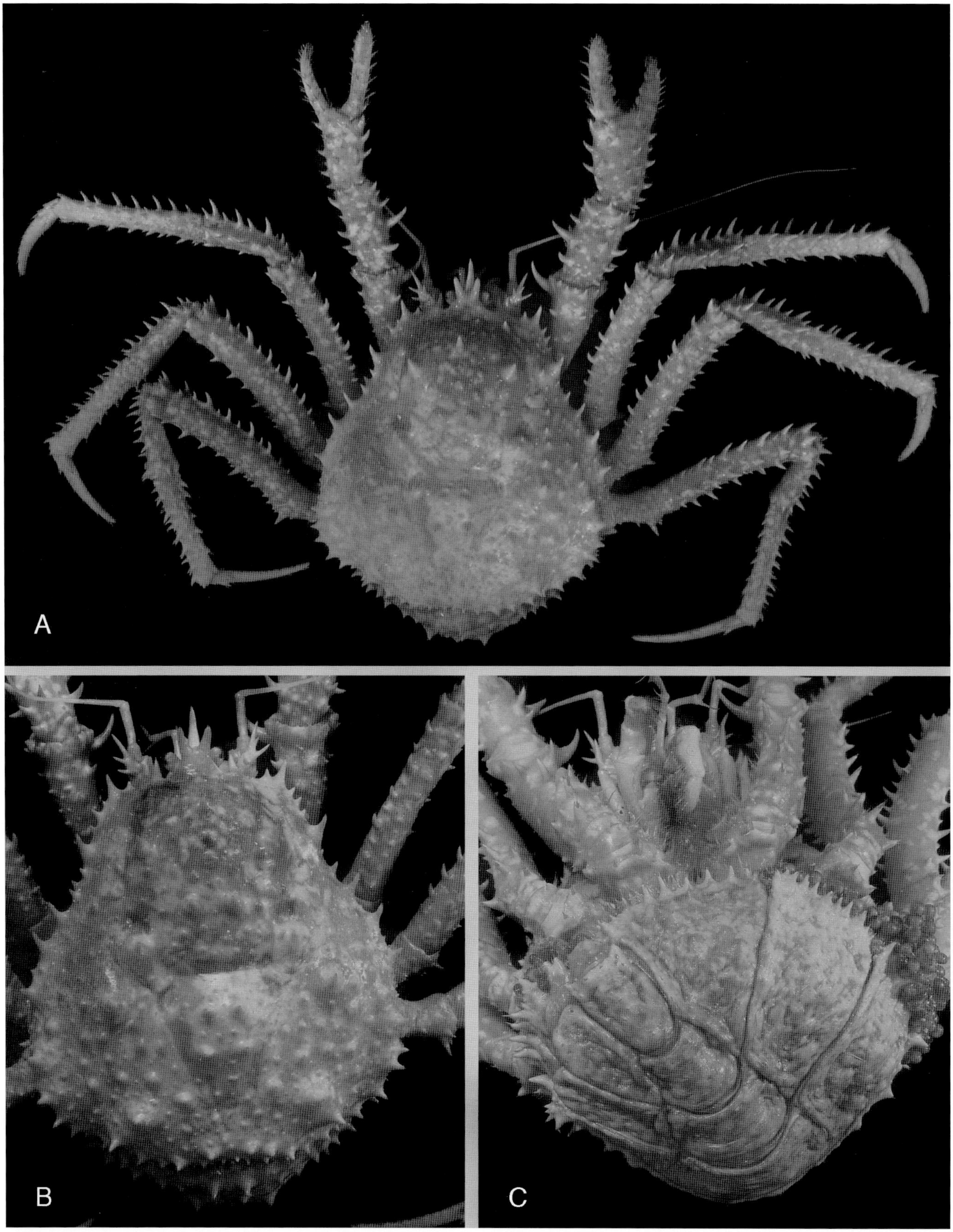

Figure 104. *Paralomis stevensi* Ahyong & Dawson, 2006, ovigerous female, pcl 70.4 mm, Ross Sea (NMNZ Cr11744). A, dorsal habitus. B, carapace. C, ventral surface and abdomen.

for slightly less than distal third, proximally crenulate; dactylus dorsal margin broadly convex, with rows of tufts of golden setae and blunt proximal tooth, 1.57–1.68 times longer than dorsal margin of palm (male), 1.55–1.88 (female).

Pereopods 2–4 (walking legs 1–3): Similar, elongate, spinose. Coxae smooth, unarmed; distal margins with dense tufts of setae. Ischiobasis with 3–5 apically setose ventral spines. Merus compressed, shorter than carapace in both sexes; extensor margin with 9–11 spines in addition to paired distal spines; dorsal surface with 5–8 spines; flexor margin with 2 rows of 4 or 5 spines; merus of pereopod 3 slightly longer than that of pereopod 2 and pereopod 4. Carpus slightly longer than half merus length, subcircular in cross section; extensor margin with 7 or 8 spines; dorsal surface with 7 or 8 spines; flexor margin unarmed. Propodus dorsoventrally flattened; slightly shorter than merus; with 11–13 spines on extensor margin; dorsal surface with small scattered spines; flexor margin with 7 or 8 spines. Dactylus broadly curved; slightly shorter than extensor margin of propodus; surface with tufts of setae; extensor margin with 3 or 4 apically corneous spines proximally; lateral proximal surfaces with short, distinct sulcus, flanked ventrally by 1 or 2 small, corneous spines; flexor margin lined with corneous spinules.

Pereopod 2 length 2.63–3.24 pcl (male), 2.06–2.10 pcl (female). Merus 0.86–1.10 pcl (male), 0.62–0.68 pcl (female); length:height ratio 5.82–7.35 (male), 4.17–4.52 (female). Carpus 0.50–0.55 merus length (male), 0.60 (female). Propodus 0.83–0.86 merus length (male), 0.89–0.94 (female); length:height ratio 7.50–9.22 (male), 6.12–7.39 (female). Dactylus 0.76–0.89 propodus length (male), 0.80–0.86 (female).

Pereopod 3 length 2.80–3.33 pcl (male), 2.09–2.23 pcl (female). Merus 0.92–1.13 pcl (male), 0.62–0.73 pcl (female); length:height ratio 6.24–7.65 (male), 4.36–4.79 (female). Carpus 0.50–0.52 merus length (male), 0.57 (female). Propodus 0.79–0.87 merus length (male), 0.88–0.93 (female); length:height ratio 8.27–10.22 (male), 6.55–7.06 (female). Dactylus 0.71–0.91 propodus length (male), 0.75–0.89 (female).

Pereopod 4 length 2.74–3.30 pcl (male), 2.12–2.17 pcl (female). Merus 0.95–1.09 pcl (male), 0.61–0.62 pcl (female); length:height ratio 6.60–7.86 (male), 4.70–5.00 (female). Carpus 0.46–0.52 merus length (male), 0.58–0.63 (female). Propodus 0.76–0.92 merus length (male), 0.93–0.99 (female); length:height ratio 8.45–11.22 (male), 6.72–7.52 (female). Dactylus 0.73–0.78 propodus length (male), 0.76–0.91 (female).

Colour in life. Deep-red overall (Pl. 3G).

Remarks. Ahyong & Dawson (2006) described *Paralomis stevensi* based on several damaged specimens from the Ross Sea, Antarctica. The availability of additional specimens, including an intact ovigerous female, has permitted more complete documentation of the species; a redescription is provided above.

Paralomis stevensi most closely resembles *P. birsteini* Macpherson, 1988a (type locality: Ross Sea) and *P. gowlettholmes* sp. nov. (type locality: Tasmania) in the combination of an armed inner margin of the scaphocerite and carapace ornamentation in which the primary spines are low and simple, and the surface between the primary spines is smooth or sparsely granular. In addition to reaching a larger overall size, *P. stevensi* can be immediately distinguished from *P. gowlettholmes* by the proportionally shorter walking leg dactyli — shorter, rather than longer than the extensor margin of the respective propodi. *Paralomis stevensi* chiefly differs from *P. birsteini* in the following features: the dactyli of the walking legs are shorter, rather than slightly longer than the extensor margin of the propodi; the right cheliped of large males (pcl 71.0–81.7 mm) is considerably more strongly inflated than the left, with the palm 1.8–1.9 times the height of the left, in contrast to 1.5–1.6 for *P. birsteini* in a similar size range (pcl 69.5–90.9 mm). The right palm of female *P. stevensi* exceeds 1.3 times the height of the left compared to less than 1.3 (1.24–1.25) in *P. birsteini*. As also indicated by Ahyong & Dawson (2006), the walking legs of male *P. stevensi* are proportionally longer and more slender than in *P. birsteini* of similar size. Comparing large, similarly sized males of *P. stevensi* (pcl 81.7 mm) and *P. birsteini* (pcl 79.4 mm), the pereopod 4 length is 3.3 versus 2.7 pcl, the pereopod 4 merus length:height ratio is 7.0 versus 4.9, and the pereopod 4 propodus length:height ratio is 10.7 versus 6.7. Note that allometric effects should be considered when using walking leg morphometrics to separate *P. stevensi* from *P. birsteini*. The proportional lengths of the walking legs of the smallest male *P. stevensi* (pcl 57.2 mm) approach those of the 79.5 mm male *P. birsteini*.

Unfortunately, most morphometric differences in the walking legs overlap between female *P. stevensi* and *P. birsteini*, so the most useful features distinguishing females are those already mentioned above — differences in dactyl length relative to the propodus, and differences in left and right cheliped height.

The smallest ovigerous *P. stevensi* and *P. birsteini* are a similar size (pcl 62.4 mm and 61.1 mm, respectively), and considerably larger than the smallest ovigerous *P. gowlettholmes* (pcl 28.8 mm).

Distribution. Antarctica, from the Ross Sea and now the Amundsen Sea; 1389–1924 m.

Paralomis taylorae sp. nov.

(Figs 105–108, Pl. 3F)

MATERIAL EXAMINED. *Holotype*: TM G3609, female (cl 49.3 mm, pcl 39.8 mm, cw 37.5 mm), Tasman Sea, Hill V, approx. 94 km SSE of Southeast Cape, Tasmania, 44°23'S, 147°09'E, 1310–1320 m, epibenthic sled, RV *Southern Surveyor*, SS01/97 stn 67, 31 Jan 1997.

DIAGNOSIS. Carapace pyriform, slightly longer than wide, margins and surface densely covered with upright spines, surfaces glabrous; outer distance between bases of anterolateral spines exceeding half carapace width; hepatic and branchial margins evenly merging, without distinct constriction. Rostrum with 2 pairs of dorsal spines; subrostral lobe absent. Distal dorsal eyespine extending anteriorly beyond cornea. Scaphocerite with 3 inner spines and 1 longer outer spine. Abdomen densely covered in short spines. Chelipeds and walking legs covered with slender spines. Walking legs slender, elongate, merus length exceeding 4 times height, about 0.9 pcl; propodus length exceeding 9 times height (exceeding 10 on walking leg 3); coxae unarmed or at most with a few blunt nodules.

DESCRIPTION. *Carapace*: Pyriform, 1.06 times longer than wide; regions distinct; surface and margins densely covered with short, slender spines of varying length, and scattered granules; surface of spines glabrous; longest spine (on hepatic margin) 0.10 pcl; cervical groove distinct. Pterygostomian region sparsely spinulose, spines widely spaced; with prominent, anterior, submarginal spine.

Rostrum 0.24 pcl; broad basally, not markedly constricted proximal to dorsal spines; median spine slender, ventral proximal margin with 4 minute spinules, subrostral lobe absent; dorsal margin with 2 pairs of laterally divergent spines proximally, directed obliquely upwards; with scattered spinules and single short spine between distal pair of dorsal spines. Posterior orbital margin concave, unarmed; outer orbital spine slender, directed anteriorly, not reaching beyond apex of cornea (when eyes directed anteriorly). Anterolateral spine shorter than outer orbital spine; outer distance between bases of anterolateral spines exceeding half carapace width.

Ocular peduncle: Longer than cornea; dorsally with scattered spines, distal spine longest, about half corneal diameter, overreaching cornea.

Antennule: Peduncle unarmed, reaching anteriorly beyond apex of antennal peduncle by quarter length of distal antennular peduncle article.

Antenna: Basal antennal article with curved anterolateral spine and distal spinule. Article 2 dorsally and laterally spinulose, with long distalolateral spine reaching apex of article 4. Article 3 with 2 distal setose tubercles. Scaphocerite a long, slender spine not overreaching distal peduncular article; with long lateral spine (one-third length of main spine), 3 shorter mesial spines and 1 or 2 scattered granules dorsally. Article 4 unarmed, about half length of article 5.

Abdomen of female: Somite 2 covered with spines of similar length to dorsal carapace spines. Somites 3–6 densely covered with short spines and acute granules, margins spinulate. Telson rounded, wider than long, with 6 blunt setose tubercles.

Pereopod 1 (chelipeds): Strongly spinose, unequal. Coxae unarmed, setose; distal margins with dense tufts of setae. Ischiobasis, merus, carpus, and propodus covered with arcuate spines and acute granules, longest distally and dorsally. Dactylus and pollex with small proximal spines, otherwise unarmed.

Major cheliped 1.78 pcl (female); upper palm length 1.59 times height (female); occlusal margins of fingers corneous for distal third, proximally with 3 calcareous nodules, proximal lowest; dactylus dorsal margin broadly convex, with tufts of golden setae and small proximal spine, 1.19 times longer than dorsal margin of palm (female).

Minor cheliped 1.85 (female); upper palm length 1.75 times height (female); occlusal margin corneous in quarter, proximally crenulate; dactylus dorsal margin broadly convex, with tufts of golden setae and 5 small proximal spines, 1.75 times longer than dorsal margin of palm.

Pereopods 2–4 (walking legs 1–3): Similar, elongate, spinose. Pereopod 2 longest. Coxae, unarmed, distal margins crenulate. Ischiobasis spinose. Merus ovate in cross section; all surfaces spinose, spines longest on extensor margins, shortest ventrally. Carpus and propodus spinose on all surfaces. Dactylus curved, laterally compressed; longer than carpus; proximal two-thirds spinose; flexor margin lined with 15–17 obliquely inclined corneous spinules; apex corneous.

Pereopod 2 length 2.68 pcl (female). Merus 0.90 pcl (female); length:height ratio 4.87 (female). Carpus 0.54 merus length (female). Propodus 0.88 merus length (female); length:height ratio 9.31 (female). Dactylus 0.71 (female).

Pereopod 3 length 2.80 pcl Merus 0.92 pcl (female); length:height ratio 4.89 (female). Carpus 0.54 merus length (female). Propodus 0.90 merus length (female); length:height ratio 10.15 (female). Dactylus 0.71 (female).

Pereopod 4 length 2.74 pcl (female). Merus 0.85 pcl (female); length:height ratio 4.35 (female). Carpus 0.57 merus length. Propodus 0.99 merus length (female); length:height ratio 10.59 (female). Dactylus 0.70 (female).

COLOUR IN LIFE. Bright red (Pl. 3F).

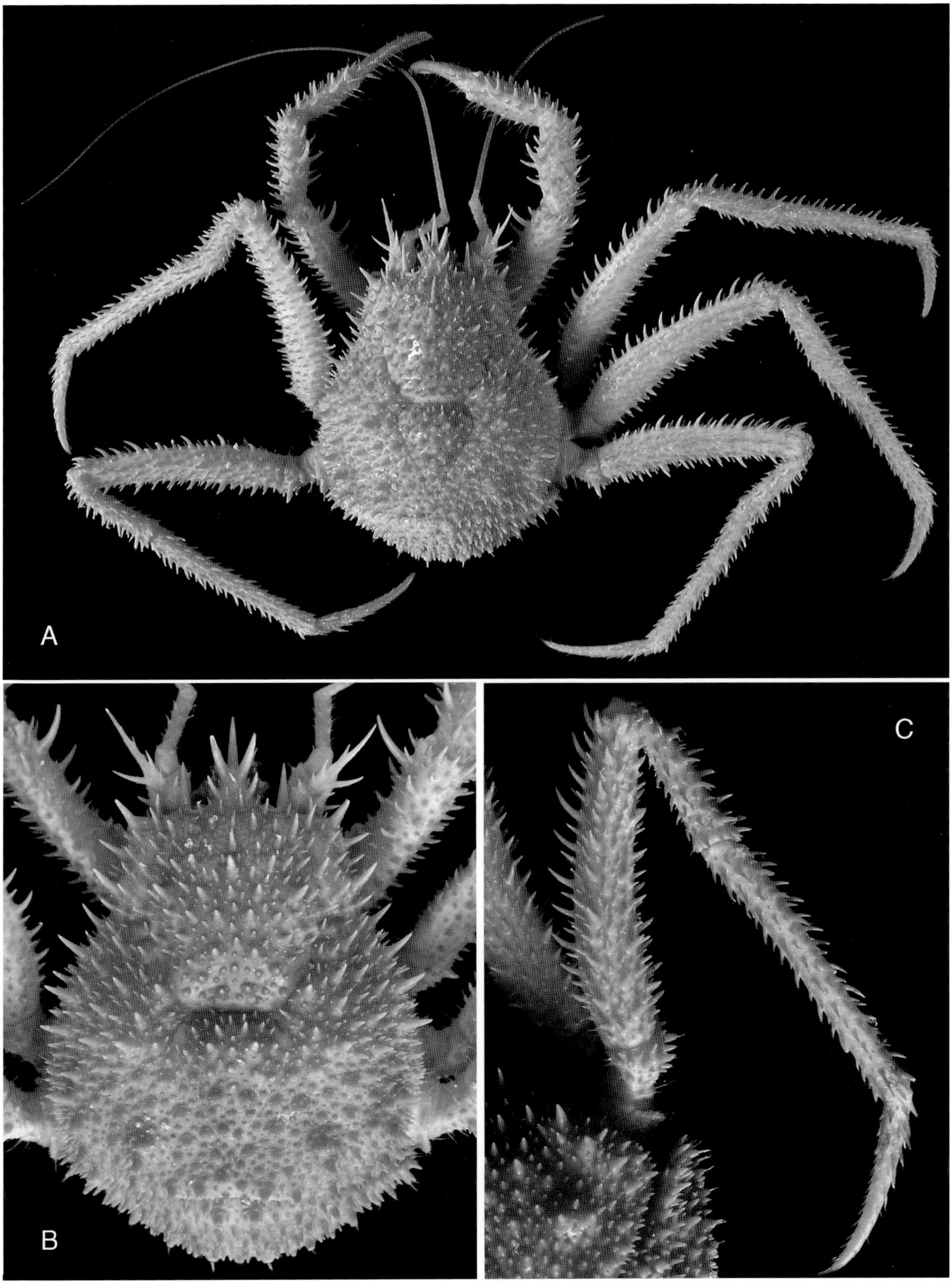

Figure 105. *Paralomis taylorae* sp. nov., female holotype, cl 49.3 mm, pcl 39.8 mm, cw 37.5 mm, Hill V, SSE of Southeast Cape, Tasmania (TM G3609). A, dorsal habitus. B, carapace. C, right pereopod 4.

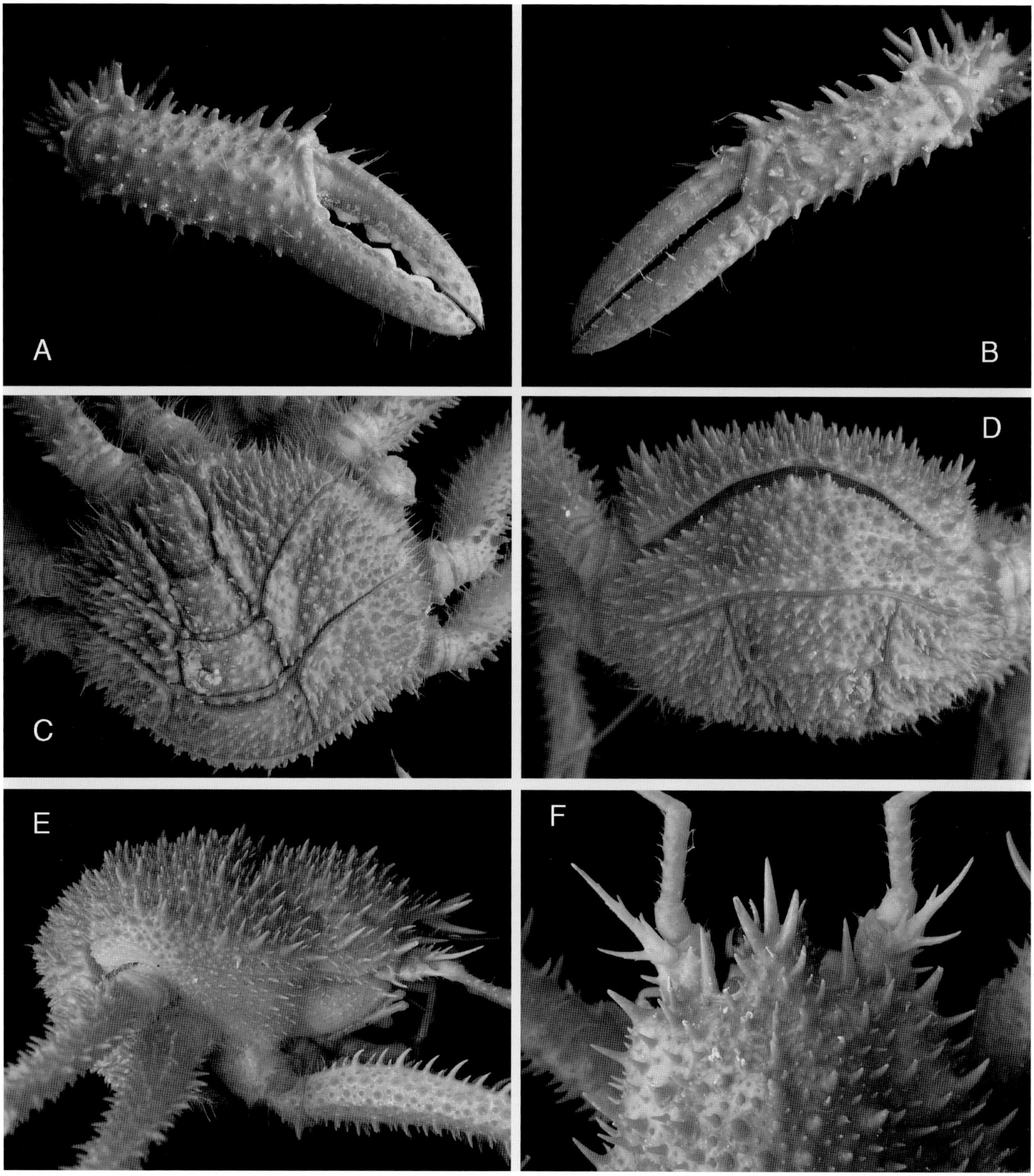

Figure 106. *Paralomis taylorae* sp. nov., female holotype, cl 49.3 mm, pcl 39.8 mm, cw 37.5 mm, Hill V, SSE of Southeast Cape, Tasmania (TM G3609). A, right chela. B, left chela. C, abdomen. D, posterior carapace and abdominal somite 2. E, carapace, right lateral view. F, anterior carapace, dorsal view.

Etymology. Named for Joanne Taylor, Museum Victoria, for her usual helpfulness and assistance with the Museum Victoria collections.

Remarks. On the basis of its densely spinose dorsum, *Paralomis taylorae* sp. nov. resembles *P. sonne* Guzmán, 2009 [type locality: Chile], *P. phrixa* Macpherson, 1992 [type locality: Peru], *P. poorei* sp. nov., *P. hystrixoides*

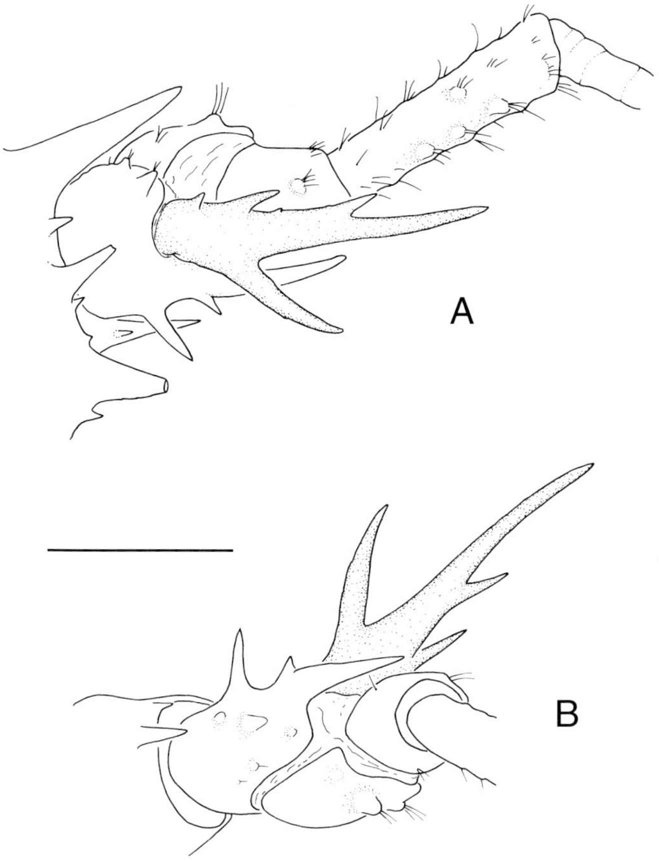

Figure 107. *Paralomis taylorae* sp. nov., female holotype, cl 49.3 mm, pcl 39.8 mm, cw 37.5 mm, Hill V, SSE of Southeast Cape, Tasmania (TM G3609). A, right antenna, dorsal view. B, right antenna, ventral view. Scale = 5.0 mm.

Sakai, 1980 [type locality: Japan], and *P. spinosissima* Birstein & Vinogradov, 1972 [type locality: off South Georgia Islands, southwestern Atlantic]. Along with *P. poorei* sp. nov., the new species differs from the remaining three species by lacking the prominent constriction in the carapace margin at the junction of the hepatic and branchial regions, and in the proportionally wider anterior carapace margin, with the distance between the bases of the anterolateral spines greater than half the maximum carapace width, rather than less than half. *Paralomis taylorae* is readily distinguished from *P. poorei* by its longer walking legs (pereopod 4 length in females exceeds 2.7 pcl versus less than 1.8 pcl) and more densely spinose carapace and walking legs (compare Figs 92 and 105; Tables 4, 5). In overall morphology, *P. taylorae* most closely resembles *P. sonne* in sharing similar carapace and antennal ornamentation, and elongate walking legs. In addition to the anterior width of the carapace and lack of a prominent carapace contriction, *P. taylorae* differs from *P. sonne* in having more slender propodi of the walking legs (pereopod 3 propodus length exceeding 10 times times height versus about 7 in *P. sonne*), in carapace proportions (slightly longer than wide rather than wider than long) and in having a sparsely, rather than a strongly and rather evenly spinose pterygostomial flap.

Paralomis taylorae further differs from *P. phrixa* and *P. hystrixoides* in having unarmed or, at most, bluntly nodular distoventral margins of the pereopodal coxae, versus spinose in the latter two species. *Paralomis taylorae* further differs from *P. spinosissima* in its more slender walking legs in which the merus length exceeds 4 times its height and the propodus exceeds 9 times its height in contrast to 3–4 and 4–6 times height, respectively.

The holotype of *P. taylorae* is a mature female, with setose ventral surfaces of the pereopodal coxae and a broad, well-developed abdomen.

Distribution. Presently known only from the type locality, off Southeast Cape, Tasmania; 1310–1320 m.

Paralomis webberi sp. nov. (Figs 81, 109–112)

Type material. *Holotype*: NMNZ Cr11141, ovigerous female (pcl 33.6 mm, cw 29.2 mm), Rumble III Seamount, Kermadec Ridge, 35°42.37′S, 178°28.69′E. 632–1255 m, trawled, 6 Sep 1998.

Diagnosis. Carapace pyriform, slightly longer than wide; surface and margins densely and uniformly covered with slender spines of similar length; outer distance between bases of anterolateral spines slightly less than half carapace width. Subrostral lobe spinulate. Ocular peduncle dorsally spinulate, overreaching cornea. Scaphocerite with 3 long lateral spines and 3 dorsal spines with longest about one-quarter length of longest lateral spine. Abdominal somites densely covered with slender spines, becoming shorter on successive somites. Chelipeds short, stout, outer and upper surfaces densely spinose; mesial surface of palm with low, scattered, setose tubercles, otherwise smooth; dactyli proximally spinose. Walking legs similar, short, stout, not exceeding 1.5 pcl; segments densely spinose except for ventral surface of meri and coxae; pereopod 4 merus length less than 2.5 times height; propodus length about 4.4 times height; dactylus shorter than carpus, surface spinose for proximal two-thirds. Somite 6 length 1.32 times width.

Description. *Carapace*: Pyriform, 1.15 times longer than wide; regions distinct; surface and margins densely and uniformly covered with slender spines of similar length; surface of spines with minute, scattered, well-spaced simple setae; longest spine (on anterior branchial margin) 0.18 pcl; cardiac region with 27

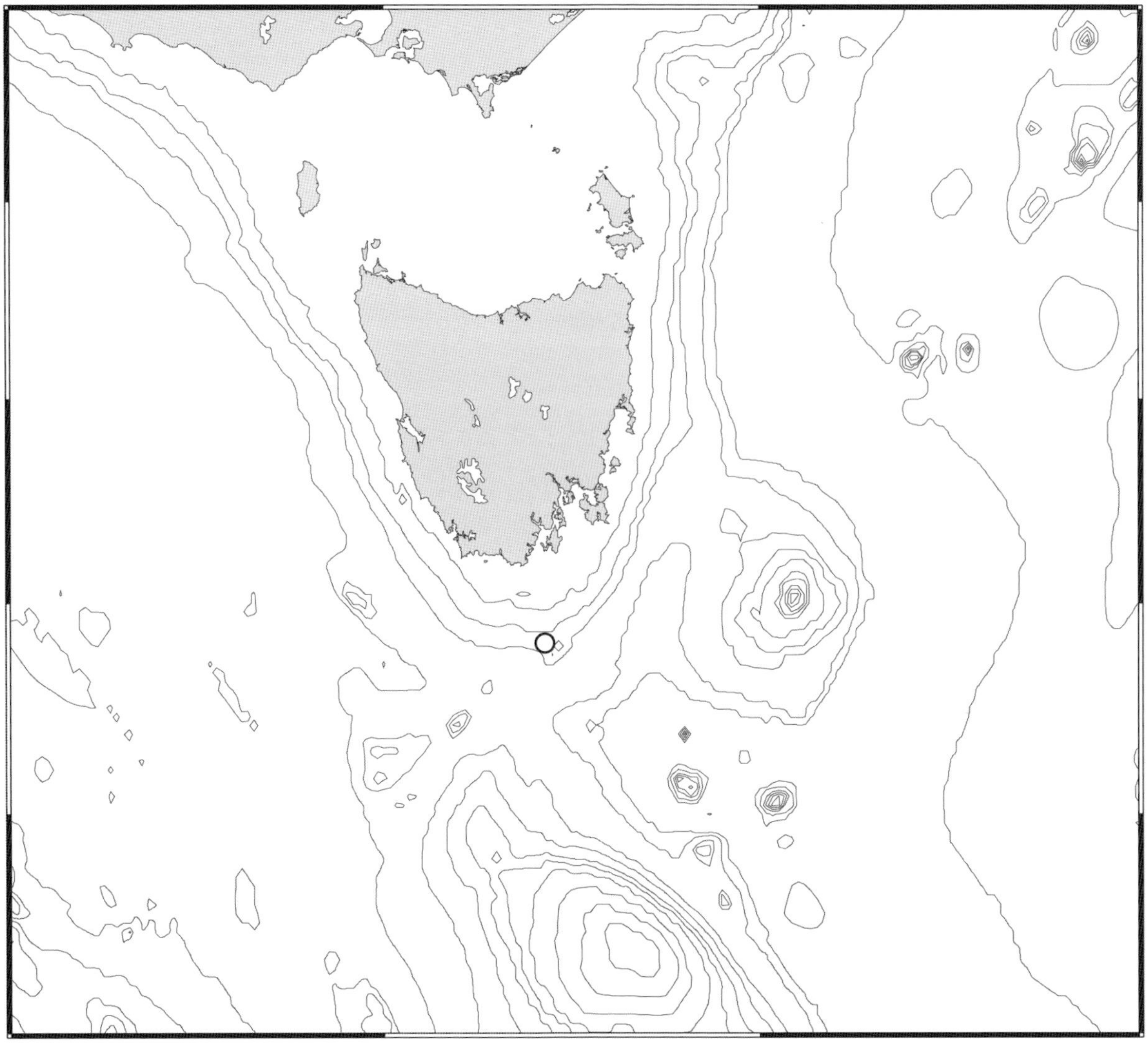

Figure 108. Geographic distribution of *Paralomis taylorae* sp. nov.

spines; cervical groove distinct. Pterygostomian region with slender anterior spine; spinulose on posterior two-thirds; surface of anterior third smooth except for scattered granules near dorsal margin and with cluster of 6 or 7 spinules anteriorly.

Rostrum 0.22 pcl; broad basally, not constricted proximal to dorsal spines; median spine slender, subrostral lobe angular, armed with 9 spinules; dorsally with 2 median spines and pair of laterally divergent spines directed obliquely upwards. Posterior orbital margin concave, unarmed; outer orbital spine slender, directed anterolaterally, reaching midlength of cornea. Anterolateral spine as long as outer orbital spine; outer distance between bases of anterolateral spines slightly less than half carapace width.

Ocular peduncle: Longer than cornea; with scattered granules and 10–12 dorsal spines, anterior 4 or 5 the longest, anteriorly directed, arranged in arcuate row above cornea; longest spine overreaching cornea by length of cornea.

Antennule: Peduncle unarmed, reaching anteriorly beyond antennal peduncle by length of distal antennular peduncle article.

Antenna: Basal antennal article with slender, curved, anterolateral spine, not reaching base of scaphocerite. Article 2 with 2 inner distal spines; outer margin with 3 or 4 slender spines and long distal spine reaching distal quarter of ultimate peduncular article. Article 3 with small inner distal tooth. Scaphocerite a long, slender spine overreaching cornea spines; lateral margins with 3 long spines; dorsally with 3 spines of about one-quarter length of longest lateral spine; inner margin unarmed. Article 4 unarmed, about one-third length of article 5.

Abdomen of female: Somites densely covered with slender spines, becoming shorter on successive somites.

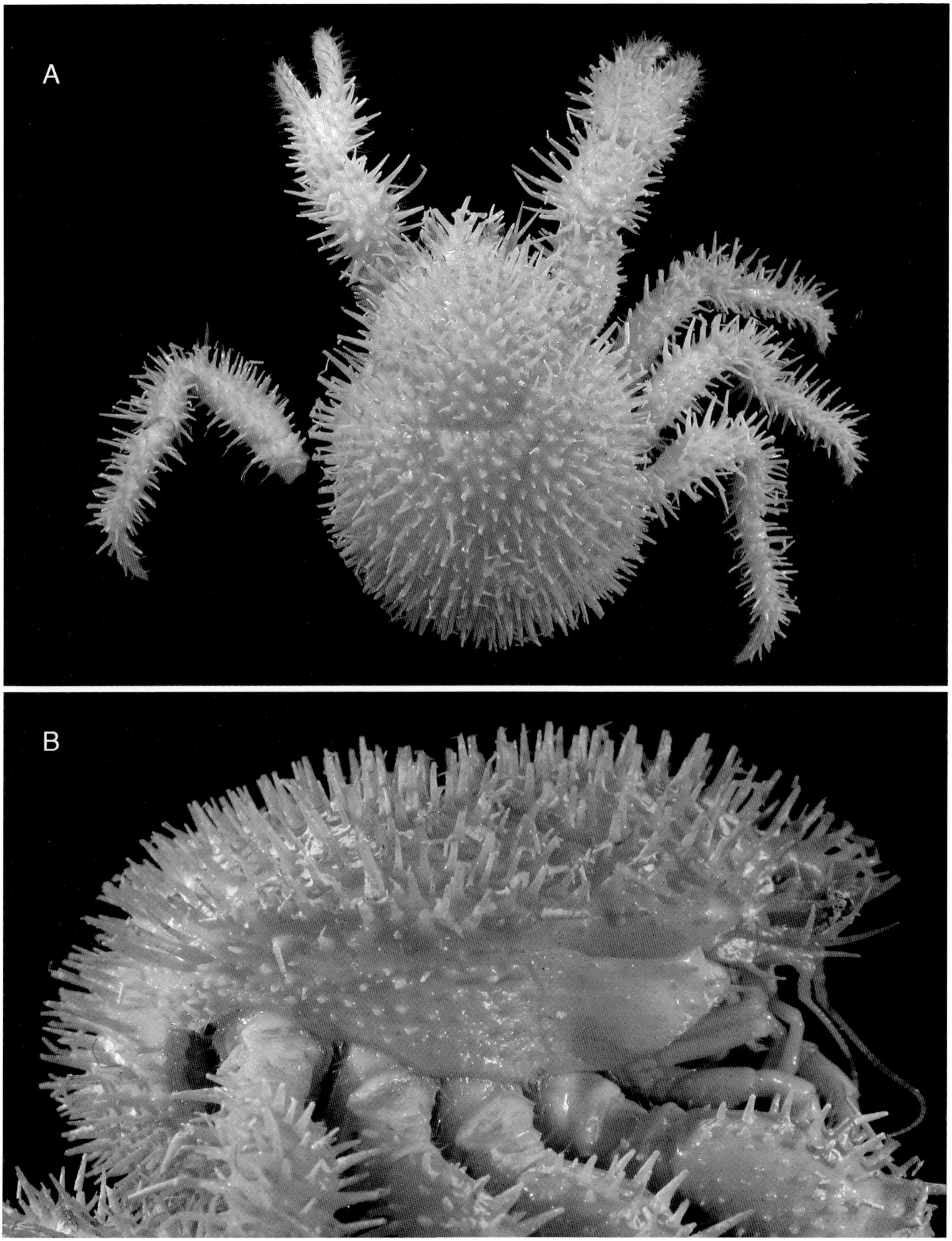

Figure 109. *Paralomis webberi* sp. nov., ovigerous female holotype, pcl 33.6 mm, cw 29.2 mm, Rumble III Seamount (NMNZ Cr11141). A, dorsal habitus. B, carapace, right lateral view.

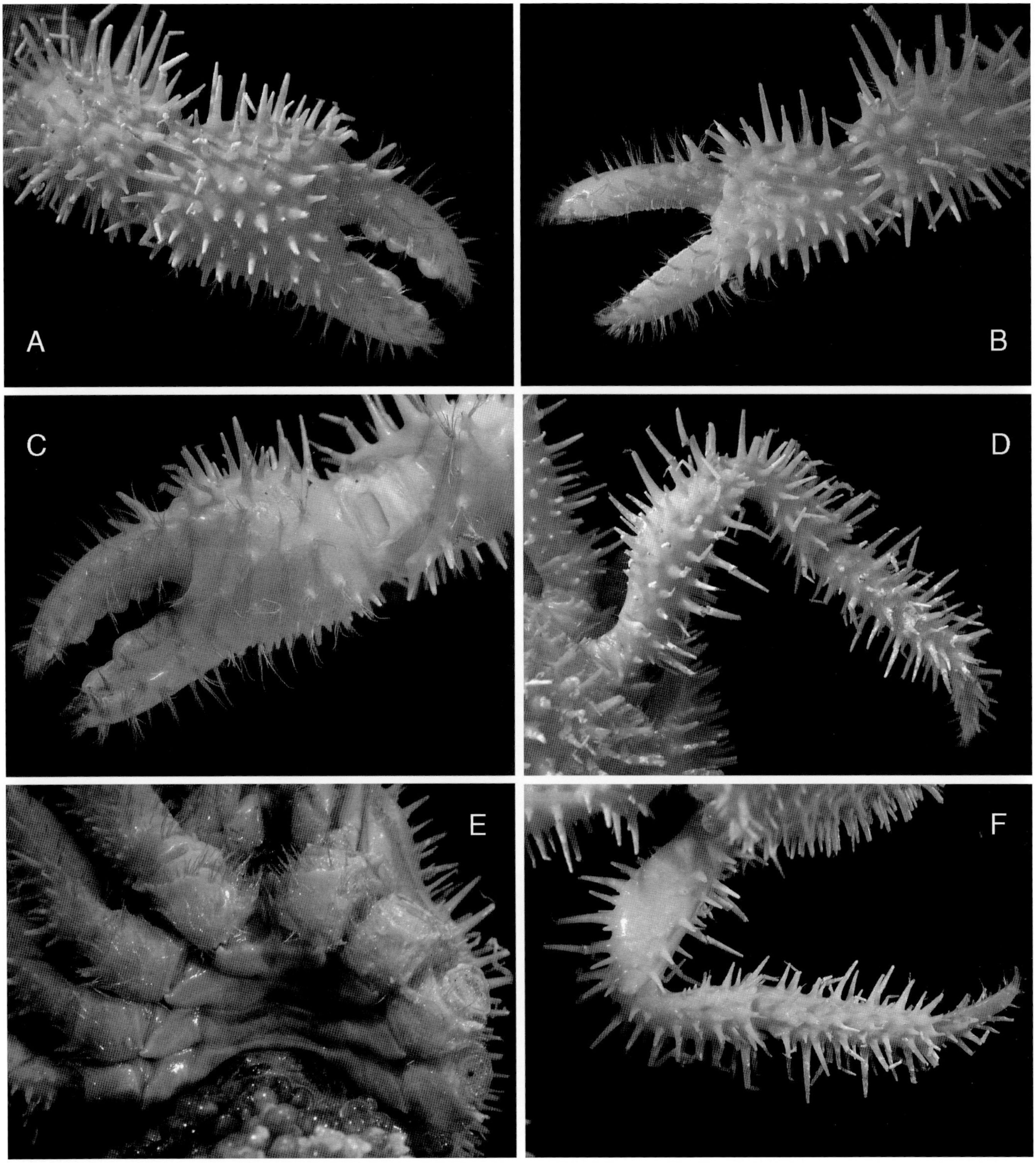

Figure 110. *Paralomis webberi* sp. nov., ovigerous female holotype, pcl 33.6 mm, cw 29.2 mm, Rumble III Seamount (NMNZ Cr11141). A, right chela outer surface. B, left chela, outer surface. C, left chela, inner surface. D, right pereopod 4, dorsal view. E, thoracic sternum and pereopodal coxa. F, right peropod 4, ventral view.

Spines on somite 2 similar to dorsal carapace spines. Spines of somites 3–6 less than half length of dorsal carapace spines. Somite 6 length 1.32 times width. Telson semicircular, with 8 blunt, apically setose tubercles.

Pereopod 1 (chelipeds): Densely spinose, unequal. Major cheliped 1.31 pcl, 1.41 times height of minor cheliped; minor cheliped 1.26 pcl. Coxae unarmed, setose; distal margins with dense tufts of setae. Ischiobasis spinose laterally, ventrally with clusters of

Figure 111. *Paralomis webberi* sp. nov., ovigerous female holotype, pcl 33.6 mm, cw 29.2 mm, Rumble III Seamount (NMNZ Cr11141). A, posterior carapace and abdominal somite 2. B, abdomen. C, anterior carapace, right lateral view. D, posterior abdominal somites and telson. E, carapace, anterior view. F, carapace, anteroventral view.

golden setae. Merus with smooth mesial surface, other surfaces spinose. Carpal surfaces densely spinose except for mesial surface with clusters of golden setae. Upper palm length 1.04 (major chela) and 1.11 times height (minor chela), surfaces densely spinose except for mesial surface with low scattered tubercles bearing tufts of golden setae. Fingers with tufts of golden setae; dactyli with proximal cluster of 5 spines. Fingers of major chela with occlusal margins corneous for distal quarter, proximally with 4 low calcareous prominences;

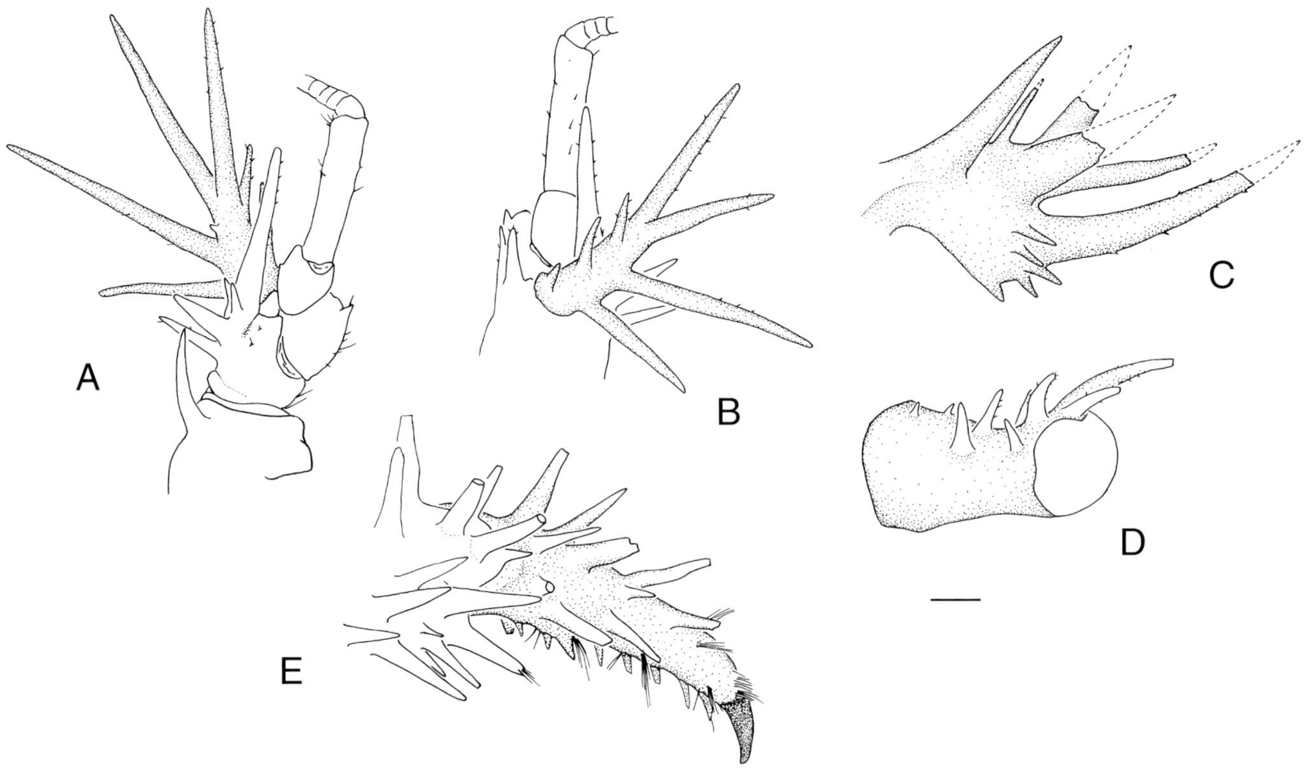

Figure 112. *Paralomis webberi* sp. nov., ovigerous female holotype, pcl 33.6 mm, cw 29.2 mm, Rumble III Seamount (NMNZ Cr11141). A, right antenna, ventral view. B, right antenna, dorsal view. C, rostrum, right lateral view. D, right eye, lateral view. E, right pereopod 4 dactylus. Scale = 1.0 mm.

dactylus 1.28 times dorsal margin of palm. Fingers of minor chela with occlusal margins corneous for distal third to half, proximally crenulate; dactylus 1.64 times upper margin of palm.

Pereopods 2–4 (walking legs 1–3): Similar, short, stout, densely spinose. Pereopod 2 longest, 1.33 pcl. Coxae unarmed, surface setose, distal margins crenulate. Ischiobasis spinose. Merus ovate in cross section, shorter than half pcl; dorsal, extensor and flexor surfaces spinose, longest extensor spines near distal third; ventral surface unarmed. Carpus spinose, longest spines as high as carpus. Propodus dorsoventrally flattened; spinose, longest spines about 1.5 times propodal height. Dactylus curved; shorter than carpus; surface spinose for proximal two-thirds; distal third with tufts of golden setae; flexor margin lined with 7 or 8 obliquely inclined corneous spinules; apex corneous.

Pereopod 2 length 1.33 pcl. Merus 0.42 pcl; length: height ratio 2.62. Carpus 0.77 merus length. Propodus 0.89 merus length; length:height ratio 4.37. Dactylus 0.65 propodus length.

Pereopod 3 length 1.26 pcl. Merus 0.39 pcl; length: height ratio 2.59. Carpus 0.80 merus length. Propodus 0.91 merus length; length:height ratio 4.25. Dactylus 0.66 propodus length.

Pereopod 4 length 1.23 pcl. Merus 0.36 pcl; length: height ratio 2.47. Carpus 0.84 merus length. Propodus 1.00 merus length; length:height ratio 4.37. Dactylus 0.65 propodus length.

Colour in life. Not known.

Etymology. Named for Rick Webber, Museum of New Zealand Te Papa Tongarewa, for his helpfulness throughout the course of this work and for making the large Te Papa lithodid collection available for study.

Remarks. *Paralomis webberi* sp. nov. closely resembles *P. histrix* (de Haan, 1844) from Japan and *P. echidna* sp. nov. from southeastern Australia in the short walking legs and densely spinose body and pereopods. As with *P. echidna*, *P. webberi* is distinguished from *P. histrix* by the pattern of carapace spination. Although spine length varies between the mid-dorsal surface and carapace margins, for example, adjacent spines are generally of uniform or near uniform length in *P. webberi* and *P. echidna*, whereas in *P. histrix*, the longer spines are interspersed with numerous shorter spines of about half the length of the former. Moreover, these shorter spines in *P. histrix* have rounded, rather than acute apices, and are often curved and inclined anteriorly. All dorsal carapace spines of *P. histrix* are glabrous, rather than minutely setose as in *P. webberi* and *P. echidna*. Distinctions between *P. webberi* and *P. echidna* are subtle, as follows:

- The density of surface spination is greater in *P. webberi* than in *P. echidna* (at least in females). For example, in *P. webberi* the cardiac region bears 27 spines versus 18–22 in *P. echidna* (in size-matched females). The differences are best observed, however, by direct comparison of specimens (Figs 74, 109). Although spine length and density varies allometrically in lithodids, the known females of *P. webberi* and *P. echidna* are of similar size, permitting direct comparison of size-related features.
- The subrostral lobe is multispinulate in *P. webberi*, unarmed in *P. echidna*.
- Abdominal somite 6 is more slender in *P. webberi* than in *P. echidna* (length 1.32 times width versus 1.08–1.15).
- The propodi of the walking legs are more slender in *P. webberi* (length:height ratio 4.37 versus 3.68–3.73 on pereopod 4 in females). Unfortunately, male *P. webberi* are not yet known.

Paralomis webberi superficially resembles juvenile *P. zealandica* in the dense dorsal spination. The most obvious distinguishing feature is size at maturity — *P. webberi* is mature by at least 33.6 mm pcl whereas *P. zealandica* matures at not less than about 60 mm pcl. *Paralomis webberi* is also readily distinguished from *P. zealandica* by its multispinulate subrostral lobe (unarmed or unispinose in *P. zealandica*), more dense dorsal spination, absence of short spines on the inner surfaces of the palms of the chelipeds, and presence of spines on the proximal two-thirds rather than proximal quarter to third of the extensor margins of the walking leg dactyli. Males of *P. webberi* are not presently known, though they can be expected to have proportionally longer pereopods than in females, as in other lithodids.

Unfortunately, the specific habitat on Rumble III seamount from which the holotype of *P. webberi* was collected is not known. Rumble III is an active hydrothermal vent site, so *P. webberi* is possibly a vent associate.

DISTRIBUTION. Presently known only from Rumble III Seamount, southern Kermadec Ridge; 532–1255 m.

Paralomis zealandica Dawson & Yaldwyn, 1971 (Figs 113–119, Pl. 2F, 4F)

Paralomis zealandica Dawson & Yaldwyn, 1971: 51–54 [type locality: Chatham Rise, 44°18′S, 174°31′E, 640 m]. — Dawson & Yaldwyn, 1985: 70. — Dawson, 1989: 318. — O'Shea *et al.*, 1999: 49, fig. 17. — Zaklan, 2002: 774, 796. — Batson, 2003: 137. — Webber & Naylor, 2004b: 62. — Naylor *et al.*, 2005: 45–42. — Ahyong *et al.*, 2007: 157. — Dawson, 2008: fig. 7.

Paralomis shinkaimaruae Takeda & Hatanaka, 1984: 14–17, fig. 3–5 [type locality: Bromley Plateau, 31°13′30″S, 34°49′W, 668 m; but see Remarks below]. — Macpherson, 1988b: 87–88, fig. 39, pl. 19B. — Zaklan, 2002: 773, 794. [New synonymy].

Acantholithus hystrix. — Dawson & Yaldwyn, 1985: 70 [misspelling; not *A. histrix* (de Haan, 1844)].

Paralomis zelandica. — McLay, 1988: 40–42, fig. 4. — Webber, 1997: 82.

Paralomis hystrix. — McLay, 1988: 44. — Takeda *in* Amaoka *et al.*, 1990: 362 [misspelling; not *P. histrix* (de Haan, 1844)].

TYPE MATERIAL. *Holotype*: NIWA 591, male (cl 111.6 mm, pcl 97.0 mm, cw 101.6 mm), Chatham Rise, 44°18′S, 174°31′E, 640 m, fine sandy mud, medium Agassiz trawl, Slope Benthos Cruise, FV *Taranui*, E0423, 15 Oct 1965.

OTHER MATERIAL EXAMINED. *Holotype* of *Paralomis shinkaimaruae* Takeda & Hatanaka, 1984: NSMT Cr8945, female (pcl 75.4 mm, cw 78.1 mm), Bromley Plateau, 31°13′30″S, 34°49′W, 668 m, RV *Shinkai-Maru*, 25 Mar 1977.

Cook Strait: NMNZ Cr11139, 1 male (cl 27.3 mm, pcl 21.5 mm, cw 21.9 mm), S of Cape Turakirae, VUZ96, 28 Aug 1957; NMNZ Cr11140, 1 male (cl 27.9 mm, pcl 23.1 mm, cw 24.4 mm), off Cape Turakirae, VUZ111, 28 Jan 1958; NMNZ Cr11142, 2 males (cl 24.9–32.9 mm, pcl 19.0–25.7 mm, cw 18.3–24.5 mm), 1 female (cl 23.7 mm, pcl 18.9 mm, cw 18.5 mm), off Cape Turakirae, 41°30.2′S, 174°52′E, 380 m, BS424, RV *Acheron*, 11 Dec 1974.

Chatham Rise: NMNZ, 1 male (cl 120.9 mm, pcl 112.7 mm, cw 119.9 mm), Independent Island, 42°59.2′S–43°09.6′S, 174°41.1–07.4′E, 693–635 m, Tow 48, trip 1591, N. Mitchell, 9 Jan 2002;

NIWA 42891, 1 female (pcl 79.4 mm, cw 78.9 mm), 43°07.29–05.71′S, 174°07.41–10.90′E, 668–690 m, TAN0701/94, 15 Jan 2007; NMNZ Cr11750, 1 male (cl 112.9 mm, pcl 100.3 mm, cw 107.4 mm), 3 ovigerous females (pcl 74.9–84.2 mm, cw 73.7–83.4 mm), Mernoo Saddle, 43°14.42′S, 173°55.63′E, 940–982 m, WIL/04/89, FV *Will Watch*, 23 Sep 1989; NIWA 42892,1 female (pcl 93.3 mm, cw 92.3 mm), 43°26.07–28.51′S, 174°14.52–16.90′E, 563–557 m, TAN0701/96, 15 Jan 2007; NIWA 61193, 1 female (cl 110.5 mm, pcl 93.7 mm, cw 95.6 mm), 43°26.68′S, 174°05.50–05.82′E, 709–739 m, TAN0101/120, 21 Jan 2001; NIWA 34927, 1 male (cl 121.0 mm, pcl 99.7 mm, cw 99.7 mm), 43°51.7–57.4′S, 179°25.38–12.50′E, 495–574 m, trip 2413/89, coll. J. Williamson, 13 May 2007; NMNZ Cr12022, 1 male (cl 87.3 mm, pcl 73.6 mm, cw 78.5 mm), 43°52.6–51.2′S, 179°14.4–17.4′W, 254–305 m, bottom temp. 4.5°C, FV *Azuchi Maru*, coll. M. Yates, 12 Oct 1987; NMNZ, 1 male (cl 137.8 mm, pcl 91.1 mm, cw 93.9 mm), 43°57.99′S to 44°00.00′S, 178°49.36–52.48′E, 638–675 m, TAN9212/074, 11 Jan 1993; NIWA 3921, 2 males (cl 46.2–58.8 mm, pcl 34.0–45.7 mm, cw 34.7–43.5 mm), 44°03.57–03.76′S,

179°20.82–22.89′E, 677–692 m, TAN9812/95, Z9384, 27 Oct 1998; NIWA 61192, 1 male (cl 101.7 mm, pcl 81.5 mm, cw 86.1 mm), 44°03.57–03.76′S, 179°20.82–22.89′E, 677–692 m, TAN9812/95, Z9384, 27 Oct 1998; NIWA, 2 females (cl 106.4–118.7 mm, pcl 90.6–101.4 mm, cw 90–103.4 mm; larger ovigerous), 44°03.57–03.85′S, 179°21.99–23.67′E, 673–688 m, TAN9812/94, Z9385, 27 Oct 1998; NIWA 61194, 1 female (pcl 87.8 mm, cw 84.1 mm), 44°09.52–10.84′S, 179°55.36–51.57′W, 627–636 m, TAN0101/67, 10 Jan 2001; NMNZ, 2 males (cl 75.9 mm, pcl 60.1 mm, cw 63.1 mm; damaged, approx. pcl 80 mm), 44°12.9′S, 179°39.30′W–178°24.73′W, 598–560 m, TAN9105/26, 1 Jan 1992; NMNZ Cr5045, 1 male (cl 117.3 mm, pcl 103.2 mm, cw 106.8 mm) W of Chatham Islands, 44°13.4–14.5′S, 178°14.4′W–177°53.1′W, 517 m, FV *Azuchi Maru*, coll. M. Yates, 21 Oct 1987; NMNZ Cr5897, 2 males (cl 118.5–120.8 mm, pcl 99.0–101.5 mm, cw 104.1–110.0 mm), 44°19.6–19.9′S, 173°36.6–37.0′E, 614–617 m, RV *James Cook*, J17/003/84, 29 Aug 1984; NMNZ, 1 male (cl 137.8 mm, pcl 117.4 mm, cw 127.2 mm), 44°20.80–21.64′S, 177°47.74′W–176°16.29′W, 626–631 m, TAN9106/39, 4 Jan 1992; NIWA 41407, 1 male (pcl 51.6 mm, cw 50.7 mm), 44°28.0′S, 175°47.0′E, 688 m, Trip 2617, FV *Amaltal Atlantis*, coll. S. Yeoman, 4 May 2008; NMNZ Cr9240, 1 male (cl 50.2 mm, pcl 38.3 mm, cw 37.4 mm), 44°28.81–29.71′S, 179°45.53–48.26′E, 1210–1212 m, TAN9104/55, 21 Oct 1991; NMNZ Cr11775, 1 male (cl 107.1 mm, pcl 86.7 mm, cw 89.4 mm), SE of Banks Peninsula, 44°31.00–36.83′S, 173°38.00–26.92′E, 761–764 m, RV *James Cook*, J15/13/83, 26 Nov 1983.

Bounty Plateau: NMNZ Cr11776, 1 female (cl 92.9 mm, pcl 77.8 mm, cw 81.5 mm), SE of Bounty Islands, 48°15.5–19.0′S, 179°48.5′E–180°00.0′E, 512–503 m, RV *Shinkai Maru*, stn 79, coll. J.G. Jones, 17 Nov 1978; NMNZ Cr11719–11720, 2 males (cl 123.6–124+ mm, pcl 106.9–114.9 mm, cw 112.3–124.6 mm), SW of Bounty Islands, 48°01–05′S, 178°32–34′E, 545–454 m, RV *Shinkai Maru*, 12/78 stn 239, 11 Jan 1979.

Campbell Plateau: NIWA 60575, 1 male (pcl 85.0 mm, cw 89.0 mm), 46°59.35′S, 169°58.25′E, 729–710 m, TAN0911/3, 27 Nov 2009; NIWA 61191, 1 male (pcl 87.9–92.5 mm, cw 90.1–90.5 mm), 1 female (pcl 87.9–92.5 mm, cw 90.1–90.5 mm), Campbell Plateau, 49°12.82–12.22′S, 167°53.74–58.24′E, 702–685 m, TAN0012/45, Z10601, 7 Dec 2000; NMNZ Cr5896, 1 male (pcl 94.2 mm, cw 96.5 mm), 2 females (larger ovigerous) (cl 98.7–114.7 mm, pcl 79.4–92.9 mm, cw 77.7–94.0 mm), 49°19.2–17.8′S, 173°38.4–39.4′E, 597–617 m, RV *James Cook*, J04/001/84, 17 Feb 1984; NMNZ Cr4857, 1 male (cl 68+ mm, pcl 60.9 mm, cw 58.5 mm), NW of Urry Bank, 49°30.6–39.54′S, 174°08.6–01.8′E, 798 m, FV *Oyang*, #7, Tow 5, coll. C.D. Roberts, 7 May 1987; NMNZ Cr11146, 1 female (pcl 50.73 mm, cw 51.29 mm), E of Auckland Islands, 50°50′S, 166°52′E, 390–400 m, FV *Peterson*, coll. D. Paterson & M. Miranovich, 11 Nov 1994; NIWA 60564, 1 spent female (cl 87.0 mm, pcl 69.4 mm, cw 70.3 mm), 51°12.33′S, 168°50.42′E, 552–555 m, TAN0911/44, 8 Dec 2009; NIWA 60572, 1 male (cl 113.3 mm, pcl 89.5 mm, cw 91.9 mm), 51°29.10′S, 173°06.03′E, 544–545 m, TAN0911/25, 3 Dec 2009; NIWA 41572, 1 male (pcl 52.5 mm, cw 52.5 mm), 1 ovigerous female (pcl 90.1 mm, cw 95.5 mm), 1 damaged female (pcl ~ 65 mm), N of Campbell Island, 51°44.04–43.00′S, 168°35.61–40.17′E, 540–517 m, TAN0317/33, 26 Nov 2003; NMNZ Cr12023, 1 ovigerous female (pcl 112.9 mm, cw 115.0 mm), NE of Campbell Island, 52°18.75′S, 171°26.10′E, 497–565 m, trip 1994, tow 21, coll. R. Hansen, 3 Oct 2004; NMNZ Cr11143, 1 female (cl 71.2 mm, pcl 54.2 mm, cw 60.3 mm), Campbell Rise, 53°00′S, 173°13′E, 508–510 m, RV *Shinkai Maru*, stn 269, 20 Jan 1977; NMNZ, 2 males (cl 94.3–116.3 mm, pcl 78.2–97.2 mm, cw 79.5–102.3 mm), SSW of Campbell Island, 53°05.29–02.46′S, 168°55.36–53.95′E, 619–641 m, TAN9105/60, 27 Nov 1991; NMNZ Cr6333, 1 male (cl +112 mm, pcl 96.3 mm, cw 106.2 mm), SE of Campbell Island, 53°14.4–08.3′S, 170°35.5–22.6′E, 400 m, FV *Ryuyu Maru*, coll. H. Kavale, 30 Aug 1987; NIWA 60571, 1 male (pcl 88.6 mm, cw 92.2 mm), 1 female (cl 117.3 mm, pcl 96.7 mm, cw 98.4 mm), 53°23.37′S, 170°46.91′S, 523–566 m, TAN0911/31, 5 Dec 2009; NIWA 61190, 2 females (pcl 87.9–92.5 mm, cw 90.1–90.5 mm), Campbell Rise, 53°28.39–31.09′S, 171°04.09–06.01′E, 558 m, TAN0012/23, Z10600, 1 Dec 2000; NIWA 60570, 1 male (cl 103.1 mm, pcl 88.3 mm, cw 88.9 mm), 53°37.16′S, 170°44.94′S, 776–778 m, TAN0911/32, 5 Dec 2009; NMNZ Cr11768–11770, 3 females (cl 79.8–113.2 mm, pcl.64.1–105.4 mm, cw 65.2–107.0 mm), Auckland to Campbell Islands, Feb 1992; NSMT Cr10117, 1 male (cl 42.9 mm, pcl 33.0 mm, cw 31.9 mm), Campbell Plateau, 542 m, T-142, 8 Nov 1983.

New Zealand, no specific data: NMNZ Cr11575, 2 males (cl 105.0–118.6 mm, pcl 89.1–100.5 mm, cw 90.8–105.8 mm), New Zealand, no other data; NMNZ Cr11576, 1 male (cl 104.0 mm, pcl 88.2 mm, cw 95.5 mm), New Zealand.

Diagnosis. Carapace pyriform, slightly longer to slightly wider than long, margins and surface densely covered with conical spines, surfaces glabrous, apices with few setae in early adults, corneous in late adults; outer distance between bases of anterolateral spines distinctly less than half carapace width. Rostrum prominently trispinous; subrostral lobe rounded or produced to small spine. Scaphocerite prominently multispinose. Abdomen densely covered with conical spines. Pereopods covered with spines, most elongate on dorsal and extensor margins; dactylus longer than carpus; spines on dactylar extensor margin restricted to proximal one-third of margin.

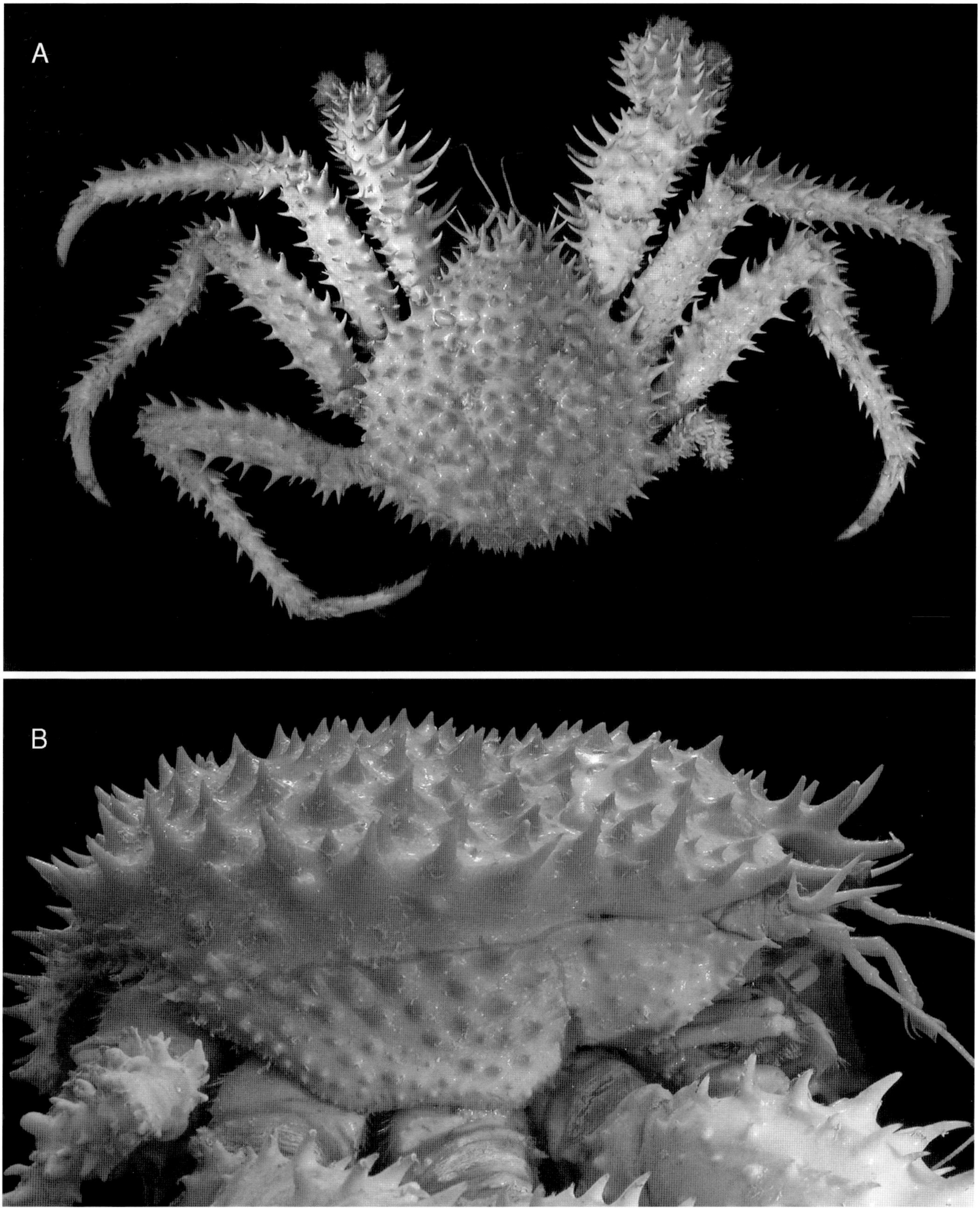

Figure 113. *Paralomis zealandica* Dawson & Yaldwyn, 1971, male, pcl 94.2 mm, Campbell Plateau (NMNZ Cr5896). A, dorsal habitus. B, carapace, right lateral view.

Figure 114. *Paralomis zealandica* Dawson & Yaldwyn, 1971, male, pcl 94.2 mm, Campbell Plateau (NMNZ Cr5896). A, carapace. B, ventral surface and abdomen. C, anterior carapace, dorsal view. D, posterior carapace and abdominal somite 2.

Description. *Carapace*: Pyriform, 0.90–1.05 times longer than wide; regions distinct; surface and margins uniformly covered with short, conical spines (longer and slender in early adults and juveniles); spines with smooth surface and several apical simple setae, which are lost in largest specimens; longest spine (on anterior branchial margin) 0.07–0.26 pcl; cardiac region with 13–15 spines; cervical groove distinct. Pterygostomian region with acute granules and small spines; with prominent anterior spine.

Rostrum 0.07–0.31 pcl; broad basally, not constricted proximal to dorsal spines; median spine slender, ventral lobe bluntly angular or with sharp apex; proximally with 2 pairs of laterally divergent spines directed obliquely upwards and single upright spine between proximal pair of dorsal spines. Posterior orbital margin concave, unarmed; outer orbital spine slender, directed anteriorly, overreaching eyes. Anterolateral spine as long as or longer than outer orbital spine; outer distance between bases of anterolateral spines distinctly less than half carapace width.

Ocular peduncle: Longer than cornea; dorsally with short, scattered granules or spines, distalmost spine longest; distal spine about half corneal diameter, overreaching cornea.

Figure 115. *Paralomis zealandica* Dawson & Yaldwyn, 1971, male, pcl 94.2 mm, Campbell Plateau (NMNZ Cr5896). A, right chela. B, left chela. C, anterior carapace, right lateral view. D, left pereopod 4 dactylus. E, left pereopod 4.

Antennule: Peduncle unarmed, reaching anteriorly slightly beyond apex of antennal peduncle by three-quarters to full length of distal antennular peduncle article.

Antenna: Basal antennal article with slender, arcuate anterolateral spine. Article 2 with slender inner spine, occasionally bifid; outer margin with 2–4 small spines proximally and long, distal spine overreaching article 3. Article 3 unarmed. Scaphocerite multispinose; primary spine long, slender, overreaching distal peduncular article; with 2 long lateral spines, first minute, second about two-thirds length of main spine; dorsomesially with 2 spines, shorter than half length of lateral spines. Article 4 unarmed, about half length of article 5.

Abdomen: Ornamentation of both sexes similar. Somites 2–6 covered with spines of similar length to dorsal carapace spines. Telson with 2 or 3 pairs of stout spines; semicircular.

Pereopod 1 (chelipeds): Strongly spinose, unequal; ornamentation similar on both sides and in both sexes.

Figure 116. *Paralomis zealandica* Dawson & Yaldwyn, 1971, ovigerous female, pcl 92.9 mm, Campbell Plateau (NMNZ Cr5896). A, dorsal habitus. B, left pereopod 4. C, carapace. D, ventral surface and abdomen.

Coxal margins spinose. Ischiobasis spinose laterally and ventrally. Merus with relatively smooth mesial surface, other surfaces spinose. Carpus prominently spinose, without setal clusters. Palm with most surfaces prominently spinose; inner (mesial surface) sparsely spinose, with some apically setose.

Major cheliped 1.51–1.68 pcl (male), 1.13–1.18 (female); upper palm length 0.94–1.02 times height (male), 0.98–1.06 (female); occlusal margins of fingers corneous for distal third, proximally with 3 calcareous nodules; dactylus dorsal margin broadly convex, with rows of tufts of golden setae and proximal cluster of 3–6 small spines, 0.93–1.10 times longer than dorsal margin of palm (male), 1.14–1.24 (female).

Minor cheliped 1.30–1.37 pcl (male), 1.07–1.11 (female); upper palm length 1.04–1.14 times height (male), 1.02–1.06 (female); occlusal margin corneous in distal half, proximally crenulate; dactylus dorsal margin

Figure 117. *Paralomis zealandica* Dawson & Yaldwyn, 1971. A–D, male holotype, cl 111.6 mm, pcl 97.0 mm, cw 101.6 mm, Chatham Rise (NIWA 591). E, female, pcl 92.5 mm, Campbell Rise (NIWA 61190). F, female, pcl 87.9 mm, Campbell Rise (NIWA 61190). A, dorsal habitus. B, ventral surface and abdomen. C–D, left pereopod 4, dorsal and ventral views. E–F, anterior carapace, right lateral view.

broadly convex, with rows of tufts of golden setae and proximal cluster of 2–4 small spines, 1.49–1.57 times longer than dorsal margin of palm (male), 1.52–1.61 (female).

Pereopods 2–4 (walking legs 1–3): Similar, elongate, spinose. Pereopod 3 longest. Coxae distally spinose; ventral surfaces with short spines or acute tubercles, becoming low and blunt in mature females and obsolete

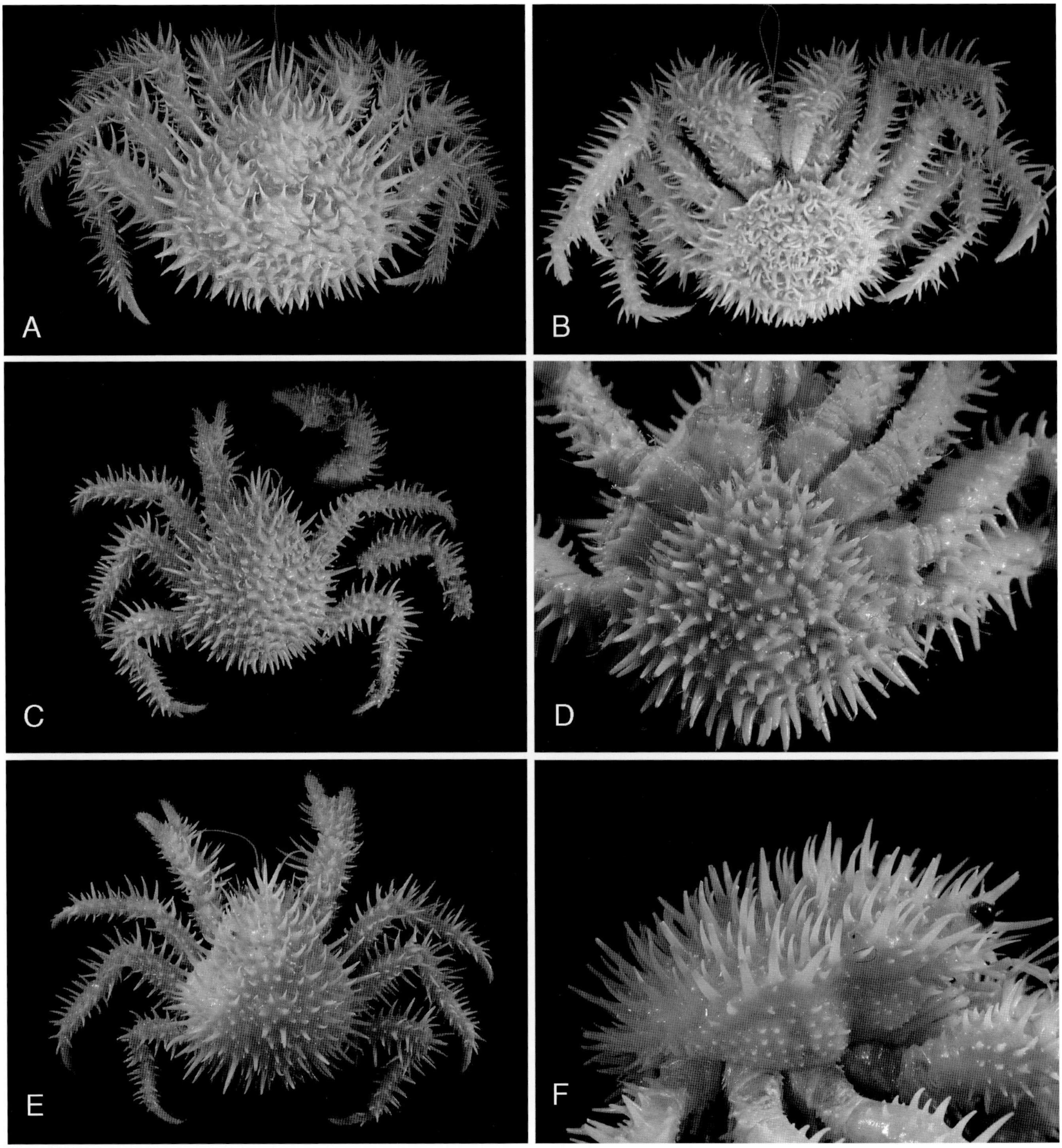

Figure 118. *Paralomis zealandica* Dawson & Yaldwyn, 1971, A–B, dorsal and ventral views, female, pcl 54.2 mm, Campbell Rise (NMNZ Cr11143). C–D, dorsal and ventral views, male, pcl 25.7 mm, Cook Strait (NMNZ Cr11142). E–F, dorsal and right lateral views, male, pcl 21.5 mm, Cook Strait (NMNZ Cr11139).

on pereopod 4. Ischiobasis spinose. Merus subovate in cross section; dorsal surface, extensor and flexor margins prominently spinose, each with relatively even rows of 7–12 spines; ventral surface with a row of small spines or acute tubercles near flexor and extensor margins. Carpus spinose on all surfaces. Propodus dorsoventrally flattened, shorter than merus; extensor and flexor margins each with row of 6–8 spines; dorsal and ventral surfaces with 2 relatively even rows of 6–8 spines. Dactylus curved, laterally compressed; longer than carpus (subequal in juvenile males < 30 mm pcl); with cluster of spines on proximal quarter to third, and scattered tufts of golden setae along length; flexor margin lined with 7–9 obliquely inclined corneous spinules; apex corneous.

Pereopod 2 length 2.31–2.34 pcl (male), 1.49–1.66 pcl (female). Merus 0.73–0.78 pcl (male), 0.47–0.51 pcl (female); length:height ratio 3.22–3.74 (male), 2.33–2.61 (female). Carpus 0.60–0.63 merus length (male), 0.68–0.76 (female). Propodus 0.77–0.81 merus length (male), 0.80–0.86 (female); length:height ratio 5.39–5.62 (male), 3.57–4.26 (female). Dactylus 0.96 propodus length (male), 0.81–0.90 (female).

Pereopod 3 length 2.44–2.53 pcl (male), 1.49–1.69 pcl (female). Merus 0.73–0.80 pcl (male), 0.45–0.53 pcl (female); length:height ratio 3.12–3.81 (male), 2.45–2.78 (female). Carpus 0.60–0.64 merus length (male), 0.67–0.74 (female). Propodus 0.79–0.84 merus length (male),

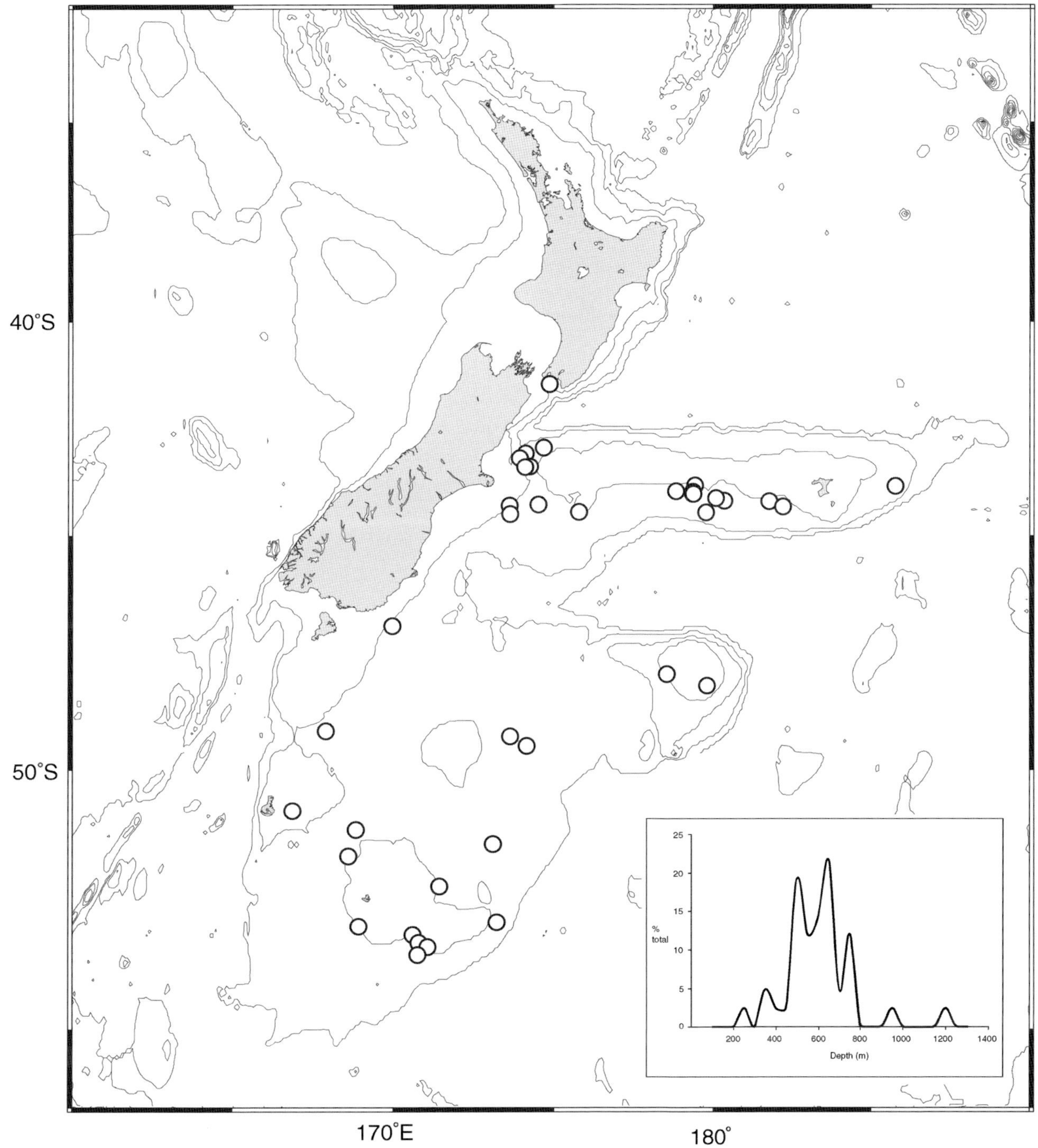

Figure 119. Geographic and bathymetric distribution of *Paralomis zealandica* Dawson & Yaldwyn, 1971.

0.81–0.87 (female); length:height ratio 5.32–6.01 (male), 3.67–4.44 (female). Dactylus 0.92–0.97 propodus length (male), 0.78–0.93 (female).

Pereopod 4 length 2.31–2.42 pcl (male), 1.46–1.68 pcl (female). Merus 0.53–0.74 pcl (male), 0.35–0.50 pcl (female); length:height ratio 2.98–4.01 (male), 2.48–2.81 (female). Carpus 0.63–0.72 merus length (male), 0.59–0.76 (female). Propodus 0.84–0.88 merus length (male), 0.89–0.92 (female); length:height ratio 5.69–6.22 (male), 3.76–4.60 (female). Dactylus 0.96–0.99 propodus length (male), 0.83–0.90 (female).

COLOUR IN LIFE. Overall ivory-white, often with pale red-orange mottling on pereopods. Carapace with spines on gastric and cardiac regions, and usually also branchial regions, red-orange. (Pl. 2F).

REMARKS. *Paralomis zealandica* Dawson & Yaldwyn, 1971, is the most widespread species of the genus in New Zealand, ranging from the Chatham Rise to the Bounty and Campbell plateaus. It attains a large size and undergoes significant allometric changes in dorsal spination. Juveniles are densely covered in slender spines, superficially resembling other small, spiny species such as the Japanese *P. histrix* (de Hann, 1844) (frequently misspelt as *P. hystrix*), *P. echidna* sp. nov. (southeastern Australia), *P. webberi* sp. nov. (Kermadec Ridge, New Zealand), and *P. poorei* sp. nov. (southern Australia and New Zealand). The dorsal spines of *P. zealandica* become proportionally shorter with increasing body size, becoming short conical tubercles in adults. These allometric changes in spination of *P. zealandica* have led to erroneous reports of the Japanese *P. histrix* from New Zealand (e.g. Dawson & Yaldwyn 1985; Takeda *in* Amaoka *et al.* 1990). The specimen on which the original New Zealand report of *P. histrix* was based (NMNZ Cr11143, Fig. 118A–B) is an immature female of *P. zealandica*. Similarly, Takeda *in* Amaoka *et al.* (1990) clearly depicts *P. zealandica* (as *P. hystrix*). As suggested by McLay (1988), New Zealand records of *P. histrix* are based on juvenile *P. zealandica*. *Paralomis zealandica* and *P. histrix* superficially resemble each other both as juveniles and adults because both species undergo similar allometric reduction in spination and attain a similar adult size. *Paralomis zealandica*, however, is readily distinguished from *P. histrix* by its more uniform length carapace spines (versus long spines interspersed with numerous, anteriorly recurved, apically rounded spines of about half the length of the former, Fig. 82), and in the extensor marginal spines of the dactyli of the walking legs being restricted to the proximal quarter to third, rather than the proximal half. Features distinguishing *P. zealandica* from *P. poorei*, *P. echidna*, and *P. webberi* are given under the respective accounts of the latter three species.

Other allometric changes in *P. zealandica* are in the proportional lengths of the walking legs, which are proportionally longest in large males, and in the proportions of the walking leg meri, which are more slender in the largest males (length:height ratio 2.7–2.8 at 19.0–25.7 mm pcl, 3.5–4.0 at 91.1–103.2 mm pcl). The walking leg dactyli are about as long as the carpus length in juveniles < 30 mm pcl, and longer than the carpus in larger specimens. The ventral surfaces of the coxae of the chelipeds and walking legs are armed with acute tubercles in juveniles of both sexes and retained in adult males. In females, however, the ventral coxal tubercles become obsolete as the abdomen expands with maturity, covering the coxae. The smallest ovigerous female is 74.9 mm pcl, although specimens appear to be mature by about 60 mm pcl.

Takeda & Hatanaka (1984) described *P. shinkaimaruae* from Bromley Plateau, southwestern Atlantic, based on an ovigerous female collected by the RV *Shinkai-Maru*, apparently in March 1977. Distinguishing *P. shinkaimaruae* from adult *P. zealandica*, however, is problematical. The holotype of *P. shinkaimaruae* differs from the holotype of *P. zealandica* in the presence of a small subrostral spine rather than a blunt lobe. This feature, however, is variable in *P. zealandica*, with the subrostral lobe often being produced into a small spine. Thus, the two species cannot be distinguished and are herein synonymised. The resulting disjunct distribution of *P. zealandica*, however, is anomalous. It is not adequately explained by sampling gaps because regular fishing effort in intermediate regions is likely to have yielded further specimens. Likewise, the prospect of artificial introduction is unlikely. A more parsimonious explanation is that the holotype of *P. shinkaimaruae* originated from New Zealand, probably the Bounty Plateau, where the *Shinkai-Maru* was operating in early 1977. The specimen was probably mislabelled or mixed-up with samples collected only one to two months later when the *Shinkai-Maru* continued operations on the Bromley Plateau, southwestern Atlantic.

Paralomis zealandica occurs only in southeastern New Zealand waters, from Cook Strait to the Campbell Plateau.

DISTRIBUTION. Southeastern New Zealand, from Cook Strait to the Campbell Plateau; 254–1212 m, usually 500–700 m.

DISCUSSION

Prior to the present study, a total of 10 extant species of king crab were believed to occur in New Zealand and Australia: *Lithodes murrayi* Henderson, 1888, *L. longispina* Sakai, 1971, *Neolithodes brodiei* Dawson & Yaldwyn, 1970, *N. flindersi* Ahyong, 2010, and six species of *Paralomis*. Seven lithodid species had been reported from New Zealand waters, and six from Australia (including Macquarie Island) of which three were believed to occur on both sides of the Tasman Sea. Three species in two genera of lithodid have been recorded from the Ross Sea, Antarctica. A total of 23 species in three genera are now known from the study area: 13 species in New Zealand, 12 in Australia (four shared by both regions) and three in the Ross Sea (wth *Paralomis birsteini* Macpherson, 1988a extending onto the Macquarie Ridge). The Macquarie Ridge record of *P. birsteini* is significant in providing a biogeographic link between Antarctic *P. birsteini* and *P. stevensi*, and the very similar *P. gowlettholmes* sp. nov. from southern Tasmania. Results of the present study demonstrate that rather than being relatively depauperate in comparison to the north Pacific, and presumed centre of origin of the group, the lithodid fauna of the southwestern Pacific is considerably richer than previously known.

That only four lithodid species are shared by Australia and New Zealand (*Lithodes macquariae* sp. nov., *Paralomis echidna* sp. nov., *P. poorei* sp. nov., *P. staplesi* sp. nov.) implies limited dispersal potential of the regional lithodids. Larvae of some lithodids, such as *Paralithodes camtschaticus*, are planktotrophic and therefore likely to disperse via ocean currents, but larvae of *Lithodes*, *Paralomis*, and probably *Neolithodes*, are lecithotrophic, limiting prospects for pelagic dispersal (Shirley & Zhou 1997; Thatje *et al.* 2005). *Neolithodes brodiei* and *L. longispina* were previously thought to occur on both sides of the Tasman Sea, but present results indicate that Australian records of *N. brodiei* are referable to *N. flindersi*, and records of *L. longispina* from New Zealand and Australia are based on a different new species in each region (*L. aotearoa*, *L. australiensis*). *Lithodes longispina*, itself, previously believed to be widespread (Japan to Guam, New Zealand and Australia) (Dawson 1989), has been shown to probably occur only in the Japanese region, following recognition of *L. aotearoa*, *L. robertsoni*, and *L. australiensis* herein, in addition to similar species in Guam (*L. paulayi* Macpherson & Chan, 2008) and Taiwan (*L. formosae* Ahyong & Chan, 2010). Likewise, the discovery that previous records of *L. murrayi* from Australia and New Zealand are referable to *L. aotearoa* and *L. macquariae* further undermines the presumed circumpolar distribution of *L. murrayi* (see also Macpherson 1988b, c). At present, *L. murrayi* is reliably known only from the southwestern Indian Ocean. Wide ranges within Lithodidae, at least in the western Pacific, appear to be the exception rather the rule, and ranges of all putatively widespread lithodids should be examined closely.

Many species of lithodid from the study area are known from few specimens, so little can generally be inferred about their distributions: *L. chaddertoni*, *L. rachelae*, *P. webberi*, and *P. taylorae* are each known only from a single locality or localities in close geographic proximity. Other species, such as *P. echidna*, *P. poorei*, and *P. staplesi*, also known from only a few sites, have been collected from widely disparate localities on both sides of the Tasman Sea, and can be regarded as relatively wide ranging. *Lithodes robertsoni* is apparently relatively widespread in New Zealand, from the Challenger Plateau, Chatham Rise south to the Campbell Plateau. *Lithodes jessica* is known from the southern Lord Howe Rise and off Hawkes Bay, New Zealand. Similarly, *L. richeri* and *P. dawsoni* have relatively wide southwestern Pacific ranges, from the Coral Sea south to southeastern Australia and New Zealand, respectively. This is probably also true for *N. bronwynae* if records of *N. vinogradovi* from New Caledonia prove referable to the former, as suggested above. Species for which the most substantial distributional data exist are *L. aotearoa*, *N. brodiei*, and *P. zealandica* from New Zealand, and *L. australiensis*, *N. flindersi*, and *P. gowlett-holmes* from Australia.

The ranges of the two most common and most widespread lithodids in New Zealand waters, *L. aotearoa* and *N. brodiei*, follow the deep shelf and slope surrounding mainland New Zealand, usually at depths of around 1000 m, principally the southern part of the West Norfolk Ridge, the Challenger Plateau, Chatham Rise and Campbell Plateau — essentially the ancient landmass of Zealandia. The two most common species of *Paralomis* in New Zealand waters, *P. dawsoni* and *P. zealandica*, apparently have an additional oceanographic component affecting their distributions. *Paralomis dawsoni* occurs along undersea ridges extending from the Coral Sea (Solomon Islands and New Caledonia) south to the Chatham Rise. *Paralomis zealandica* ranges from Cook Strait and the Chatham Rise south to the Campbell Plateau. The Chatham Rise acts as the intersection between the warm, south-flowing Tasman Front originating in the Coral Sea, and the colder north-flowing subantarctic inflow. This intersection of water masses, the Subtropical Convergence, corresponds to the distributional overlap of *P. dawsoni* and *P. zealandica*. Moreover, this distributional 'boundary' between *P. dawsoni* and *P. zealandica* also corresponds to broader patterns of benthic diversity and may be regarded as a general biogeographic boundary (Nodder *et al.* 2003).

The ranges of the two most common Australian lithodids, *L. australiensis* and *N. flindersi*, essentially follow the outer shelf and continental slope off southeastern Australia. *Lithodes australiensis* also occurs along the southeastern coast from New South Wales south to eastern Tasmania and onto the South Tasman Rise. *Neolithodes flindersi* ranges from New South Wales around Tasmania and into the Great Australian Bight. *Paralomis gowlettholmes* is presently known only from the southern Tasmanian seamounts, and appears to have its nearest relatives in Antarctica and on the Macquarie Ridge (*P. birsteini*).

Because most species of *Neolithodes* and *Lithodes* in New Zealand and Australia do not occur beyond slope depths, their geographic distributions are related to the extent of contiguous continental shelf and slope habitat. Consequently, the Australian and New Zealand species of *Neolithodes* and *Lithodes* are isolated and endemic to their respective regions (though it is conceivable that some southern overlap might occasionally occur via the South Tasman Rise and Macquarie Ridge, and northern communication might occur via the various ridge systems in the northern Tasman Sea). In contrast, species that occur down to abyssal depths can be expected to range more widely, as observed in *Paralomis staplesi* and *P. birsteini*, which occur at depths approaching or exceeding 2000 m and have trans-continental ranges.

The lithodid fauna of New Zealand and Australia is now known to be about twice as large as previous estimates. In the present study, emphasis on limb morphometrics has proven particularly useful in distinguishing species when considered in the context of allometry and sexual dimorphism. Also, for *Lithodes*, the spination of the walking leg dactyli, spination of the posterior branchial margin, and fusion of the plates of abdominal somite 2 have also proven to be taxonomically useful. These featutres should be considered in any future studies of *Lithodes*. The large collections available for study have permitted most species to be documented from both sexes over a wide size range. Some species, however, remain to be more adequately characterised, such as *Lithodes robertsoni* and *L. chaddertoni*, being known only from very few specimens. Males of *Paralomis taylorae* and *webberi* are not presently known, nor are females of *Neolithodes bronwynae* and *Paralomis staplesi*. Clearly, further research into the regional lithodid fauna is still required. It is also likely that more species will be discovered from the region, especially in northern New Zealand waters and southwestern Australia. Northern Australian waters remain unstudied for Lithodidae.

ACKNOWLEDGMENTS

This work was largely supported by the Foundation for Research Science and Technology (contract C01X0502) and the NIWA Capability Fund. I wish to thank Rick Webber (NMNZ), Stephen Keable (AM), Joanne Taylor and Gary Poore (NMV), Tin-Yam Chan (NTOU), Thierry Laperousaz (SAM), Roger Buttermore (TM), Peter Ng (ZRC), Hironori Komatsu (NSMT), Rafael Lemaitre (USNM), and Régis Cleva (MNHN) for making their collections available for study. Many of the specimens studied here were collected under numerous research programmes funded by the New Zealand Ministry of Fisheries and Foundation for Research Science and Technology, including the following programmes: Ocean Ecosystems (C01X0223, C01X027), Seamounts: their importance for fisheries and marine ecosystems (C01X0224), Biodiversity of the western Ross Sea (ZBD200303), IPY-CAML project (LIN08302), Oceans Survey 2020, and various trawl surveys (project MDT2007-01C). Malcom Clark, Karen Gowlett-Holmes, Sadie Mills, Reyn Naylor, Steve O'Shea, Ashley Rowden, Kareen Schnabel, Darren Stevens, Dean Stotter, and Di Tracey are thanked for their help in numerous ways, including provision of colour images of selected species. Enrique Macpherson, Colin McLay, and Pat Mclaughlin are also thanked for valuable comments and useful discussions, as are Dennis Gordon and Tracy Farr for their careful copy-editing. Lastly, special thanks to Rachel for her tireless support.

REFERENCES

AHYONG, S.T. 2009. New species and new records of hydrothermal vent shrimps from New Zealand (Caridea: Alvinocarididae, Hippolytidae). *Crustaceana* 82(7): 775–794.

AHYONG, S.T. 2010. *Neolithodes flindersi*, a new species of king crab from southeastern Australia (Crustacea: Decapoda: Lithodidae). *Zootaxa* 2362: 55–62.

AHYONG, S.T.; BOOTH, J.; BRUCE, N.L.; LOERZ, A.-N.; NAYLOR, J.R.; SCHNABEL, K.; WEBBER, W.R. 2007. Phylum Arthropoda. *In*: Tracey, D.M.; Anderson, O.F.; Naylor, J.R. (eds). A guide to common deepsea invertebrates in New Zealand waters, pp. 135–195. New Zealand Ministry of Fisheries, Wellington.

AHYONG, S.T.; CHAN, T.Y. 2010. *Lithodes formosae*, a new species of king crab from Taiwan (Crustacea: Decapoda: Lithodidae). *Zootaxa* 2332: 61–68.

AHYONG, S T.; DAWSON, E.W. 2006. Lithodidae from the Ross Sea, Antarctica, with descriptions of two new species (Crustacea: Decapoda: Anomura). *Zootaxa* 1303: 45–68.

AHYONG, S.T.; MACPHERSON, E.; CHAN, T.Y. 2010. Lithodoidea (King crabs). *In*: Chan T.Y. (ed.). Crustacean fauna of Taiwan: crab-like anomurans (Hippoidea, Lithodoidea, Porcellanidae), pp. 42–66. National Taiwan Ocean University, Keelung.

AHYONG, S.T.; O'MEALLY, D. 2004. Phylogeny of the Decapoda Reptantia: resolution using three molecular loci and morphology. *Raffles Bulletin of Zoology*, 52(2): 673–693.

AHYONG, S.T.; SCHNABEL, K.E.; MAAS, E. 2009. Anomuran phylogeny: new insights from molecular data. *In*: Martin, J.W.; Crandall, K.A.; Felder, D.L. (eds). Decapod Crustacean Phylogenetics. *Crustacean Issues 18*, pp. 399–414. CRC Press, Boca Raton.

ALCOCK, A. 1901. A descriptive catalogue of the Indian deep-sea Crustacea Decapoda Macrura and Anomala, in the Indian Museum. Being a revised account of the deep-sea species collected by the Royal Indian Marine Survey Ship *Investigator*. Trustees of the Indian Museum, Calcutta, iv + 286 pp., 3 pls.

ALCOCK, A.; ANDERSON, A.R.S. 1899. Natural history notes from H.M. Royal Indian marine survey ship *Investigator*, commander T.H. Heming, R.N., commanding. — Series III, No. 2. An account of the deep-sea Crustacea dredged during the surveying season of 1897–98. *Annals and Magazine of Natural History* (7) 3: 1–27.

AMAOKA, K.; MATSUURA, K.; INADA, T.; TAKEDA, M.; HATANAKA, H.; OKADA, K. 1990. Fishes collected by the R/V *Shinkai Maru* around New Zealand. Japan marine Fishery Resource Research Center, Tokyo. 410 pp.

ANDERSON, A.R.S. 1896. Natural history notes from R.I.M. Survey Steamer *Investigator*, Commander C.F. Oldham, R.N., commanding. Series II, no. 21. An account of the deep sea Crustacea collected during the season 1894–95. *Journal of the Asiatic Society of Bengal (2, Natural History)* 65: 88–106.

ANDRADE, H. 1980. Nueva especie de *Paralomis* en aguas de Chile: *Paralomis chilensis* n. sp. (Crustacea, Anomura, Lithodidae). *Boletín del Museo Nacional de Historia Natural de Chile* 37:269–273.

ARANA, P.M.; RETAMAL, M. 1999. Nueva distribución de *Paralomis birsteini* Macpherson, 1988 en agues antárticas (Anomura, Lithodidae, Lithodinae). *Investigaciones Marinas, Valparaiso* 27: 101–110.

ARNAUD, P.M.; DO-CHI, T. 1977. Données biologiques et biométriques sur les lithodes *Lithodes murrayi* (Crustacea: Decapoda: Anomura) des îles Crozet (SW Océan Indien). *Marine Biology* 39: 147–159.

BALSS, H. 1911. Neue Paguriden aus den Ausbeuten der Deutschen Tiefsse-Expedition *Valdivia* und der japanischen Expedition Prof. Dofleins. *Zoologischer Anzeiger* 38: 1–9.

BARNARD, K.H. 1947. Descriptions of new species of South African decapod crustaceans, with notes on synonymy and new records. *Annals and Magazine of Natural History* (11) 19: 361–392. [Dated 1946, published 1947]

BATSON, P. 2003. Deep New Zealand: Blue Water, Black Abyss. Canterbury University Press, Christchurch, 240 pp.

BENEDICT, J.E. 1895. Descriptions of new genera and species of crabs of the family Lithodidae, with notes on the young of *Lithodes camtschaticus* and *Lithodes brevipes*. *Proceedings of the United States National Museum* 17: 479–488.

BIRSTEIN, Y.A.; VINOGRADOV, L.G. 1967. Nakhozdeniye *Paralomis spectabilis* Hansen (Crustacea, Decapoda, Anomura) v Antarktike. Issledpvaniye Fauny Morei IV (XII). *Resultaty Biologicheskikh Issledovaniye Sovetskoy Antarkticheskoy Ekspeditsii 1955–1958* 3: 381–388. [English translation, 1968 as: Occurrence of *Paralomis spectabilis* Hansen (Crustacea, Decapoda, Anomura) in the Antarctic. Explorations of the fauna of the sea. IV (XII). *Biological Results of the Soviet Antarctic Expedition (1955–1958)* 3: 390–398. Jerusalem: Israel Program for Scientific Translation. Published for the National Science Foundation and Smithsonian Institution, Washington D.C.]

BIRSTEIN, Y.A.; VINOGRADOV, L.G. 1972. Crustacea (Decapoda, Anomura, Lithodidae) of the Atlantic sector of the Antarctic, South America and South Africa. *Zoologichesky Zhurnal* 51: 351–363. [In Russian]

BOUVIER, E.L. 1894a. Sur la signification de *Hapalogaster* dans l'évolution des Lithodinés. *Compte Rendu Sommaire des Séances de la Société Philomathique de Paris* 18: 1–5.

BOUVIER, E.L. 1894b. Sur la transformation des Paguriens en crabes anomures de la sous-famille des Lithodinés.

Comptes Rendus Hebdomadaires des Séances de l'Académie des Sciences 119: 350–352.

BOUVIER, E.L. 1895. Recherches sur les affinités des *Lithodes* et les *Lomis* avec les Pagurides. *Annales des Sciences Naturelles, Zoologie et Paléontologie* (7) 18: 157–213.

BOUVIER, E.L. 1896. Sur la classification des Lithodinés et sur leur distribution dans les oceans. *Annales des Sciences Naturelles, Zoologie et Paléontologie* (8) 1: 1–46.

BOUVIER, E.L. 1897. La transformation des Bernards l'Ermite en *Lithodes*. *Naturaliste* (2) 19(230): 41–43.

BRACKEN, H.D.; TOON, A.; FELDER, D.L.; MARTIN, J.W.; FINLEY, M.; RASMUSSEN, J.; PALERO, F.; CRANDALL, K.A. 2009. The decapod tree of life: compiling the data and moving toward a consensus of decapod evolution. *Arthropod Systematics and Phylogeny* 67(1): 99–116.

BRANDT, J.F. 1848. Die Gattung *Lithodes* Latreille nebst vier neuer ihr verwandten von Wosnessenski entdeckten, als Typen einer besondern Unterabtheilung (Tribus Lithodes) der Edward'shen Anomuren. *Bulletin de la Classe physico-mathématique de l'Académie Impériale des Sciences de Saint Pétersbourg* 7: 171–176.

BRANDT, J.F. 1850. Vorläufige Bermerkungen über eine aus zwei noch unbeschreibenen Gattungen und Arten gebildete Unterabtheilung (Hapalogastrica) der Tribus Lithodina, begleitet von einer Charakteristik der eben genannten Tribus der Anomuren. *Bulletin de la Classe physico-mathématique de l'Académie Impériale des Sciences de Saint Pétersbourg* 8: 266–269.

CHACE, F.A. 1939. Reports on the scientific results of the first *Atlantis* Expedition to the West Indies, under the joint auspices of the University of Havana and Harvard University. Preliminary descriptions of one new genus and seventeen new species of decapod and stomatopod Crustacea. *Memorias de la Sociedad Cubana de Historia Natural* 13(1): 31–54.

CUNNINGHAM, C.W.; BLACKSTONE, N.W.; BUSS, L.W. 1992. Evolution of the king crabs from hermit crab ancestors. *Nature* 355: 539–542.

DAVIE, P.J.F. 2002. Crustacea: Malacostraca: Eucarida (Part 2): Decapoda — Anomura. Brachyura. *Zoological Catalogue of Australia*, 19.3B. CSIRO Publishing, Melbourne, xiv + 641 pp.

DAWSON, E.W. 1988. The offshore fauna of Macquarie Island: history and biogeography—results from New Zealand and United States research cruises. *Papers and Proceedings of the Royal Society of Tasmania* 122(1): 219–229.

DAWSON, E.W. 1989. King crabs of the world (Crustacea: Lithodidae) and their fisheries: a comprehensive bibliography. *New Zealand Oceanographic Institute Miscellaneous Publication* 101: 1–338.

DAWSON, E.W. 2008. *Paralomis hirtella* Saint Laurent & Macpherson: the first hydrothermal vent-associated stone crab (Crustacea: Decapoda: Anomura: Lithodidae) in New Zealand waters. *Occasional Papers of the Hutton Foundation New Zealand*, 12: 1–17.

DAWSON, E.W.; YALDWYN, J.C. 1970. Diagnosis of a new species of *Neolithodes* (Crustacea: Anomura: Lithodidae) from New Zealand. *New Zealand Journal of Marine and Freshwater Research*, 4(2): 227–228.

DAWSON, E.W.; YALDWYN, J.C. 1971. Diagnosis of a new species of *Paralomis* (Crustacea: Anomura: Lithodidae) from New Zealand. *Records of the Dominion Museum*, 7(7): 51–54.

DAWSON, E.W.; YALDWYN, J.C. 1985. King crabs of the world or the world of king crabs: an overview of identity and distribution—with illustrated diagnostic keys to the genera of the Lithodidae and to the species of *Lithodes*. Proceedings of the International King Crab Symposium, Anchorage, Alaska, USA, January 22–24, 1985. Alaska Sea Grant Report No. 85-12: 69–106.

DE GRAVE, S.; PENTCHEFF, N.D.; AHYONG, S.T.; CHAN, T.Y.; CRANDALL, K.A.; DWORSCHAK, P.C.; FELDER, D.L.; FELDMANN, R.M.; FRANSEN, C.H.J.M.; GOULDING, L.Y.D.; LEMAITRE, R.; LOW, M.E.Y.; MARTIN, J.W.; NG, P.K.L.; SCHWEITZER, C.E.; TAN, S.H.; TSHUDY, D.; WETZER, R. 2009. A classification of living and fossil genera of decapod Crustaceans. *Raffles Bulletin of Zoology*, Supplement 21: 1–109.

DELL, R. K. 1963. Native Crabs. Reed, Auckland. 64 pp.

DESBRUYÈRES, D.; SEGONZAC, M.; BRIGHT, M. 2006. Handbook of Deep-Sea Hydrothermal Vent Fauna. *Denisia* 18: 1–544.

ELDREDGE, L.C. 1976. Two new species of lithodid (Anomura, Paguridea, Lithodidae) crabs from Guam. *Micronesica* 12: 309–315.

FAXON, W. 1893. Reports on the dredging operations off the west coast of Central America to the Galapagos to the west coast of Mexico and in the Gulf of California — by the U.S. Fish Commission steamer *Albatross*, during 1891. VI. Preliminary description of new species of Crustacea. *Bulletin of the Museum of Comparative Zoölogy, Harvard College* 24: 149–220.

FAXON, W. 1895. Reports on an exploration off the west coasts of Mexico, Central and South America, and off the Galapagos Islands, in charge of Alexander Agassiz, by the U.S. Fish Commission Steamer *Albatross* during 1891, Lieut.-Commander Z.L. Tanner, U.S.N., commanding. XV. The stalk-eyed Crustacea. *Memoirs of the Museum of Comparative Zoölogy at Harvard College* 18: 1–292, pls 1–56.

FELDMANN, R.M. 1998. *Paralomis debodeorum*, a new species of decapod crustacean from the Miocene of New Zealand: first notice of the Lithodidae in the fossil record. *New Zealand Journal of Geology and Geophysics* 41: 35–38.

FILHOL, H. 1883. Note sur quelques espèces nouvelles d'*Eupagurus* recueilles en Nouvelle-Zélande. *Bulletin de la Société Philomathique de Paris* (7) 8: 66–68.

FILHOL, H. 1884. Explorations sous-marines. Voyage du *Talisman*. *La Nature, Revue des sciences et de leurs applications aux arts et à l'industrie (Paris)* 12 (568): 326–330.

FILHOL, H. 1885a. Considérations relatives à la faune des Crustacés de la Nouvelle-Zélande. *Bibliothèque de l'École des Hautes Études, Section des Sciences Naturelles* 30: 3–60.

FILHOL, H. 1885b. La vie au fond des mers. *La Nature, Revue des sciences et de leurs applications aux arts et à l'industrie (Paris)* 13 (2)630: 55–58.

FRANKHAM, J. 2009. Aliens: benthic fauna of the Ross Sea. *New Zealand Geographic* 98: 64–71.

GARCÍA RASO, J.E.; MANJÓN-CABEZA, M.E.; RAMOS, A.; ILASO, I. 2005. New record of Lithodidae (Crustacea, Decapoda, Anomura) from the Antarctic (Bellingshausen Sea). *Polar Biology* 28: 642–646.

GUZMÁN, G. 2009. *Paralomis sonne* nueva especie de Litodido (Decapoda, Paguroidea, Lithodidae) en aguas profundas frente a Antofagasta, Chile. *Latin American Journal of Aquatic Research* 37(2): 253–257.

HAAN, W. DE 1833–1850. Crustacea. *In*: von Siebold, Ph. F. Fauna Japonica sive descriptio animalium, quae in itinere per Japoniam, jusse et auspiciis superiorum, qui summum in India Batavia Imperium tenent, suscepto, annis 1823–1830 collegit, notis observationibus et adumbrationibus illustravit. Lugdunum Batavorum: A. Arnz. 243 pp.

HAIG, J. 1974. Observations on the lithodid crabs of Peru, with description of two new species. *Bulletin of the Southern California Academy of Sciences,* 73(3): 152–164.

HALE, H.M. 1941. Decapod Crustacea. *Reports of the B.A.N.Z. Antarctic Research Expedition, 1929–1931*, 48: 257–285.

HALL, S.; THATJE, S. 2009a. Global bottlenecks in the distribution of marine Crustacea: temperature constraints in the family Lithodidae. *Journal of Biogeography* 36(11): 2125–2135.

HALL, S.; THATJE, S. 2009b. Four new species of the family Lithodidae (Decapoda: Anomura) from the collections of the National Museum of Natural History, Smithsonian Institution. *Zootaxa* 2302: 31–47.

HANSEN, H.J. 1908. Crustacea Malacostraca. I. *Danish Ingolf Expedition* 3(2): 1–20.

HASWELL, W.A. 1882. *Catalogue of the Australian Stalk- and Sessile-eyed Crustacea.* The Australian Museum, Sydney, xxiv + 324 pp.

HEALY, A.; YALDWYN, J. 1970. *Australian Crustaceans in Colour*. Reed, Sydney. 112 pp.

HELD, C.; WÄGELE, J.W. 2005. Cryptic speciation in the giant Antarctic isopod *Glyptonotus antarcticus* (Isopoda: Valvifera: Chaetiliidae). *Scientia Marina* 69: 175–181.

HELLER, C. 1865. Crustaceen. In: *Reise der österreichischen Fregatte Novara um die Erde in den Jahren 1857, 1858, 1859 unter den Befehlen des Commodors B. von Wüllerstorf-Urbair.* Kaiserlich-königlichen Hof-und Staatsdruckerei, Wien. Zoologischer Theil. 2(3): 1–280, pls. 1–25.

HENDERSON, J.R. 1888. Report on the Anomura collected by H.M.S. *Challenger* during the years 1873–76. Report on the Scientific Results of the Voyage of H.M.S. *Challenger* during the years 1873–76. *Zoology* 27(69): 1–221, 21 pls.

HESS, W. 1865. Beitrage zur Kenntniss der Decapoden-Krebse Ost-Australiens. *Archiv für Naturgeschichte* 31(1): 127–173, pls. 6–7.

HOMBRON, J.B.; JACQUINOT, H. 1842–1854. Crustacés. *Atlas d'Histoire Naturelle. Zoologie. Voyage au Pôle Sud et dans l'Océanie sur les corvettes l'Astrolabe et la Zélée pendant les années 1837–1838–1839–1840, Crustacés,* pls. 1–9.

HUTTON, F.W. 1882. The stalk-eyed Crustacea of New Zealand. *New Zealand Journal of Science* 1: 263–264.

ICZN. 1958. Opinion 511. Validation under the Plenary Powers of the generic name *Maja* Lamarck, 1801 (Class Crustacea, Order Decapoda) and designation under the same Powers of a type species for that genus in harmony with established practise. *Opinions and Declarations of the International Commission on Zoological Nomenclature, 18(15)*: 257–272.

IKEDA, H. 1998. The deep-sea crabs of Sagami Bay. Hayama Shiosai Museum, Kanagawa, Japan. 180 pp.

KENSLEY, B. 1980. Decapod and isopod Crustaceans from the west coast of southern Africa, including seamounts Vema and Tripp. *Annals of the South African Museum* 83(2): 13–32.

KIKUCHI, T.; OHTA, S. 1995. Two caridean shrimps of the families Bresiliidae and Hippolytidae from a hydrothermal field on the Iheya Ridge, of the Ryukyu Islands, Japan. *Journal of Crustacean Biology* 15(4): 771–785.

KING, M.D. 1958. Close-up photography of small plants and animals. *Tuatara* 7(2): 63–70.

KIRKWOOD, J.M. 1984. A guide to the Decapoda of the Southern Ocean. *ANARE Research Notes* 11: 1–47.

KLAGES, M.; GUTT, J.; STARMANS, A.; BRUNS, T. 1995. Stone crabs close to the Antarctic Continent: *Lithodes murrayi* Henderson, 1888 (Crustacea; Decapoda; Anomura) off Peter I Island (68°51'S, 90°51'W). *Polar Biology* 15: 73–75.

KOMAI, T.; SEGONZAC, M. 2004. A new genus and species of alvinocaridid shrimp (Crustacea: Decapoda: Caridea) from hydrothermal vents on the North Fiji and Lau Basins, south-western Pacific. *Journal of the Marine Biological Association of the United Kingdom*, 84: 1179–1188.

KOSLOW, J.A; GOWLETT-HOLMES, K. 1998. The seamount fauna off southern Tasmania: benthic communities, their conservation and impacts of trawling: final report to Environment Australia & The Fisheries Research Development Corporation. CSIRO Marine Research. 104 pp.

LATREILLE, P.A. 1806. Genera crustaceorum et insectorum secundum ordinem naturalem in familias disposita, iconibus exemplisque plurimus explicata. Amand Koenig, Paris, 280 pp.

LINNAEUS, C. 1758. Systema Naturae per Regna Tria Naturae, Secundum Classes, Ordines, Genera, Species, cum

Characteribus, Differentiis, Synonymis Locis. Edition 10, volume 1: iii + 824 pp. Holmiae.

LÖRZ, A.-N.; GLENNER, H.; LÜTZEN, J. 2008. New records of Rhizocephala (Cirripedia) from New Zealand, including the first rhizocephalan records from hot vents and cold seeps. *Crustaceana*, 81(8): 1013–1019.

LÖRZ, A.-N.; MAAS, E.W.; LINSE, K.; COLEMAN, C.O. 2009. Do circum-Antarctic species exist in peracrid Amphipoda? A case study in the genus *Epimeria* Costa, 1851 (Crustacea, Peracarida, Epimeriidae). *Zookeys* 18: 91–128.

LOVRICH, G.A.; PERRONI, M.; VINUESA, J.H.; TAPELLA, F.; CHIZZINI, A.; ROMERO, M.C. 2002. Occurrence of *Lithodes confundens* (Decapoda: Anomura) in the intertidal of the southwestern Atlantic. *Journal of Crustacean Biology* 22: 894–902.

MACPHERSON, E. 1982. A new species of *Paralomis* (Decapoda Anomura) from the southeastern Atlantic. *Crustaceana* 43(2): 142–146.

MACPHERSON, E. 1984. Crustáceos Decápodes del Banco Valdivia (Atlántico sudoriental). *Resultados Expediciones Cientificas* 12: 39–105.

MACPHERSON, E. 1988a. Three new species of *Paralomis* (Crustacea, Decapoda, Anomura, Lithodidae) from the Pacific and Antarctic Oceans. *Zoologica Scripta* 17(1): 69–75.

MACPHERSON, E. 1988b. Revision of the Family Lithodidae Samouelle, 1819 (Crustacea, Decapoda, Anomura) in the Atlantic Ocean. *Monografías de Zoología Marina* 2: 9–153.

MACPHERSON, E. 1988c. Lithodid crabs (Crustacea, Decapoda, Lithodidae) from Madagascar and La Réunion (SW Indian Ocean) with descriptions of two new species. *Bulletin du Muséum national d'Histoire naturelle, Paris 4(A)* 10(1): 117–133, pls. 1–3.

MACPHERSON, M. 1989. A new species of the genus *Paralomis* (Crustacea, Decapoda, Anomura, Lithodidae) from the Indian Ocean. *Scientia Marina* 53: 117–120.

MACPHERSON, E. 1990. Crustacea Decapoda: On some species of Lithodidae from the western Pacific. *In:* Crosnier, A. (ed.) Résultats des Campagnes MUSORSTOM, Volume 6. *Mémoires du Muséum national d'Histoire naturelle, Paris 4(A)* 145: 217–226.

MACPHERSON, E. 1991. A new species of the genus *Lithodes* (Crustacea, Decapoda, Lithodidae) from French Polynesia. *Bulletin du Museum National d'Histoire Naturelle, Paris 4(A)* 13(1–2): 153–158.

MACPHERSON, E. 1992. *Paralomis phrixa* (Decapoda, Anomura, Lithodidae), a new species from northern Peru, and a key to the eastern Pacific species of the genus. *Crustaceana* 63(3): 313–317.

MACPHERSON, E. 1994. Occurrence of two lithodid crabs (Crustacea: Decapoda: Lithodidae) in the cold seep zone of the South Barbados accretionary prism. *Proceedings of the Biological Society of Washington* 107: 465–468.

MACPHERSON, E. 2001. New species and new records of lithodid crabs (Crustacea, Decapoda) from the southwestern and central Pacific Ocean. *Zoosystema* 23(4): 797–805.

MACPHERSON, E. 2003. Some lithodids (Crustacea: Decapoda: Lithodidae) from the Solomon Islands (SW Pacific Ocean), with the description of a new species. *Scientia Marina* 67(4): 413–418.

MACPHERSON, E. 2004. A new species and new records of lithodid crabs (Crustacea: Decapoda: Lithodidae) from the Crozet and Kerguelen Islands area (Subantarctica). *Polar Biology* 27: 418–422.

MACPHERSON, E.; CHAN, T.Y. 2008. Some lithodid crabs (Crustacea: Decapoda: Lithodidae) from Taiwan and adjacent waters, with the description of one new species from Guam. *Zootaxa* 1924: 43–52.

MAKAROV, V.V. 1934. Eine neue Gattung der Familie Lithodidae aus dem japanischen Meer. *Zoologischer Anzeiger* 108: 251–254.

MAKAROV, V.V. 1938 Rakoobraznyey. Anomura. [Crustacés Décapodes anomures]. *In*: Shtakel'berg, A.A. (ed.). Fauna SSSR, (n. ser.) 16, (10) (3): i–x, 1–324, pls. 1–5. Academii Nauk, Moscow and Leningrad [English translation, 1962. Anomura. Fauna of the USSR. Crustacea, Anomura. X(3): 1–278, pls 1–5. Jerusalem: Israel Program for Scientific Translation. Published for the National Science Foundation and Smithsonian Institution, Washington D.C.]

MARTIN, J.W.; DAVIS, G.E. 2001. An updated classification of the Recent Crustacea. *Natural History Museum LA County Science Series* 39: 1–124.

McLAUGHLIN, P.A.; LEMAITRE, R. 1997. Carcinization in the Anomura — fact or fiction? I. Evidence from adult morphology. *Contributions to Zoology* 67(2): 79–123.

McLAUGHLIN, P.A.; LEMAITRE, R.; TUDGE, C.C. 2004. Carcinization in the Anomura — fact or fiction? II. Evidence from larval, megalopal and early juvenile morphology. *Contributions to Zoology* 73(3): 165–205.

McLAUGHLIN, P.A.; LEMAITRE, R.; SORHANNUS, U. 2007. Hermit crab phylogeny: a reappraisal and its «fall out». *Journal of Crustacean Biology* 21(1): 97–115.

McLAY, C.L. 1988. Crabs of New Zealand. *Leigh Laboratory Bulletin* 22: iv + 1–463.

McLAY, C.L. 2007. New crabs from hydrothermal vents of the Kermadec Ridge submarine volcanoes, New Zealand: *Gandalfus* gen. nov. (Bythograeidae) and *Xenograpsus* (Varunidae) (Decapoda: Brachyura). *Zootaxa* 1524: 1–22.

MIERS, E. J. 1876. Catalogue of the stalk- and sessile-eyed Crustacea of New Zealand. Colonial Museum and Geological Survey Department London. xii + 136p., 3 pls.

MILNE-EDWARDS, A.; BOUVIER, E.L. 1894a. Crustacés décapodes provenant des campagnes du yacht *l'Hirondelle* (1886, 1887, 1888). I. Brachyures et Anomoures. *Résultats des Campagnes Scientifiques accompliés sur son Yacht par Albert Ier Prince Souverain de Monaco* 7: 3–112, pls. 1–11.

MILNE-EDWARDS, A.; BOUVIER, E.L. 1894b. Troisième campagne du yacht *l'Hirondelle*, 1887. *Neolithodes*, genre nouveau de la sous-famille des Lithodinés. *Bulletin de la Société Zoologique de France* 19: 120–122.

MOLINA, G.I. 1782. *Saggio sulla storia natural del Chili*. Bologna, 367 pp.

MORRISON, C.L.; HARVEY, A.W.; LAVERY, S.; TIEU, K.; HUANG, Y.; CUNNINGHAM, C.W. 2002. Mitochondrial gene rearrangements confirm the parallel evolution of the crab-like form. *Proceedings of the Royal Society of London B* 269(1489): 345–350.

NAYLOR, J. R.; WEBBER, W.R.; BOOTH, J. 2005. A guide to the common offshore crabs in New Zealand waters. *New Zealand Aquatic Environment and Biodiversity Report* 2:1–47.

NODDER, S.C.; PILDITCH, C.A.; PROBERT, P.K.; HALL, J.A. 2003. Variability in benthic biomass and activity beneath the Subtropical Front, Chatham Rise, SW Pacific Ocean. *Deep-Sea Research I* 50: 959–985.

O'DRISCOLL, R.L.; BOOTH, J.D.; BAGLEY, N.W.; ANDERSON, O.F.; GRIGGS, L.H.; STEVENSON, M.L.; FRANCIS, M.P. 2003. Areas of importance for spawning, pupping or egg-laying, and juveniles of New Zealand deepwater fish, pelagic fish, and invertebrates. *NIWA Technical Report* 119: 1–377.

ORTMANN, A.E. 1892. Die Decapoden-Krebse des Strassburger Museums, mit besonderer Berücksichtigung der von Herrn Dr. Döderlein bei Japan und bei den Liu-Kiu-Inseln gesammelten und zur Zeit im Strassburger Museum aufbewahrten Formen. IV. Die Abtheilungen Galatheidea und Paguridea. *Zoologische Jahrbücher, Abtheilung für Systematik, Geographie und Biologie der Thiere* 6: 241–326.

O'SHEA, S.; McKNIGHT, D.; CLARK, M. 1999. Bycatch – the common, unique, and bizarre. *Seafood New Zealand* 7(5): 45–51.

POORE, G.C.B. 2004. *Marine Decapod Crustacea of Southern Australia: a guide to identification*. CSIRO Publishing, Melbourne. 574 pp.

POORE, G.C.B.; HART, S.; TAYLOR, J.; TUDGE, C. 1998. Decapod Crustaceans from Tasmanian seamounts. *In*: Koslow, J.A.; Gowlett-Holmes, K. (eds). The seamount fauna off southern Tasmania: benthic communities, their conservation and impacts of trawling, pp 65–78. CSIRO Marine Research.

POORE, G.C.B.; McCALLUM, A.W.; TAYLOR, J. 2008. Decapod Crustacea of the continental margin of southwestern and central Western Australia: preliminary identifications of 524 species from FRV *Southern Surveyor* voyage SS10-2005. *Museum Victoria Science Report* 11: 1–106.

RICHTER, S.; SCHOLTZ, G. 1994. Morphological evidence for a hermit crab ancestry of lithodids (Crustacea, Decapoda, Anomala, Paguroidea) *Zoologischer Anzeiger* 233: 187–210.

SAINT LAURENT, M. DE; MACPHERSON, E. 1997. Une nouvelle espèce du genre *Paralomis* White, 1856, des sources hydrothermales du Sud-ouest Pacifique (Crustacea, Decapoda, Lithodidae). *Zoosystema* 19(4): 721–727.

SAKAI, T. 1971. Illustrations of 15 species of crabs of the family Lithodidae, two of which are new to science. *Researches on Crustacea* 4–5: 1–49, figs. 1–3, pls. 1–21, maps 1–3.

SAKAI, T. 1978. Decapod Crustacea from the Emperor Seamount Chain. *Researches on Crustacea* Supplement 8: 1–39.

SAKAI, T. 1980. New species of crabs of the families Lithodidae and Calappidae. *Researches on Crustacea* 10: 1–11, pl. 1, frontispiece 1, 2.

SAMOUELLE, G. 1819. The Entomologist's Useful Compendium; or an introduction to the knowledge of British Insects, comprising the best means of obtaining and preserving them, and a description of the apparatus generally used; together with the genera of Linné, and modern methods of arranging the Classes Crustacea, Myriapoda, spiders, mites and insects, from the affinities and structure, according to the views of Dr. Leach. Also an explanation of the terms used in entomology; a calender of the times of appearance and usual situations of near 3,000 species of British Insects; with instructions for collecting and fitting up objects for the microscope. Thomas Boys, London. 496 pp., 12 pls.

SCHMITT, W.L. 1921. The marine decapod Crustacea of California with special reference to the decapod Crustacea collected by the United States Bureau of Fisheries Steamer "*Albatross*" in connection with the biological survey of San Francisco Bay during the years 1912–1913. *University of California Publications in Zoology* 23: 1–359, pls 1–50.

SHIRLEY, T.C.; ZHOU, S.J. 1997. Lecithotrophic development of the golden king crab *Lithodes aequispinus* (Anomura: Lithodidae). *Journal of Crustacean Biology* 17: 207–216.

SMITH, S. I. 1882. Reports on the results of dredging under the supervision of Alexander Agassiz, on the east coast of the United States during the summer of 1880, by the U.S. Coast Survey Steamer "Blake", Commander J.R. Bartlett, U.S.N., commanding. Report on the Crustacea, Part 1. Decapoda. *Bulletin of the Museum of Comparative Zoölogy, at Harvard College* 10: 1–108.

SPIRIDONOV, V.A.; TÜRKAY, M.; ARNTZ, W.E.; THATJE, S. 2006. A new species of the genus *Paralomis* (Crustacea: Decapoda: Lithodidae) from the Spiess seamount near Bouvet Island (Southern Ocean), with notes on habitat and ecology. *Polar Biology* 29: 137–146.

STEBBING, T.R.R. 1905. South African Crustacea. Part 3. *Marine Investigations in South Africa* 4(3): 21–123, pls 17–26.

STEVENS, D.W. 2004. Stomach contents of the Antarctic toothfish (*Dissostichus mawsoni*) from the western Ross Sea, Antarctica. *CCAMLR Research Report WG-FSA*, Hobart. 16 pp.

STIMPSON, W. 1858. Crustacea Anomura. Prodromus descriptionis animalium evertebratorum, quae in Expeditione ad Oceanum Pacificum Septentrionalem a Republica Federata missa, Cadwaladaro Ringgold et Johanne Rodgers Ducibus, observavit et descripsit. Pars

7. *Proceedings of the Academy of Natural Sciences of Philadelphia* 10: 225–252.

TAKEDA, M. 1974. On three species of the Lithodidae (Crustacea, Anomura) from the Central Pacific. *Bulletin of the National Science Museum, Tokyo, series A* 17(3): 205–212, pls 1–3.

TAKEDA, M. 1985. A new species of *Paralomis*, the crab-shaped Anomura (Crustacea, Decapoda), from the Kyushu-Palau submarine ridge. *Bulletin of the National Science Museum, Tokyo, series A* 11(3): 137–140.

TAKEDA, M.; S. BUSSARAWIT, S. 2007. A new species of the genus *Paralomis* White, 1856 (Crustacea, Decapoda, Anomura, Lithodidae) from the Andaman Sea. *Bulletin of the National Science Museum, Tokyo* 33(2): 51–59.

TAKEDA, M.; HASHIMOTO, J. 1990. A new species of the genus *Paralomis* (Crustacea, Decapoda, Lithodidae) from the Minami-Ensei Knoll in the Mid-Okinawa Trough. *Bulletin of the National Science Museum, Tokyo, series A* 16(2): 79–88.

TAKEDA, M.; HATANAKA, H. 1984. Records of decapod Crustaceans from the southwestern Atlantic collected by the Japanese fisheries research trawlers. *Bulletin of the National Science Museum, Tokyo, series A* 10(1): 7–24.

TAKEDA, M.; MIYAKE, S. 1980. A new species of *Paralomis* (Crustacea, Anomura) from the East China Sea. *Annotationes Zoologicae Japonenses* 53(1): 42–45.

TAKEDA, M.; NAGAI, S. 2004. Four species of giant crustaceans from the Indonesian depths, with description of a new species of the family Lithodidae. *Bulletin of the National Science Museum, Tokyo, series A* 30(1): 9–21.

TAKEDA, M.; OHTA, S. 1979. A new species of the Lithodidae (Crustacea, Anomura) from Suruga Bay, Central Japan. *Bulletin of the National Science Museum, Tokyo, series A* 5(3): 195–200, pls 1–3.

THATJE, S.; ANGER, K.; CALCAGNO, J.A.; LOVRICH, G.A.; PÖRTNER, H.O.; ARNTZ, W.E. 2005. Challenging the cold: crabs reconquer the Antarctic. *Ecology* 86: 619–625.

THATJE, S.; HALL, S.; HAUTON, C.; HELD, C.; TYLER, P. 2008. Encounter of lithodid crab *Paralomis birsteini* on the continental slope off Antarctica, sampled by ROV. *Polar Biology* 31: 1143–1148.

THATJE, S.; LÖRZ, A.-N. 2005. First record of lithodid crabs from Antarctic waters off the Balleny Islands. *Polar Biology* 28, 334–337.

TKACHUK, L.P. 1985. *Paralomis spectabilis* (Decapoda) as new host for *Briarosaccus callosus* (Rhizocephala). *Zoologichesky Zhurnal* 64(11): 1733–1734. [In Russian]

TILESIUS, W.C. 1815. De Cancris Camtschaticis, Oniscis, Entomostracis et Cancellus marinis microscopicis noctilucentibus cum tabulis IV: Aenaeis et appendice adnexo de Acaris et Ricinis Camtschaticis. Auctore Tilesio. *Mémoires de l'Académie Impériale de Sciences de St Pétersbourg*, 5: 331–405, pl. 5–8.

TSANG, L.M.; MA, K.Y.; AHYONG, S.T.; CHAN, T.Y.; CHU, K.H. 2008. Phylogeny of Decapoda using two nuclear protein-coding genes: origin and evolution of the Reptantia. *Molecular Phylogenetics and Evolution* 48: 359–368.

TUNG, Y.M.; WANG, F.Z.; LI, Z.C. 1984. A new species of Lithodidae (Crustacea, Anomura) from East China Sea. *Zoological Research* 5: 329–332. [In Chinese]

WEBBER, R. 1997. The Royal Family: king crabs at home and abroad. *Seafood New Zealand* 5(4): 81–84.

WEBBER, W.R. 2004. A new species of *Alvinocaris* (Crustacea: Decapoda: Alvinocarididae) and new records of alvinocaridids from hydrothermal vents north of New Zealand. *Zootaxa* 444: 1–26.

WEBBER, R.; NAYLOR, R. 2004a. King Crabs I: the three big reds. *Seafood New Zealand* 12(1): 78–79.

WEBBER, R.; NAYLOR, R. 2004b. King Crabs II: *Paralomis* species, warts and all. *Seafood New Zealand* 12(10): 62–63.

WHITE, A. 1848. Description of *Echinocerus cibarius* a new species and subgenus of Crustacea. *Proceedings of the Zoological Society of London* 1848, 16: 47–49, pls. 2–3.

WHITE, A. 1856. Some remarks on Crustacea of the genus *Lithodes* with a brief description of a species apparently hitherto unrecorded. *Proceedings of the Zoological Society of London* 1856: 132–135, pl. 42.

WILLIAMS, A.B.; SMITH, C.R.; BACO, A.R. 2000. New species of *Paralomis* (Crustacea, Anomura, Lithodidae) from a sunken whale carcass in the San Clemente Basin off southern California. *Journal of Crustacean Biology* Special no. 2: 281–285.

WILSON, R. 1990. *Paralomis otsuae*, a new species of Decapoda Anomura from deep water off the Chilean coast. *Crustaceana* 58(2): 130–135.

YALDWYN, J.C.; DAWSON, E.W. 1970. The stone crab *Lithodes murrayi* Henderson: the first New Zealand record. *Records of the Dominion Museum*, 6(17): 275–284.

ZAKLAN, S.D. 2002. Review of the family Lithodidae (Crustacea: Anomura: Paguroidea): distribution, biology and fisheries. *In*: Paul, A.J.; Dawe, E.G.; Elner, R.; Jamieson, G.S.; Kruse, G.H.; Otto, R.S.; Sainte-Marie, B.; Shirley, T.C.; Woodby, D. (eds). Crabs in Cold Water Regions: biology, management, and economics, Pp. 751–845. University of Alaska Sea Grant Program. Fairbanks, Alaska.

SPECIES INDEX

Principal taxonomic account is in bold font; species illustrations in bold italic. Synonymic entries and material examined are not included.